DEVELOPMENTS IN
GEOPHYSICAL EXPLORATION METHODS—3

CONTENTS OF VOLUMES 1 AND 2

Volume 1

Volume 2

DEVELOPMENTS IN GEOPHYSICAL EXPLORATION METHODS—3

Edited by

A. A. FITCH

Consultant, Formerly of Seismograph Service (England) Limited, Keston, Kent, UK

APPLIED SCIENCE PUBLISHERS

LONDON and NEW YORK

APPLIED SCIENCE PUBLISHERS LTD
Ripple Road, Barking, Essex, England
Sole Distributor in the USA and Canada
ELSEVIER SCIENCE PUBLISHING CO., INC.
52 Vanderbilt Avenue, New York, NY 10017, USA

British Library Cataloguing in Publication Data

Developments in geophysical exploration methods—
3—(The Developments series)
1. Prospecting—Geophysical methods
I. Series
622'.15   TN269

ISBN-13:978-94-009-7351-0   e-ISBN-13:978-94-009-7349-7
DOI: 10.1007/978-94-009-7349-7

WITH 10 TABLES AND 164 ILLUSTRATIONS

The selection and presentation of material and the opinions expressed in this publication
are the sole responsibility of the authors concerned.

PREFACE

One of the themes in current geophysical development is the bringing together of the results of observations made on the surface and those made in the subsurface. Several benefits result from this association.

The detailed geological knowledge obtained in the subsurface can be extrapolated for short distances with more confidence when the geological detail has been related to well-integrated subsurface and surface geophysical data. This is of value when assessing the characteristics of a partially developed petroleum reservoir. Interpretation of geophysical data is generally improved by the experience of seeing the surface and subsurface geophysical expression of a known geological configuration. On the theoretical side, the understanding of the geophysical processes themselves is furthered by the study of the phenomena in depth. As an example, the study of the progress of seismic wave trains downwards and upwards within the earth has proved most instructive.

This set of original papers deals with some of the more vigorous developments in subsurface geophysics: and it is hoped that it will contribute to the understanding of geophysical phenomena in the solid.

The editor thanks the busy workers in the several fields who have made time to produce these contributions.

A. A. FITCH

CONTENTS

LIST OF CONTRIBUTORS

A. K. BOOER

Senior Systems Analyst, Plessey Marine Research Unit, Templecombe, Somerset, U K.

R. C. CARLSON

Applied Geophysics Group, Lawrence Livermore National Laboratory, PO Box 808, Livermore, California 94550, USA.

S. GIANZERO

Gearhart Industries, Inc., 2525 Wallingwood, Austin, Texas 78746, USA. Formerly, Schlumberger-Doll Research Center, Ridgefield, Connecticut, USA.

J. R. HEARST

Applied Geophysics Group, Lawrence Livermore National Laboratory, PO Box 808, Livermore, California 94550, USA.

W. S. KEYS..

Chief Borehole Geophysics Research Project, US Geological Survey, Denver, Colorado 80225, USA.

P. G. KILLEEN

Head, Borehole Geophysics Section, Geological Survey of Canada, 601 Booth Street, Ottawa, Canada K1A OE8.

J. H. MORAN

Consultant, Briarcliff 840, Spicewood, Texas 78669, USA.

D. RADER

Director, Advanced Development, Teleco Oilfield Services Inc., 105 Pondview Drive, Meriden, Connecticut 06450, USA.

A. ROY

National Geophysical Research Institute, Hyderabad 500 007, A.P., India.
Present address: Department of Geology, University of Ibadan, Ibadan, Nigeria.

A. J. RUDMAN

Professor of Geophysics, Indiana University, 1005 East Tenth Street, Bloomington, Indiana 47405, USA.

Chapter 1

UNDERGROUND GEOPHYSICS OF
COAL SEAMS

A. K. BOOER

Plessey Marine, Templecombe, Somerset, UK

SUMMARY

The economic extraction of coal using modern, highly mechanised techniques relies on the existence of large, relatively undisturbed areas of coal. The planning of the mining operation is based on the knowledge of geological discontinuities which may impede the progress of a coal face. Any technique which enables the structure of the coal seam to be clarified may be of great benefit. In-seam seismic surveying is one such technique, and the quality of result which may be obtained from it is heavily dependent on an understanding of the propagation of seismic waves in the underground environment. The study of geological structures using guided waves relies on techniques of traditional seismic surveying, suitably modified to account for phenomena which occur in the rather special environment of a layered coal seam. The scale of geological features which may be detected by such methods is sufficiently small to be of great potential use to mining engineers.

1. INTRODUCTION

Investigations into techniques which might reveal the geophysical structure of coal seams are primarily justified on economic grounds. It might be argued that there are considerable benefits to be gained in the areas of safety and other aspects of the mining operation by having detailed knowledge of the geophysics of an area to be mined, but it is undeniable that the major impact would be on the productivity of a mine. To understand the engineering problems of this operation, and to appreciate the methods and solutions which underground geophysics may offer, it is necessary to investigate modern mining techniques and their reliance on a continuous yield from a face throughout its planned lifetime.

Modern coal mining is extremely capital intensive and the investment in terms of time and equipment required to set up a new coal face may amount to several millions of pounds. Before investing such a sum it is necessary to ensure that the initial outlay will be adequately repaid by coal won from the face. What then are the circumstances which may arise to prevent this objective from being achieved? A brief survey of the statistics of some recent coal faces is enough to conclude that a significant proportion are abandoned prematurely through the discovery of hitherto unknown geological faults, which may be quite small (displacing the seam vertically on either side by only 1 or 2 m), which would make the continued extraction of the coal uneconomic because of the effort involved in 'carrying' the fault and realigning the cutting equipment to follow the seam's direction.

The impact of losing a face on the output of the mine may to some extent be lessened by the provision of 'standby' faces, set up in advance, which may be worked for the duration of the time which it takes to dismantle the equipment at the abandoned face and set it up once more at a new site. There is, of course, some cost involved in having suitable standby faces lying idle if production is proceeding without difficulty, as planned, at the other faces in the mine. A more desirable solution would be to carry out a detailed survey of the geology of the area to be mined in advance of setting up a face, and this is, of course, already done using established techniques. The evidence produced by such a survey may not be clear cut in identifying the existence of an undisturbed area in which to drive a face, relying as it does on the interpretation of ambiguous data. What might also be useful is some advance warning that a coal face is approaching some discontinuity in the seam which may impede its progress, in order that some sensible estimate of its remaining life may be made, to assist in planning the switch to another face.

Underground seismic survey may provide some of the desirable facilities outlined above. To appreciate the part that it may play in planning and monitoring the progress of an advancing face it is necessary to describe more fully the method of 'longwall mining' which is used in the majority of coal faces in the UK, and to consider present techniques for assessing the geological structure of a seam in advance of its extraction.

Coal mining today is not the labour intensive activity that it once was. A high degree of mechanisation has reduced the number of men required to work at the actual site of a coal face to a handful. It is unlikely, however, that technology will advance sufficiently rapidly for this number to be reduced significantly in the near future. Situations will always arise where human intervention is the fastest and best solution to the problem.

This being the case, a necessary environmental limitation to the depth at which coal may be mined economically is about 1500 m with a minimum seam thickness of a little less than 1 m, due to increasing temperature with depth and the increased costs involved in ventilating the mine adequately. In the UK it is unusual for a seam to exceed a couple of metres in thickness, and the coal cutting equipment used at the face is designed to accommodate this range of seam thicknesses.

In the UK, 80% of the faces are worked using the longwall mining method. Two parallel tunnels, or 'roadways', are driven horizontally into a panel of coal to be extracted. These two roadways are joined by another tunnel which forms the working face, where the coal cutting equipment is installed (see Fig. 1). This machinery consists of a track,

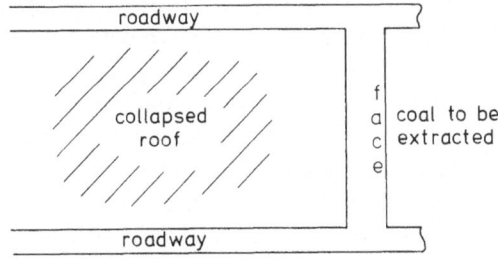

FIG. 1. Sketch plan of an advance longwall coal face.

laid along the face, supporting a moving coal shearer with a rotating drum of cutting edges which slice off a thickness of about 50 cm of coal on each pass of the shearer. The coal falls on to a conveyor belt which runs along the track and is carried to one of the roadways and back to a main shaft, where it is brought to the surface. The roof of rock in the region of the face is supported by massive hydraulic props, several hundreds of them spaced along the length of the face, which push forward after every pass of the shearer carrying the track and conveyer with them. The roof behind the advancing face is encouraged to collapse to relieve the build up of strain on the face itself. It is this equipment which is so costly to install in terms of money and time before any coal may be won from the new face.

The vulnerability of the mining operation to the presence of actually rather small faults on a geological scale is readily appreciated, since a fault with a vertical throw of only the seam's thickness will result in the apparent disappearance of the coal from the vicinity of the face. In addition, the fault may lie at any angle to the line of advance of the face, and may therefore take an unknown period of time to carry if work is continued. Considerable effort has been directed to finding methods of averting this situation, and techniques span a very wide range of fields. The technique which is the topic of this discussion is that of inferring the geological structure of the coal seam from measurements taken in underground seismic surveys at the seam itself. To place this technique in context, there follows a brief discussion of other methods which may be used to the same, or similar, ends.

A very practical solution to the problem of advancing the face into relatively unknown areas of the mine, and the risk that involves, is to drive the lateral roadways, which will be joined by the face, into the area before setting up the face across the far end of the roadways and mining the coal 'in retreat' as it is termed. If some unexpected fault is encountered whilst the initial roadways are being driven into the coal then a more accurate assessment of the impact that this would have on the face may be made at that time, resulting perhaps in the face not being worked at all. Since the equipment has not yet been set up at the face, the cost of abandoning the working at this stage is not nearly as great as if the fault had not been discovered. Whilst giving some assurance that the area to be worked is relatively free of faults, retreat mining does not ensure that the face will be free of unexpected geological features in the course of its extraction. For this reason, and also because of the higher initial cost, retreat mining is not used for the majority of coal faces. More commonly, several surveying techniques may be used to determine the suitability of an area for mining.

Underground electromagnetic techniques, essentially an attempt to apply the technology of radar in an underground environment, suffer from the extremely severe attenuation experienced by propagating waves, particularly in the presence of water. The range over which information may be extracted from the underground strata by this method is very limited and would not serve to give adequate warning of any discontinuity in the path of the face.

Surface magnetic and gravimetric surveys have been used to detect old workings, another possible source of disruption to the operation of a face and one which is known to have been the cause of major mining accidents in the past. These methods generally rely on the location of surface dumps of spoil produced by mining and the detection of old buildings and shafts, not on the geological structure of the seam itself.

Seismic techniques in general seem to have a number of desirable characteristics for their application to this problem. Seismic waves are known to travel very long distances through rock. They have, for a long time, been used for the exploration of geological features in the search for oil and gas. For this reason, efficient transducers of seismic waves are available, as is equipment for capturing and recording large amounts of data. In conventional seismic exploration, the rock strata of interest are very much deeper than are coal seams, but an overall scaling down of the standard seismic method would appear to be applicable to the surveying of shallow coal seams. There are, however, particular problems which arise when trying to assess the structure of coal seams using surface seismic techniques (Ziolkowski, 1979). Although these may, in general, be overcome, the resolution required is right at the upper limit of that which may be achieved from the surface. Data may also be expensive to acquire in this manner, or even entirely precluded by surface environmental conditions.

Another application of seismic techniques has been found by gathering data underground, using seismic waves travelling within the coal strata. Such waves are very susceptible to changes in the geological structure of the regions through which they travel, and may therefore be an important source of information about the condition of the coal seam.

In-seam seismic techniques are based on the fact that acoustic energy may be trapped within rock strata of particular characteristics. This phenomenon is essentially due to the process of total internal reflection, so that sound propagating in a low velocity layer will be reflected at the upper and lower interface with higher velocity layers. Seismic waves constrained in such a manner are effectively limited to a two-dimensional plane and may potentially travel over large distances.

It has long been known that rock strata, in general, may act as a guide for acoustic energy, and an early observation that this occurred in coal seams in particular was reported by Evison (1955). Evison, working in New Zealand, recorded vibrations due to a small explosive shot at a distance of about 150 m in a 5 m thick seam, and obtained good agreement with the theoretical characteristics of a Love wave (a particular class of guided waves). He noted that guided waves might find useful applications in mining, particularly for the location of faults.

Subsequent observations of guiding in coal seams were made by Krey
(1963). Since that time, a number of groups have been actively engaged in
underground seismic trials, and supporting work has been done in model
experiments and theoretical studies in order to investigate some of the
problems which arise in the collection and processing of underground
seismic data.

Underground experimentation has been complicated by the need for
extensive precautions in fire-proofing of the equipment which is to be
used in the potentially explosive atmosphere of a mine. Apart from this,
however, the equipment required for underground seismic work is essentially the same as that used for small-scale surface seismic work. Digital
recording is generally favoured since subsequent data processing is
carried out on a digital computer. The processing which is performed on
the data is somewhat different from conventional seismic stacking due to
the rather special geometry underground and the possible existence of a
fault at any range and orientation to the face. The unusual propagation
characteristics of acoustic waves in this environment also has a significant impact on the processing.

The end product of this processing is a representation of the disposition of geological discontinuities within the seam. This map of the seam
is, inevitably, affected by the geometry of the receiving array and shot
points used in the gathering of data. These effects are apparent as
ambiguities or noise in the reconstructed map and complicate the
interpretation of the underlying geological structure of the seam. To
assist in the interpretation of results it has been necessary to identify the
characteristics of the data collection and signal processing system as a
whole and to attempt to overcome some of their limitations.

The techniques discussed here have gone some way towards the goal
of inferring the geophysical structure of coal seams using data derived
from underground experiments. The overall problem is quite complex, as
are all remote sensing and identification problems, and the methods used
constitute rather more an art than a science. However, the following
discussion presents those aspects which may be reasonably quantified
and yield, at least to some extent, to scientific methods.

2. DATA COLLECTION

The investigation of the structure of a seam by guided acoustic waves
presents some unique problems in the gathering of data. The detection of

channelled waves in the seam requires the placement of geophone sensors within the seam itself. Discarding, as uneconomic and impractical, the approach of drilling into the seam from above or below, the remaining solution is to position the geophones at points which are easily accessible from existing roadways. In the confines of a mine, these points of access are even more limited than might be imagined, a typical roadway being cluttered with such items as roof supports, shuttering, conveyors and a variety of other equipment. The underground seismologist has therefore very much less choice concerning the positioning of the geophone array and its overall dimensions than that available to the designer of a surface array.

Many of the techniques which have been learnt to counter problems in the recording of surface seismic data are simply not possible to apply to the underground array. Many of the classical problems of seismic arrays, however, still exist; problems such as ground roll, inconsistency of geophone coupling, electrical interference, measurement of geophone position, and so on. The whole problem of array design is coloured by the limitation in the vertical positioning of a geophone within the seam, which may be a few metres thick. The array geometry is constrained to one dimension, and the overall array length is effectively limited by the distance over which the roadway runs along the seam.

The utilisation of clusters of detectors along the array to overcome poor coupling is not helpful for several reasons. On a practical front, it is extremely difficult to plant even one geophone at any desired position in the array, for reasons already given. On theoretical grounds, the frequencies involved are relatively high and a grouping of detectors would rapidly degrade the high frequency response of the array. To reconstruct a reflection map of the seam it would be desirable to sample the seismic energy at intervals of less than one-half the wavelength of its maximum frequency, in order to have sampled the phase fronts adequately in a spatial sense. Again, this is often not possible in practice and may not even be approached. This has profound effects on the methods used to produce a map from the seismic data, a point which will be returned to later.

The size of the array, in terms of the number of geophones, is very small indeed, and whereas in surface seismic work the volume of recorded data may almost be an embarrassment because of the time required to process it all, the same is not true in this situation. It is necessary to squeeze as much information as possible from the data by extensive processing, in order that it may be utilised to the full. The relatively modest amount of data coming from the array does enable recording to be a quite straightforward process. This is not to say that the recording equipment may be in any way less sophisticated than in other work, having the same requirement for automatic gain control, large dynamic range, digital recording, and so on.

Environmental considerations also have a significant impact on underground coal seam exploration in the choice of seismic source which may be used. In the confines of a mine it is clearly inadvisable to use large explosive charges, and heavy equipment for generating complex seismic waveforms is also out of the question. The sources available are therefore limited to small charges, noise produced by standard mining equipment and noise generated from within the coal itself due to micro-faulting relieving internal stresses.

Research into listening in a purely passive way to the noises generated within the coal has been done, but not with the intention of determining the overall geophysical structure of the seam (Salamon and Wiebols, 1974). Noise generated by the coal cutting machinery has been used in some experiments and this might be a useful source for the everyday monitoring of the progress of an existing face, gathering data all the time that the face advances to gain more and more certainty of the predicted structure in front of the face. It does, however, complicate the signal processing required to extract information from the data, and to some extent the data recording, since it may be necessary to record the noise from the cutter at some point close to it in order that a reference signal may be obtained and used to correlate with the signals recorded elsewhere.

By far the most convenient source is a small explosive charge buried some small distance into the coal seam. The advantages that this source provides are many. The onset of the explosion is easily measured, as it is triggered electrically; the frequency range of the signal energy is wide, being initially impulsive, and offers the opportunity in theory to implement high resolution mapping of the seam; it is readily available, as is the expertise required to handle it. Assuming that the explosion is initially a narrow, broadband pulse, there is no requirement to measure its actual form. The errors introduced into the signal processing by this assumption are minor compared with the drastic alteration of the signal as it propagates within the confines of the seam. This filtering effect of the seam is the dominating factor in underground (in-seam) seismic work and will be considered in more detail later.

The duration of signals which need to be recorded after the shot is fired is a little longer than the two-way travel time of the energy over the range which is of interest. With velocities in the region of 2000 m/s and a maximum range of perhaps 1 km, a little over one second's worth of data should suffice. The frequency range in question extends up to about 500 Hz, which determines the sample rate for digitising the recorded signal as being 1 kHz. With a necessarily modest array, perhaps ten detectors, the overall amount of data recorded totals not much more than 10 000 data points, a minute amount of data in comparison to seismic surveys undertaken in, for example, the oil industry. Nevertheless, an accurate representation of the structure of the seam, containing that number of information points over an area of about 1 km², would be of enormous value to the mining geologist.

The final resolution of such a map could be no higher than the wavelength of the energy launched into the seam, a few tens of metres, but it must be remembered that this figure relates to the lateral resolution of the imaged scene. Faulting of the seam by as little as 10 cm or so in the vertical direction is enough to cause a significant perturbation to the propagating waves and constitute a measurable amount of reflected energy which appears in the recorded signal.

The most convenient underground locations at which to test in-seam seismic surveying have been at retreat faces, since access to the panel of coal is available on all four sides and the roadways give some indication as to the condition of the coal. Such a site gives a much freer hand in the geometry of the shot points and geophone array, and in particular allows the measurement of the propagation characteristic through the panel by means of a transmission shot (Fig. 2). The source and detector positioning may be arranged such that any reflections from the other roadways are well separated in time from the direct path. Alternatively, the reflected paths may be used to image one of the roadways, giving an

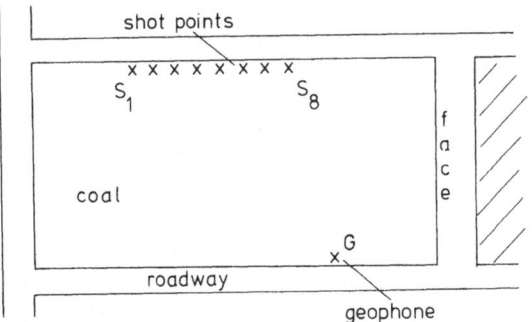

FIG. 2. Transmission shot configuration on a longwall retreat face.

unusual opportunity to reconstruct a map of a scatterer which is at an easily measured location in relation to the shot and geophone array.

Reference has already been made to the difficulty of planting a geophone array with consistently reliable coupling throughout all the geophones. The geophones themselves may be placed in small holes drilled 1 or 2 m into the seam and then packed with sand or some other material. Coupling may be checked by recording the geophones' response to a nearby blow on the coal with a hammer, and it is likely that the response along the array will vary considerably. In the event of having only one or two good geophones in the array, it is possible to 'synthesise' the results of a larger array by using multiple shot points. Referring to Fig. 2, the signals received at a single geophone G in a series of shots S_1–S_8 may also be considered as the signals which would have been recorded at locations S_1–S_8 due to a single shot at G. This is entirely analogous to the conventional surface seismic technique of extending the coverage of an array by using an array of shot points.

The more normal situation of an advancing face, or single roadway, affords only limited access to the coal seam (Fig. 3). In this case, there is

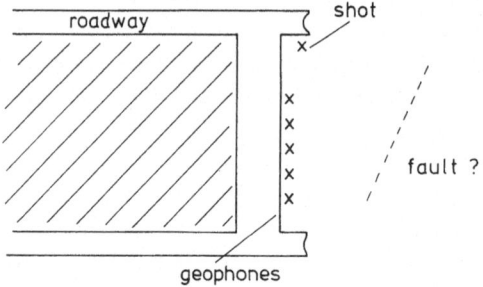

FIG. 3. Reflection shot configuration on a longwall advance face.

no opportunity for a transmission shot to calibrate the seam and recorded signals will be the result of reflections from scatterers within the seam and possibly the direct path between the shot point and geophone array, the underground equivalent of ground roll. The possibilities for reducing the interference by this roadway mode are discussed later.

The techniques for gathering underground seismic data are not, then, very different from those of surface work, although there are some additional difficulties to be overcome in the underground environment. What really distinguishes this work from conventional seismology is the nature of the recorded data due to its transmission and guiding by the coal seam.

3. GUIDED WAVE PROPAGATION

Considerable attention has been paid, over the last 100 years or more, to the characteristics and effects of waves (both acoustic and electro-magnetic) travelling in layered media. An early notable in this field, as in many others, was Lord Rayleigh, whose name is given to a class of waves guided by a single surface layer. A common characteristic shared by guided waves in many forms is that their velocity of propagation is a function of their frequency. This effect is known as dispersion and arises from the propagating waves being constrained in one or more dimension by the geometry of the guiding structure, and having to satisfy certain boundary conditions which the guide might impose. This variation of velocity means that there is not the usual simple inverse relationship between frequency and wavelength.

Dispersion is often represented as a dispersion characteristic, which shows the dependence of wavenumber (inversely proportional to wave-length) and frequency (Fig. 4). For a non-dispersive medium this disper-sion characteristic is a straight line through the origin. The velocity of propagation of a particular frequency (the phase velocity) is taken from the dispersion characteristic by dividing its frequency by its wavenumber. Graphically, this may be represented by the gradient of a line joining

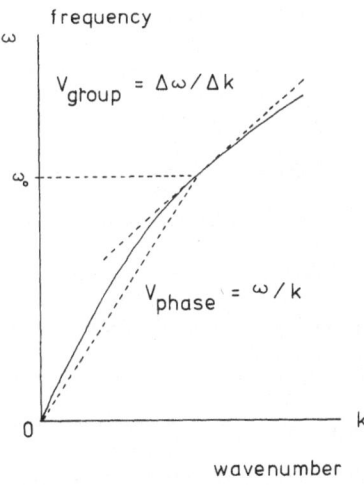

FIG. 4. A dispersion characteristic in wavenumber–frequency space, showing the graphical construction of the phase and group velocities of a particular frequency.

that point on the curve to the origin. It may be shown that a narrow band pulse travels at a velocity (the group velocity) which depends on the slope of the dispersion characteristic at the centre frequency of the pulse, that is, on the gradient of the tangent to the curve at that frequency.

In general, for a dispersive medium, the phase and group velocities are not equal to one another. This means that as a pulse centred on a particular frequency propagates, the phase of the signal within the envelope of the pulse will vary as a function of distance. This effect must be accounted for if dispersed arrivals are to be combined coherently in the signal processing. If the dispersion characteristic is not a straight line, then different groups of frequencies in a wide band pulse will travel with different group velocities. An initially impulsive signal spreads to become a long pulse which contains low frequencies which arrive first and higher frequencies which follow (or high then low frequencies, depending on the type of dispersion). The amount of spreading of the pulse is a function of the distance travelled, or the length of time for which the pulse has been propagating (Fig. 5). It is exactly this situation which complicates the interpretation of underground seismic data.

However, the complexities of the propagation of seismic waves in a seam are not confined to the effect of dispersion. The layered structure of

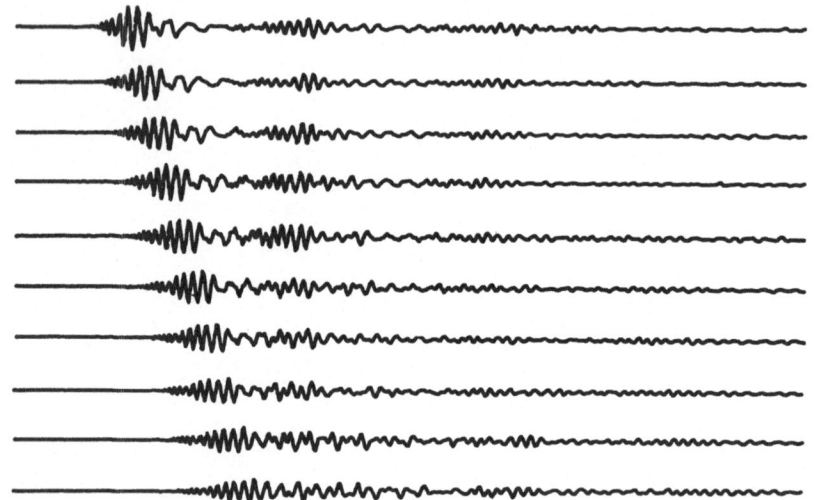

FIG. 5. Dispersed arrivals from an impulsive source and a number of reflectors recorded at several detector positions. (Model experiment on a metal plate.)

rock and coal may support more than one mode of propagation, each with its own dispersion characteristic and each contributing to the received signal. Thus a single reflector in the seam may give rise to more than one echo in the recorded wave train. This effect may confuse the issue enormously, many of the signal processing methods which could be applied to correct the dispersion being valid only in the case of a single propagating mode. The easiest solution to this problem is to suppress, as far as possible, the effects of all but a single mode at the time of recording the signals. This may be done using the knowledge that theoretical studies of wave propagation have produced (Lagasse and Mason, 1975; Buchanan, 1978). A geophone's orientation may be chosen to receive maximum signal strength in the direction of interest and discriminate against other forms of motion. Alternatively, a three-component detector may be used and its various components combined in the processing to produce even greater discrimination against unwanted modes (Millahn and Arnetzl, 1979; Beresford-Smith and Mason, 1980).

Another mode of propagation, which is the equivalent of ground roll, may exist (Lagasse and Mason, 1975). In configurations such as that shown in Fig. 3, a single length of roadway wall links the explosive shot location to the detectors. As well as generating energy into the seam, a mode may also be launched along the roadway. As previously discussed, it is often not possible to place the detectors in the array at a spacing which would reduce signals from this direction, so the careful combination of the three components of the signal may be the most convenient way to reject this mode.

Useful information concerning dispersive wave propagation may also be obtained from model experiments, which have the merit of providing 'real' data (in the sense that it is not generated synthetically, and that it contains noise) at a cost considerably lower than that of data gathered in the field. Additional advantages of model experiments are that the source and scatterer geometry may be varied at will and accurately measured.

Dresen and Freystätter (1976) have reported an extensive series of model experiments to determine the reflection coefficient of scatterers and the detectability of reflected signals. The models used for layers of rock and coal are usually metal or plastics, and acoustic waves are generated and recorded using piezo-electric transducers. The frequencies used are typically 1000 times higher than those in an actual coal seam, and wavelengths are correspondingly shorter.

4. DATA PROCESSING

The identification of a reflecting object in a dispersive medium is complicated because the received waveform is at best a distorted replica of the transmitted pulse; at worst it may be completely unrecognisable, especially if there are a number of reflectors in the field. The signal processing required to reconstitute the originally impulsive signal is not a trivial matter. With a number of reflected signals from different distances, the received signal is a combination of all these reflections which are all dispersed by varying amounts (see Fig. 5). The complications do not end there, since most of the processing methods which could be applied to the dispersed signals require absolute knowledge of the dispersion characteristic of the medium through which they have passed. This information is simply not available to the underground geophysicist. The conventional approach to acquiring this vital data is limited to either measuring it experimentally or calculating it theoretically.

The experimental approach would require a known area of the seam of interest to be free of any embedded discontinuities and for access to be available at two widely separated points. A transmission shot would then provide a model pulse, suitably dispersed, from which the dispersion characteristic might be derived. The major problem of this approach is that a suitable transmission site is not normally available, and if it is, then there is no assurance that it will not contain some feature, such as a small fault, which would completely disturb the transmission path.

The objections to deriving the dispersion characteristic theoretically are equally practical. Well-established analytical methods, such as finite element analysis, are available for calculating the characteristics of many layered structures, but all assume extensive knowledge of the material parameters and dimensions of the structure (Lagasse and Mason, 1975). The elastic properties of coal and its surrounding rock beds are not readily measured and may change significantly in a specimen which is removed from the pressure applied to it 1000 m or so beneath the surface of the earth. Also, the demarcation between the various layers (and there may be very many) of a particular coal seam are often not well pronounced, one layer gradually merging into the next.

These difficulties lead to the abandonment of any direct measurement of dispersion characteristic and the adoption of a rather more empirical approach in which, after a number of approximations, a parametric description of the dispersion law may be derived from the recorded data itself, and used to compress the dispersed waveforms.

The techniques used for removing dispersion from recorded data have varied between different groups working on the problem but have generally become more sophisticated as experience and understanding of the problems have improved. Arnetzl (1969) used a method which accounted for the difference between phase and group velocity but did not attempt to correct for pulse spreading. The calculations were done by hand, in which '. . . the travel path for the central point of the observation line is obtained by multiplying the travel time with the group velocity, while the differences of travel paths for the various geophones are given by multiplication of travel time with the phase velocity'. Paths calculated from the various arrivals were plotted on a map, using ray optics to locate specular reflectors (faults).

By 1971, Krey and Arnetzl (1976) recognised that 'because we are dealing with highly dispersive wave trains, guided by the seam, it is not possible to determine a precise beginning of the reflected seam wave train'. Their method involved the use of a set of measured signals over known paths in the seam, which they used to cross-correlate with reflection data in order to remove the majority of dispersion effects at one particular range of interest, and then some further calculation on the cross-correlation function which gave an estimate of the length of the transmission path.

An obvious extension to this method is the replacement of the measured transmission signals by synthetic signals, calculated using the available estimates of phase and group velocities and other parameters. Transmission path lengths of many values may then be synthesised and the recorded signals cross-correlated with them. Evaluation of the cross-correlation function at zero time difference would give an indication of the amplitude of a reflection (if any) at that particular range. This method has been applied to underground seismic signals (Booer *et al.*, 1976) but suffers from the disadvantage that a synthetic signal has to be calculated for every range cell investigated. More recently, an efficient technique (Booer *et al.*, 1977) for correcting most of the dispersion effects in a signal has been used to process data in such a manner that the resulting data appears similar to that which would have been recorded from the seam were it not dispersive (Mason *et al.*, 1980). This overcomes the difficulty described by Krey and Arnetzl (1976) in which '. . . there are interferences where several reflection signals follow each other'.

The theoretical justification for this method of removing dispersion is outside the scope of this discussion. The process is shown diagrammatically in Fig. 6, and is an operation performed on the data in the

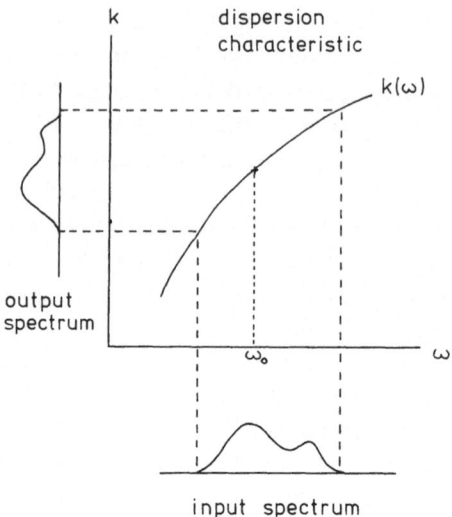

FIG. 6. Diagrammatic representation of a technique to compress dispersed signals by frequency warping.

frequency domain, which may conveniently be calculated for each geophone signal using the Fast Fourier Transform (FFT) algorithm. Each frequency component of the signal spectrum is mapped on to a new frequency axis, the mapping being defined by the dispersion characteristic of the seam. The spectrum so constructed is then inverse transformed (using the FFT) to return it to the time domain, and this results in the required non-dispersed time signals for that geophone.

The dramatic effect of this process on a set of dispersed signals may be seen in Fig. 7, which shows signals recorded from a model experiment on a metal plate. This data contains a direct path from the source to the detectors and also a reflection from one edge of the plate. The signals are considerably confused where these two arrivals are close together. In the corresponding processed data set, the pulse sharpening which occurs on the reflected signals is particularly effective. Similar improvements are possible using data acquired underground as shown in Fig. 8. The first arrival shown in this figure is the direct path between the explosive source and a geophone 250 m away in a seam about 2 m thick. The reflection comes from a roadway to one side of the direct path.

The preceding techniques have all relied to some extent on a knowledge of the dispersion characteristic of the coal seam. Although forming a necessary part of the processing strategy for dispersed seismic data, a

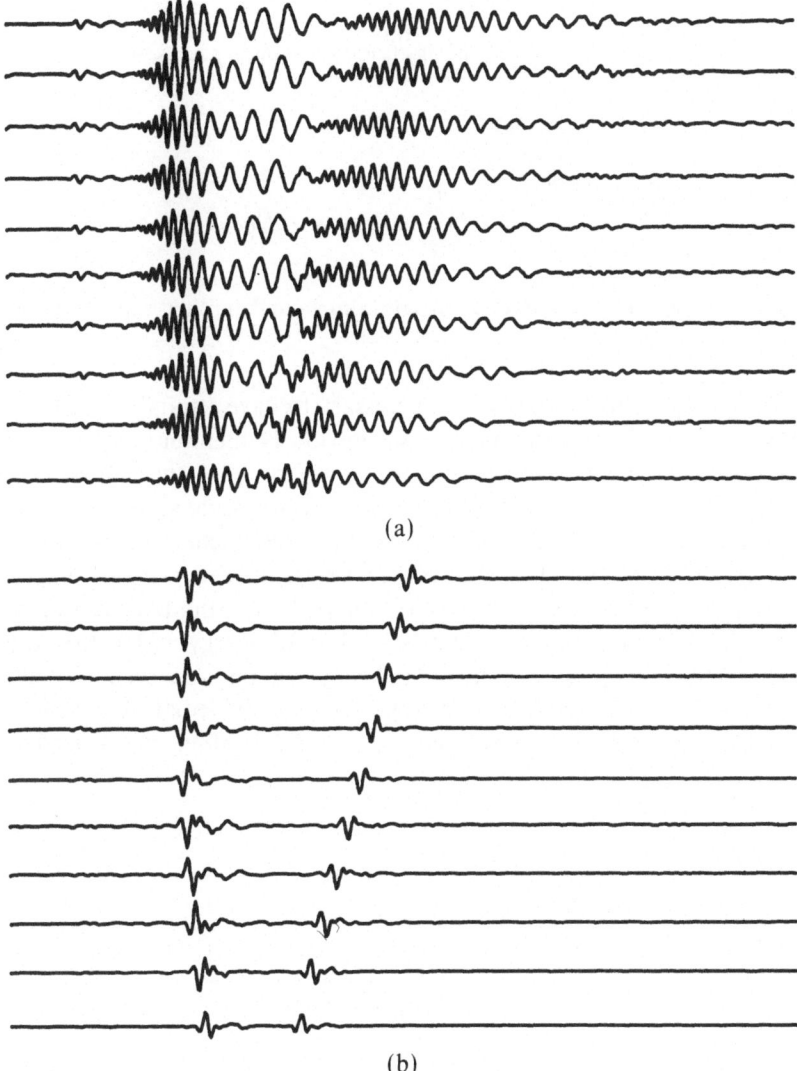

(a)

(b)

FIG. 7. (a) Dispersed and (b) compressed signals from a model experiment.

dispersion correction method in itself is not a sufficient tool to allow successful manipulation of the data. One useful source of information concerning the seam's propagation characteristic is the seismic data itself. Using a technique similar to a velocity filter, as used on conventional seismic data, it is possible to extract the form of the dispersion character-

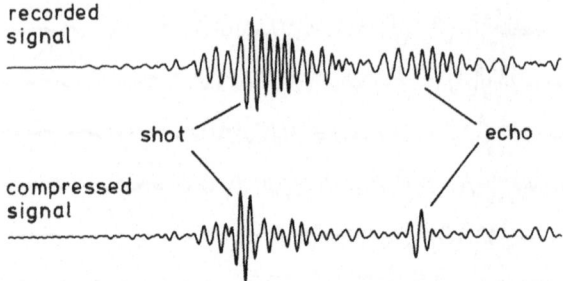

FIG. 8. Dispersed and compressed signals from an underground experiment on a coal seam.

istic. To apply this technique, it is necessary to have a suitable set of data from a transmission experiment. As previously mentioned, this is not always available, but in the situations where it is then it allows a comparison to be made with the results of other techniques.

Figure 9(a) shows a typical transmission experiment configuration, in which the response of a number of different length paths may be recorded using shots at successive locations. Neglecting the effect of loss in signal amplitude due to lateral spreading of energy within the seam, these signals may be regarded as being the impulse response of a dispersive line measured at different points along its length (Fig. 9(b)). In this one-dimensional equivalent geometry, the spreading energy has been

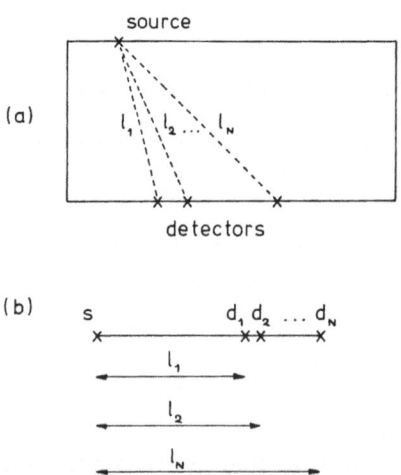

FIG. 9. (a) Configuration of source and detectors for a transmission shot. (b) 'Equivalent' one-dimensional geometry.

sampled in time and at various points in space. This space–time data set may be transformed (using a two-dimensional FFT) into wavenumber–frequency space. It is in this space that dispersion characteristics are plotted, and the transform of a dispersed set of data will result in energy only existing at points on the dispersion characteristic. In the case of a set of data containing more than one mode of propagation, all dispersion characteristics will appear in the frequency–wavenumber space as the transform is a linear operation. This gives rise to the possibility of filtering out just one mode of propagation from the set to use in the mapping of discontinuities in the seam.

Figure 10 shows the result of applying this two-dimensional transform to a set of signals recorded from a seam. The energy of the signal in the wavenumber–frequency plane is plotted as a series of equi-spaced contours. The contours fall into two groups, representing two distinct modes of propagation. Information about the dispersion of the seam can, of course, only be derived from those frequencies in the signal which contain a significant amount of energy; in this case, the information covers nearly an octave from 80 Hz to 160 Hz.

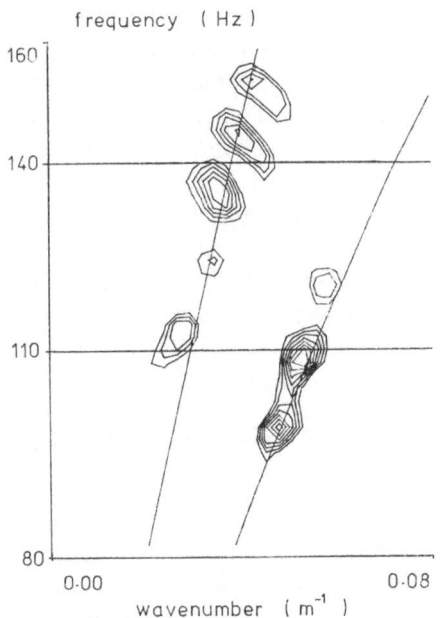

FIG. 10. Two-dimensional FT of multi-mode data.

In an ideal situation, these contours would form very well defined dispersion characteristics, but in a real situation the resolution with which they represent the characteristic is affected by the finite duration of the time history recording, the limited distance over which the waves have propagated, the uniformity of the seam, the number and spacing of the geophones, noise and other minor influences. Taking these factors into account, the results shown in Fig. 10 do indicate that seismic waves propagating a significant distance through a coal seam may be of sufficient fidelity to extract meaningful information about the medium through which they have travelled.

From similar frequency–wavenumber plots it would be possible to estimate a dispersion characteristic over the frequency band of interest to be used in the removal of dispersion. In many situations we do not have the luxury of a suitable transmission data set and it is necessary to consider other methods for extracting the dispersion characteristic from the data. Because of the nature of the seismic source and of the medium through which the signals are passed, the frequency components of the data are limited, as already described, to some finite band. This being so, it is only necessary to derive a dispersion relation which is valid over the same bandwidth.

Early attempts to correct for dispersion only utilised the phase velocity, related to the position of a point on the dispersion characteristic, and group velocity, related to the gradient of the curve at that point. It may be shown that the curvature, or second derivative, of a point on the curve affects the rate at which a band of frequencies will spread into a linear frequency-swept pulse. Such a signal is, to a good approximation, a model of the type of dispersed pulse which is recorded underground. It is reasonable, therefore, to suppose that a quadratic approximation to the dispersion relation over a limited band would be an adequate representation of a characteristic with which to compress the dispersed signals.

A suitable quadratic approximation may be written as

$$k(\Omega) = \mu\Omega^2 + \Omega + \omega_0/p$$

where

$$\Omega = (\omega - \omega_0)/c$$

ω is frequency, ω_0 is the centre of the band of interest, k is wavenumber, μ is 'chirp rate' and p and c are phase and group velocity respectively.

The suitability of this representation is verified by Figs. 7 and 8 in which just such an approximation was used. The problem of defining the dispersion of the seam has now been reduced to evaluating the three parameters phase velocity p, group velocity c and chirp rate μ, so called because it affects the rate at which the frequency sweeps (or chirps) through the travelling pulse. In fact, the problem is somewhat simpler than that, because the phase and group velocities do not affect the overall shape of the pulse group as it travels and may be accounted for in a separate process after the quadratic dispersion has been removed.

It is a straightforward process to select a particular recording of a shot received by one geophone, to identify one arrival group in the recording, and to monitor the behaviour of that arrival as the chirp rate parameter is varied. A particularly simple measurement of the amount by which the dispersed arrival has been compressed may be obtained from the pulse width. This process has been carried out interactively at a computer terminal but could if necessary be automated. Other measurements could be applied to the signal to monitor successful removal of dispersion; in particular 'entropy', a quantifiable measurement of the 'spikiness' of the signal, could be used. Having identified the correct chirp rate, the complete set of data from a seam may be processed to remove dispersion. This processed data may then be further manipulated by one of a number of techniques to produce a representation of the discontinuities within the seam.

Amongst other operations which may be applied to the data are a few cosmetic items which nevertheless contribute to the quality of the final result. Before the removal of dispersion it may be necessary to apply some sort of swept gain to the signal to accommodate the loss in signal strength which occurs as a result of cylindrical spreading within the seam. Other processing which may occur in the time domain is the selection of geophone components to assist in discriminating against a particular mode. Since the process described for removing dispersion from the signals involves a transformation to and from the frequency domain, that is also an appropriate time to apply any conventional filter characteristic to discriminate against unwanted noise, other modes, or to modify the spectrum of the seismic source.

5. PRESENTATION OF DATA

The ultimate goal of in-seam seismic surveying is the presentation of the geophysical structure of the coal seam in a form which may be readily understood by those whose job it is to control the development of the mining operation. The geology inferred from many sources of information is generally coordinated on a plan of the mine at different levels. At present, it requires a high degree of skill from the mine geologist to interpret information available to him. It is desirable to present the additional information supplied by underground seismic tests in a manner which is familiar. It may appear that this requirement demands a presentation which is no different from that of the seismograms produced in surface seismic work, but this is not so.

In conventional seismograms, it is traditional to assume that, to a first approximation, reflections in the data come from a point directly beneath the surface geophone. Subsequent corrections in the data such as migration and stacking are applied to take account of inaccuracies in this initial assumption. Due to the fact that this assumption is often not far from the truth, and also that there is a large amount of data which helps to average out statistical errors, this technique works very well. The modest amount of data collected underground would not provide an adequate representation if treated in the same manner and a different approach to producing a map of an area has to be adopted.

The problems faced in transforming a limited set of time histories recorded at discrete points in space into a representation of a spatial distribution of scatterers are shared with several other areas of current interest. Medical imaging, using acoustic or X-ray radiation, non-destructive testing, multi-static radar, and radio astronomy, to name but a few, have very similar difficulties despite the orders of magnitude which separate the scales of these various problems. It is natural, therefore, that the techniques for mapping underground discontinuities should be based on methods pioneered in these other fields.

The particular characteristics of underground configurations which must be considered in the selection of an imaging technique include the essentially two-dimensional nature of wave propagation, which may actually simplify some aspects of imaging; the dispersive nature of wave propagation, which has to a large extent been removed by pre-processing the data but the difference between phase and group velocity must still be taken into account; and the severe limitation on the number of points in space which have been sampled, which is bound to degrade the fidelity of the map produced. A particularly simple technique which encompasses all these features and has been used to some effect in the reconstruction of underground maps (Mason *et al.*, 1980) is now described.

Consider the situation of a plane containing a single omnidirectional detector and a point source which transmits a pulse at a known time. The pulse, propagating cylindrically from the source, will arrive some time later at the receiver. Using the information of travel time and velocity of propagation, the received pulse may be said to have come from some point which lies on a circle, centred on the detector. Introducing another detector at some different location in the plane will enable the position of the source to be measured uniquely (neglecting a half-plane ambiguity as to which side of the line joining the detectors the source lies) as the intersection of the two circles.

Adding another source, which transmits at the same time as the first, introduces an ambiguity which may be resolved by adding another detector. Identical in effect to adding another source is the presence of a specular, or line, reflector which produces an image of the source at a position which depends on the location and orientation of the reflector. Geological faults in a coal seam have the characteristic of lying along relatively straight lines and may behave as specular reflectors of the explosive source. The complexity of source distribution which may be interpreted correctly depends on the amount of information available about the propagating signals, specifically on the number of detectors. This intuitive method of signal reconstruction may be formalised in the following expression:

$$I(x,y) = |\sum_{n=1}^{N} S_n(d/c)|^2$$

where

$$d^2 = (x - x_n)^2 + (y - y_n)^2$$

I is the intensity distribution of the reconstructed field in the (x, y) plane, $S_n(t)$ is the time signal received by the nth detector at location (x_n, y_n) and c is the group velocity of propagation.

The reconstructed image of a source is therefore made up of a number of intersecting circles, whose amplitude is dependent on the amplitude of each of the signals at the detectors at the corresponding arrival times, as shown in Fig. 11. This is an isometric representation of the intensity distribution $I(x, y)$ due to a single pulse source as seen by six detectors.

In the case of dispersive signals which have been compressed by the technique described previously, there remains a residual correction to apply to account for the difference in phase and group velocities. The expression for $I(x, y)$ may be modified as shown to apply this correction:

$$I(x, y) = |\sum_{n=1}^{N} S_n(d/c) \exp\left\{-j\omega_0 d\left(\frac{1}{c} - \frac{1}{p}\right)\right\}|^2$$

A. K. BOOER

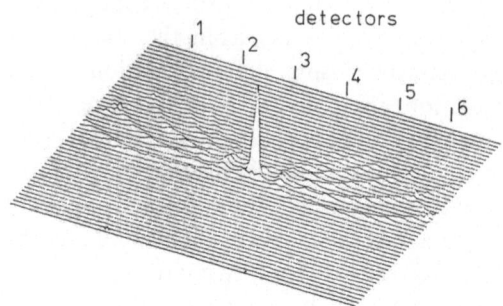

FIG. 11. Reconstruction of an impulsive source as seen by six detectors (synthetic data).

where p is the phase velocity and ω_0 the centre of the frequency band as used in the course of dispersion removal.

This reconstruction technique is useful to apply in the case of a transmission experiment in which the source location is known. The source position may be used to derive the missing phase and group velocity parameters which complete the description of the seam's dispersion characteristic in terms of its approximation as a quadratic function.

In the more realistic case of a signal being due to a single source and its reflection from a number of point, or diffuse, reflectors in the plane, then a simple modification to the above procedure will yield an alternative representation of the distribution of scatterers. A particular time slot in the received signal will contain contributions from all points which give rise to that value of source to reflector plus reflector to detector travel time. The locus of possible points for which this condition holds is in fact an ellipse with foci at the source and detector locations.

Using this reconstruction technique, the source itself will appear as a line joining the source and each detector position. The modification to the above formula required to effect this reconstruction is simply a redefinition of the parameter d in the following manner:

$$d = d_s + d_n$$

where

$$d_s^2 = (x - x_s)^2 + (y - y_s)^2$$

and

$$d_n^2 = (x - x_n)^2 + (y - y_n)^2$$

(x_s, y_s) being the coordinates of the source point in the plane.

Figure 12 shows the appearance of $I(x, y)$ using this reconstruction method applied to the data obtained from a model experiment using a metal plate. The point reflector in this case was a hole drilled through the plate. The reflector is well resolved and imaged at the correct position in the plate. Note the extended appearance of the source, as expected.

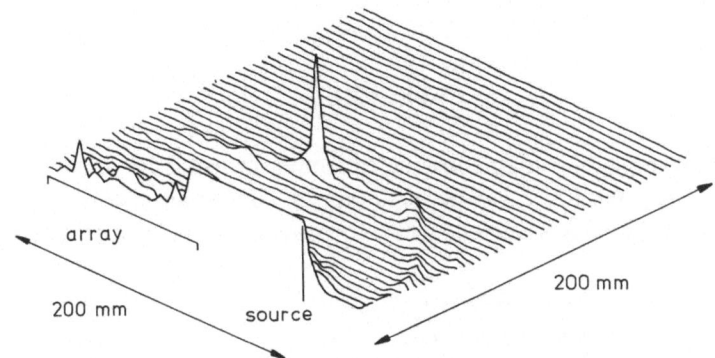

FIG. 12. Reconstruction of source and point reflector (model experiment).

As described, the fundamental limitation to the quality of such a reconstruction is the number of detectors in the receiving array. This is because, in the examples so far presented, the array has been under-sampled spatially. It is not possible to increase the amount of information from a seismic survey to an arbitrarily high level simply by increasing the number of detectors if the array is already sufficiently filled with detectors. Maximum information about the propagating field is, of course, obtained when the geophone spacing is equal to half the wavelength of the highest frequency in the data.

In the reconstruction of data from an undersampled array, ambiguities arise in the form of rings around each of the detectors, as already seen and, dependent on the bandwidth of the signals, diffraction orders which are misplaced images of the real source and scatterers. The resolution with which a scatterer may be imaged is a complex function of the aperture of the array, the source location, the signal bandwidth and geophone positions. The relative weightings which are applied to individual geophones (all assumed equal in the above description) affect the structure of sidelobes which appear in the reconstructions close to the associated objects. It is possible to form a 'super-directive' array in order to improve the image in certain parts. This method is not generally adopted though, since it is very susceptible to noise in the data and small differences in gain in the geophone channels.

The basic limitations, which are common to many applications of imaging techniques, must always be considered in the interpretation of any underground seismic map. Careful study of the map is required to avoid confusion by these artefacts, in order to determine correctly the geological structure portrayed.

6. INTERPRETATION

As an example of a map constructed from a transmission experiment on a panel of coal, Fig. 13 is perhaps typical. This is an instance of a map produced by a synthesised array constructed from a sequence of ten

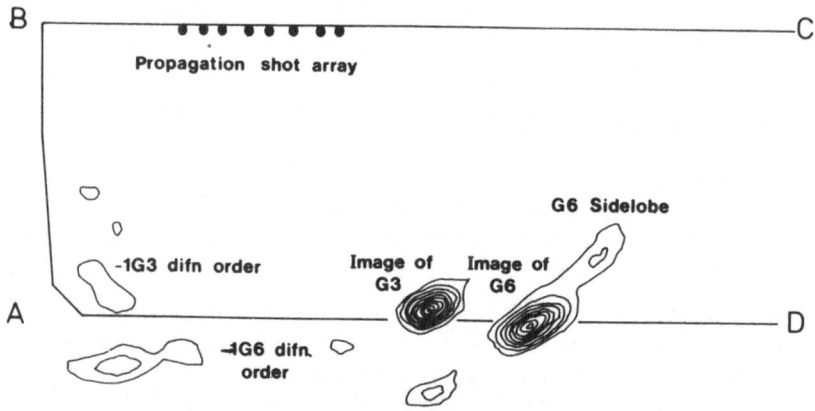

FIG. 13. Transmission map showing the reconstruction of two geophones across a 220 m wide panel of coal.

shots and recordings made by two geophones. Prior to signal processing, the data was grouped into ten traces, being the sum of the two geophone outputs for each of the individual shots. The purpose of the experiment was to examine the homogeneity of the seam, and to verify that the two geophones, spaced 90 m apart (equivalents, now, of two shots at that spacing received by a ten-element array), could be imaged with a resolution comparable to that expected if dispersion had been successfully removed.

The map is presented as a series of contours, spaced equally in intensity, overlayed on a sketch of the roadways which form the edge of the panel of coal. The distance across the panel is about 220 m and the array of shot points forms an aperture of about 160 m. The spectrum of the signal received from the shots, 227 g charges, was centred on about 110 Hz. A quadratic dispersion law was used to compress the dispersed signals and the phase and group velocities were estimated by optimising the intensity of the reconstruction ín the region of one of the geophone images.

For this particular seam, the phase and group velocities, at the nominal centre frequency, were found to be 2025 and 1720 m/s, respectively. These figures indicate the importance of accounting correctly for the difference in these two velocities, since an error of 15% (introduced by assuming that phase velocity is equal to group velocity) would give rise to signals which had travelled over paths differing by only 50 m being combined destructively rather than both contributing to the intensity of the reconstruction.

The resolution predicted for this configuration is about 25 m in range and 35 m in azimuth. These theoretical figures, based on non-dispersive propagation of signals of similar bandwidth to those measured, compare favourably with the reconstructions of the two geophones, using the standard convention of measuring resolution as the width between half intensity points of the image. The positional accuracy of the images is also excellent, the peaks being within 5 m of their measured positions on the roadway. The overall clarity of the reconstruction provides evidence as to the homogeneity of the seam and the absence of discontinuities in the zone covered by the transmission paths. The other low level peaks which are visible may be identified with one or other of the ambiguities which would be expected in an image of this type. The peaks close to the geophone images are sidelobes due to the limited aperture of the array, and those on the opposite side of the array from the geophones are first-order diffraction images due to the shot spacing (about 20 m).

The particular site characterised by the transmission experiment described above was also the location of a more realistic (in an operational sense) reflection experiment. The seam in question actually supported two propagation modes, which could be well separated by frequency domain filtering and geophone orientation. The slow mode as utilised in the transmission map of Fig. 13, contains a significant proportion of its energy within the layer of coal itself. Its low velocity is essentially determined by the physical characteristics of the coal, a relatively soft material. The other mode identified, although guided by the coal seam, contains rather more energy in the surrounding rock layers. This is to say that the particle displacement of the wave as it travels extends some distance into the rock, and the physical properties of the rock play an important part in the mode's behaviour. The velocity of the mode tends towards that of freely propagating waves in the more rigid surrounding medium. Modal behaviour of waves in coal seams has been characterised by Lagasse and Mason (1975), amongst others, and a knowledge of the elastic displacements of a mode is necessary for the correct interpretation of details in maps such as those shown here.

A reflection map of the area, produced using this faster mode, is shown in Fig. 14. The array consisted of only six geophones, spaced at 30 m

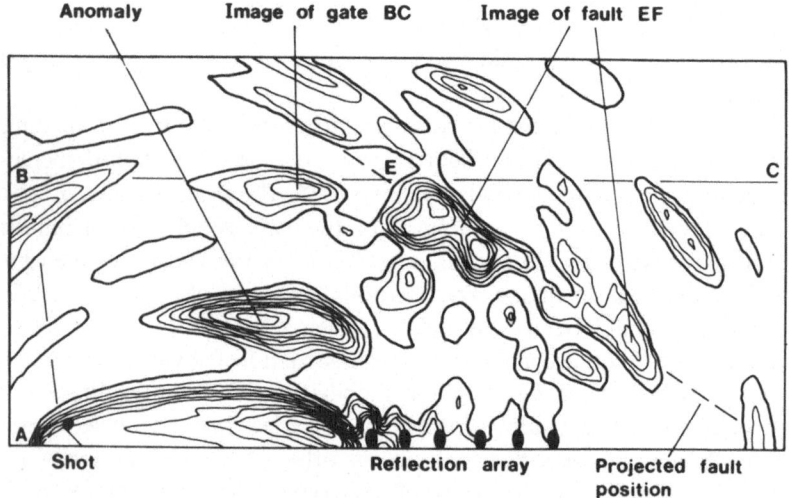

FIG. 14. Reflection map showing the reconstruction of the source point, a roadway, a fault and an anomaly.

intervals, and the data was gathered from a single shot at a corner of the panel of coal as shown in the diagram. The band centre was at 140 Hz and the phase and group velocities of this mode were found to be 3650 and 3200 m/s, respectively. Due to the rather fewer number of geophones in this array, compared to the ten used in the transmission map, the level of ambiguities in the reconstruction is inevitably greater. Since the map is based on travel times from the source to the reflection point and back to the receiving array, the appearance of the ambiguities is also somewhat different.

As expected, the source appears as an extended ridge joining the source location to the receiving array. For configurations such as this, in which the source and array lie approximately along a line, the signal received directly from the shot is inevitably the strongest component of the received waveform. It does not interfere unduly with the signals received from reflections in the seam, being spatially well separated from the rest of the reconstruction. It is possible, as a purely cosmetic measure, to remove the shot from the map by applying a time window to the recorded data. This will introduce a 'blind' area within an ellipse centred on the shot and geophone positions.

This particular map is of special interest for several reasons. The panel of coal was suspected to have a fault, of about 1 m vertical throw, cutting diagonally across the panel as indicated by the broken line on the diagram. The grounds for this belief were that the fault was observed as cutting across the two roadways either side of the panel and was also known to exist for some distance as the panels of coal either side had been previously extracted and found to contain this fault line. The severity of the fault was such that it was not considered to be a sufficient hazard to warrant abandoning the face.

The location of the fault is clearly indicated in the reconstruction. Its position in the map is in agreement with its suspected location, lying almost directly in front of the geophone array. The orientation of the fault is such that it is clearly illuminated by the shot and it is well imaged by the array. Despite the relatively high level of ambiguities it is, perhaps, not too fanciful to assume that the image of the roadway opposite the array also appears in the map. If this is so, it is reasonable to question why this well defined boundary between air and rock does not appear so strongly as the rock – rock boundary of the fault. A possible explanation of this feature of the reflection map lies in the energy distribution of the mode. This high speed mode is, as described, supported mainly by the rock around the seam. The roadway, cut alongside the coal, extends only a small distance into the rock above the seam. The fault, on the other hand, may extend some considerable distance vertically above and below the seam. The discontinuity presented by the fault to the propagating wave may therefore be significantly greater than that shown by the roadway.

Imaged with the slower mode, bounded more closely by the coal seam, the roadway might be expected to appear stronger than the fault. The choice of mode is, then, an important consideration in the mapping of faults.

Another point of interest in the map is the presence of an unexpected feature, marked as 'anomaly' on the diagram. This major feature, approximately in the centre of the panel, cannot be identified as the image of any known geological structure in the seam. It is interesting to note that it did not appear in any form on the transmission map of Fig. 13, although it might not be expected to appear strongly if it does not correspond to a specular reflection of the shot point; neither is it readily identifiable with any ambiguity of the imaging process. A possible explanation of this feature is that it is due to a propagation anomaly of the mode in that area of the seam. Its absence from the transmission map may well be due to the different effect of such a region of stress on the other mode. The most attractive explanation is, of course, that this anomaly is indeed a small, hitherto unexpected, fault in the coal panel, although it may be premature to suggest this on the evidence of such a small amount of data. It is unfortunate that there is no easy way to verify this explanation. The ultimate proof would naturally lie in the extraction of this panel and a discovery of a geological feature at the predicted position. The presence of a fault of such limited extent would not be reason enough to avoid mining this area, and the panel is planned to be worked. Time will tell if this last explanation is the correct one.

CONCLUSIONS

In summary, the examination of the geophysical structure of coal seams, using guided seismic waves, presents several problems in the areas of data collection, processing and presentation. The major difference between in-seam and conventional seismic data is dispersion, which spreads broadband pulses into long, frequency-swept arrivals. Restricted access to the coal seam limits the quantity of data available for processing.

Dispersive propagation is well understood, and methods for removing its effects from recorded data may be applied once the dispersion characteristic of the mode has been derived. Limited sets of data have required special data stacking methods to produce maps of geological discontinuities within the seam. The susceptibility of guided waves to variations in seam characteristics have, at the same time, made them suitable tools for probing seam structure but difficult to interpret unambiguously.

The difficulties of interpretation may to some extent be overcome as experience in examining reflection maps increases. This application of seismic techniques is relatively new and promises to be a useful additional source of information about the geophysical structure of coal seams.

ACKNOWLEDGEMENTS

For my introduction to, and subsequent education in, this subject whilst working at University College, London and the Department of Engineering Science at the University of Oxford, I have to thank Dr I. M. Mason of St. John's College Oxford and Dr D. J. Buchanan of the National Coal Board.

I am especially indebted to F. F. Evison of the Institute of Geophysics at the Victoria University of Wellington for drawing my attention to his early work in the field.

Figures 13 and 14 are taken from Mason *et al.* (1980) with permission of the Society of Exploration Geophysicists.

This contribution is published with the kind permission of my current employer, the Plessey Company Ltd.

REFERENCES

ARNETZL, H. (1969). Seismos GmbH Report, Hannover.

BERESFORD-SMITH, G. and MASON, I. M. (1980) Seismic imaging of faults in multi-moded coal seams, *Proc. IEEE Ultrasonics Symp.*, CH1559–4/80, pp. 107–10.

BOOER, A. K., CHAMBERS, J. and MASON, I. M. (1977). Fast numerical algorithm for the recompression of dispersed time signals, *Electronics Letters*, 13, 453–5.

BOOER, A. K., CHAMBERS, J., MASON, I. M. and LAGASSE, P. E. (1976). Broadband wavefront reconstruction in two-dimensional dispersive space, *Proc. IEEE Ultrasonics Symp.*, pp. 160–2.

BUCHANAN, D. J. (1978). The propagation of attenuated SH channel waves, *Geophys. Prospecting*, 26, 16–28.

DRESEN, L. and FREYSTÄTTER, S, (1976). Rayleigh channel waves for the in-seam seismic detection of discontinuities, *J. Geophysics*, 42, 111–29.

EVISON, F. F. (1955). A coal seam as a guide for seismic energy, *Nature*, 176, 1224–5.

KREY, T. C. (1963). Channel waves as a tool of applied geophysics in coal mining, *Geophysics*, 28, 701–14.

KREY, T. C. and ARNETZL, H. (1976). Paper at EAEG meeting, Hannover.

LAGASSE, P. E. and MASON, I. M. (1975). Guided modes in coal seams and their application to underground seismic surveying, *Proc. IEEE Ultrasonics Symp.*, CH0994–4SU, pp. 64–7.

MASON, I. M., BUCHANAN, D. J. and BOOER, A. K. (1980). Channel wave mapping of coal seams in the United Kingdom, *Geophysics*, **45**, 1131–43.

MILLAHN, K. D. and ARNETZL, H. (1979). Analysis of digital in-seam reflection and transmission surveys using two components. Paper at 41st EAEG meeting, Hamburg.

SALAMON, K. D. G. and WIEBOLS, G. A. (1974). Digital location of seismic events by an underground network of seismometers using the arrival times of compressional waves, *Rock Mechanics*, **6**, 141–66.

ZIOLKOWSKI, A. (1979). Seismic profiling for coal on land, in *Developments in Geophysical Exploration Methods—1*, Ed. A. A. Fitch. Applied Science Publishers, London, pp. 271–306.

Chapter 2

INTERRELATIONSHIP OF RESISTIVITY AND VELOCITY LOGS

ALBERT J. RUDMAN

Indiana University, Bloomington, In., USA

SUMMARY

The wide range of velocities associated with earth materials makes it difficult to relate such data to specific lithologies. A review of geologic parameters shows that depth, lithology and age are not as specifically related to velocities as are the physical parameters of density, elastic constants, pressure, temperature and porosity. The Nafe and Drake curves relating velocity to density and porosity are seen as the most effective theoretical–empirical equations in the present literature.

Although conventional resistivity data has often been related to lithology, depth and age, its dependence on porosity is quantitatively more significant. Resistivity is related to porosity through Archie's law; however, several other variations of this relation now are available in the literature. Log analysts, for example, often prefer working with the formation factor (the ratio of bulk resistivity to fluid resistivity) when studying porosity.

Equations relating velocity and resistivity to porosity form a basis for representing transit times in terms of apparent resistivity (Kim's scale function). Application of this predictive relation is highly successful in generating pseudovelocity logs from readily available resistivity log data.

1. INTRODUCTION

Seismic exploration today far exceeds all other geophysical techniques in terms of financial expenditure and numbers of people involved. Its popularity is linked to its high resolution and great depth of penetration, despite the fact that seismology only reveals how velocity varies with depth. Upon completion of any seismic survey there remains the problem of interpreting the velocity variations in terms of the underlying geologic structure. It follows that an understanding of velocities is essential to seismic exploration.

A common method of obtaining velocity information about the sub-surface involves utilising the seismic data itself. Modern exploration seismology involves common depth point (CDP) methods, where the same subsurface point is sampled many times. Such multiple coverage, under known geometric conditions, permits the calculation of CDP-derived velocities (Al-Chalabi, 1979). Without reviewing the method, it should be intuitively clear that CDP velocities cannot match the ac-curacy and resolution of velocities obtained by direct measurements in wells; hence, there remains a need for obtaining velocity information from well logging techniques.

Many, but not all, exploration wells being drilled today include some kind of continuous velocity logging (CVL). In addition to the basic velocity data, the log expert can extract from the data information about lithology, fluid content, porosity and over-pressured zones (Kennett, 1979). Although velocity information has been the primary goal for geophysicists, the primary geologic application of the CVL has been the determination of porosity of limestone and sandstone sections. In gas storage studies, for example, the CVL has been vital in providing porosity control.

Unfortunately the CVL is not obtained conventionally even today. This has been even more true in the past. In Indiana there are less than 1000 CVL logs available out of a total of 15 000 wells on file at the Geological Survey. Almost all of these wells, however, do have some type of electrical resistivity log, most often the conventional (16 in) short normal type. It follows that there are obvious needs for a method to extract velocity information from resistivity logs. It is the purpose of this chapter to examine how one might generate a pseudovelocity log from resistivity logs.

The close relationship of CVL logs and resistivity logs is usually obvious with only the briefest visual comparison. As an example, we compared two logs from a basement test well in Indiana (Fig. 1). The well penetrated 4500 ft of limestone, sandstone and shale ranging in age from Devonian to Precambrian. The two logs are similar, although there are some differences on careful study. At 3800 ft there is a marked change in the resistivity amplitudes, but only a slight change in the CVL. Nevertheless, the close correlation between these well logs encourages one to consider the viability of relating the two kinds of logs.

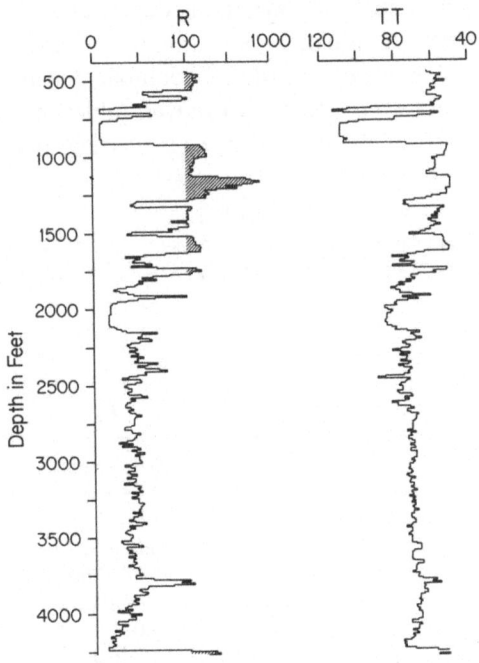

FIG. 1. Comparison of short normal (16 in) resistivity log with continuous velocity log (CVL). Resistivity R is in ohm-m and transit time TT is in μs/ft. (From Rudman et al., 1975.)

The comparative study of velocity versus resistivity presented in this chapter follows two separate approaches. First, the theoretical basis is examined to determine the parameters that control the observed resistivity and velocity. Some mathematical expressions are developed to relate the two quantities. Second, a summary of experimental evidence is presented in order to relate the data in an empirical fashion, independent of any theoretical considerations.

In all of the following discussion it is presumed that the reader is familiar with the basic logging methods, including their limitations and objectives (see, for example, Chapters 5 and 6).

2. VELOCITY

This review of velocity of earth materials emphasises the interrelationships of velocity with such parameters as depth, age of rocks, lithology, density and porosity. A similar review follows for electrical resistivity.

Any study of velocities is immediately complicated by the wide range of overlapping values encountered. Moreover, tables of average velocities (Table 1) do not reflect the dependence on physical parameters such as

TABLE 1

COMPRESSIONAL VELOCITY (V_p) AND SHEAR VELOCITY (V_s) AT LOW PRESSURES AND NORMAL TEMPERATURES OF NEAR-SURFACE ROCKS (SUMMARISED FROM PRESS, 1966)

Material	Velocity (km/s)	
	V_p	V_s
Syenite, trachyte	5·41–5·53	3·05
Granite	4·8–5·88	2·94–3·23
Monzonite, granodiorite, diorite, andesite, gabbro, diabase, basalt	4·6–6·69	2·72–3·56
Anorthosite, norite, eclogite, dunite	6·18–8·60	3·24–4·37
Sandstone, shale	1·4–4·4	2·75–3·59
Anhydrite, salt, gypsum, chalk	2·0–5·0	1·07–2·99
Limestone, dolomite	1·7–6·9	2·75–3·59

mineralogic composition, fluid content, temperature, pressure, grain size, cementation, anisotropy with respect to bedding or foliation, and alteration (Press, 1966). Another complication involves the differences that are observed between *in situ*, well log and laboratory measurements. Field measurements are generally based on explosive sources, with typical wave frequencies of 10–100 cycles/s. In contrast, laboratory and borehole measurements use resonance and pulse techniques and are based on higher frequencies (10^2–10^7 cycles/s). Details of log measurements and how one adjusts velocities to *in situ* values are discussed in Chapter 6.

2.1. Velocity Versus Depth, Lithology and Age

Interest in seismic velocities began as early as 1935 when Weatherby and Faust (1935) studied the relationship of velocity to geologic age and depth. Faust (1951) studied data from 500 wells, primarily sandstones and shales, and established a quantitative relationship of velocity V to depth Z and age T:

$$V = 125 \cdot 3 \ (ZT)^{1/6} \tag{1}$$

where V is in ft/s, Z is in ft and T is in years. The appearance of the 1/6 power as a function of depth Z or pressure difference Δp has theoretical basis (Gassman, 1951; Acheson, 1963) and is demonstrable in hexagonal packing of spheres to simulate porous media (Fig. 2). It was apparent that such geologic factors as age and depth were not sufficient to predict detailed velocity variations. Therefore, researchers also investigated the role of such physical factors as density, pressure, temperature and porosity.

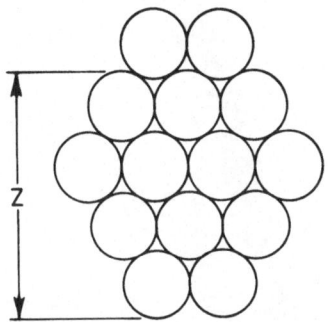

FIG. 2. Model granular aggregate, used by Gassmann (1951) in mathematical demonstrations of the formula for the velocity of sound in porous media: $V = CZ^{1/6}$. (Modified from Acheson, 1963.)

2.2. Velocity Versus Density, Pressure and Temperature

Ludwig *et al.* (1970) plotted experimental data for compressional velocities V_p and shear velocities V_s against observed densities ρ for unconsolidated sediments, sedimentary, metamorphic and igneous rocks (Fig. 3). Although these data are in wide use today, the solid lines simply define the general trend of the sample set and are not considered a functional relationship for velocity–density. The straight line fit at the high end of the plot (mantle rocks) is derived from Birch's (1964)

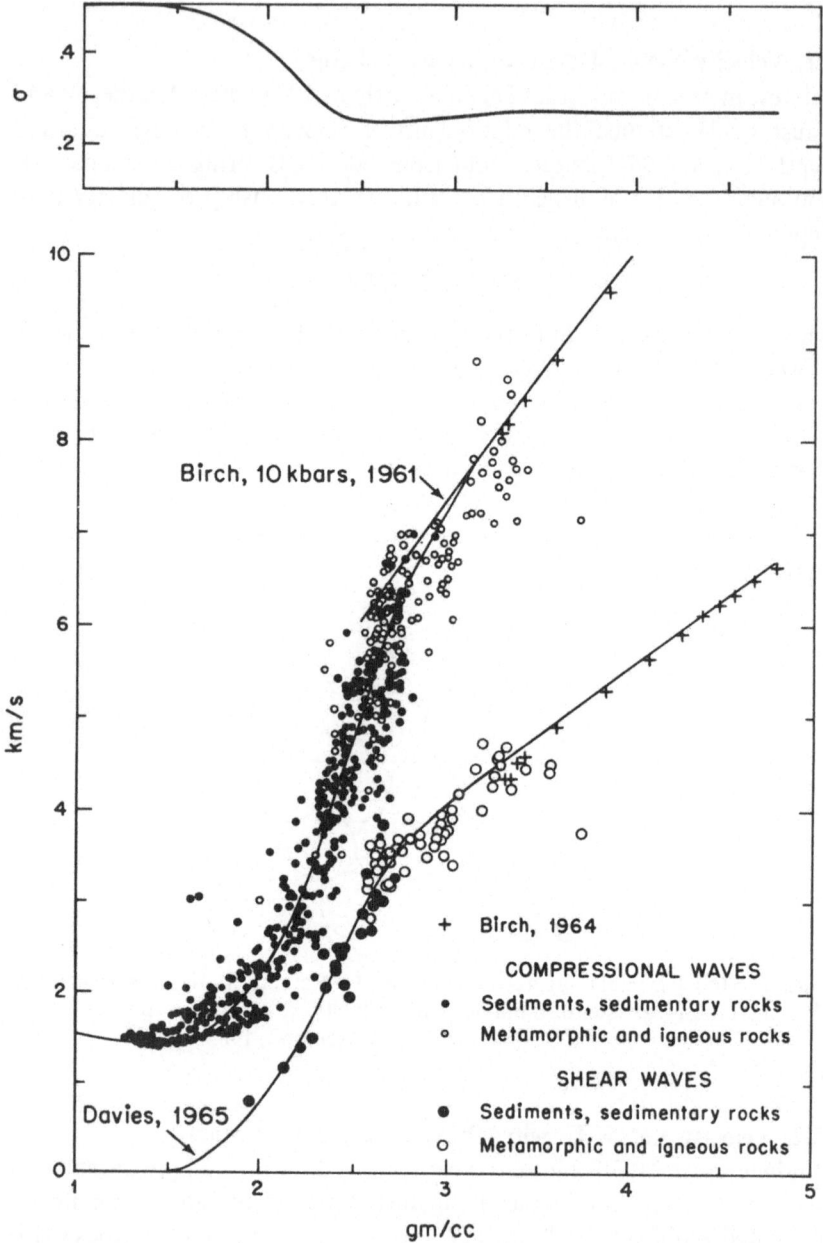

FIG. 3. Experimental data for compressional velocity V_p and shear velocity V_s in unconsolidated sediments, sedimentary, metamorphic and igneous rocks plotted against the observed densities. The curve of Poisson's ratio σ was drawn from the smooth curves of V_p and V_s. (From Ludwig *et al.*, 1970.)

relationship:

$$V_p = -0.98 + 2.76\rho \tag{2}$$

The plot of Poisson's ratio σ at the top of Fig. 3 is computed from the ratio V_p/V_s according to the relation

$$V_p^2/V_s^2 = 2(1-\sigma)/(1-2\sigma) \tag{3}$$

Hamilton (1980) summarised the available data for velocity–density relations in the sea floor (Fig. 4) and noted that compressional velocity does not usually identify rock type. For example, at 4·0 km/s the rock could be any of the six types listed in the caption. However, at higher pressures (above a few kilobars) systematic correlations become pronounced. Birch (1961) has shown that velocity is linearly related to density for materials with a common mean atomic weight (Fig. 5). At those higher pressures the porosity is sufficiently reduced that one may observe the intrinsic velocity of the material.

The problem of adjusting velocities observed in the laboratory to those expected *in situ* has been considered by Boyce (1976). His equation compensates for hydrostatic pressure and temperature of a mixture of seawater and calcite:

$$V_{III} = [(V_3 - V_2)(V_I - V_{II})/(V_1 - V_2)] + V_{II} \tag{4}$$

where V_1 = velocity of pore water at laboratory conditions; V_2 = velocity of calcite (or other mineral) at laboratory conditions; V_3 = velocity of sample at laboratory conditions; V_I = velocity of pore water at *in situ* temperature and hydrostatic pressure; V_{II} = velocity of calcite (or other mineral) at *in situ* temperature and hydrostatic pressure; V_{III} = velocity of sample at *in situ* temperature and hydrostatic pressure.

2.3. Velocity Versus Porosity

The previous paragraphs have demonstrated the dependence of velocity on density. Density, in turn, is linearly related to porosity:

$$\rho = \rho_1\phi + \rho_2\phi_2 \tag{5}$$

where the subscripts 1 and 2 refer to fluid and particles respectively and ρ = wet bulk density, ϕ = porosity and $\phi = 1 - \phi_2$. A plot of densities versus porosities (Fig. 6) shows that a variety of rock types and sediments all lie along the same line given by

$$\rho = 1.0\phi + 2.65\phi_2 \tag{6}$$

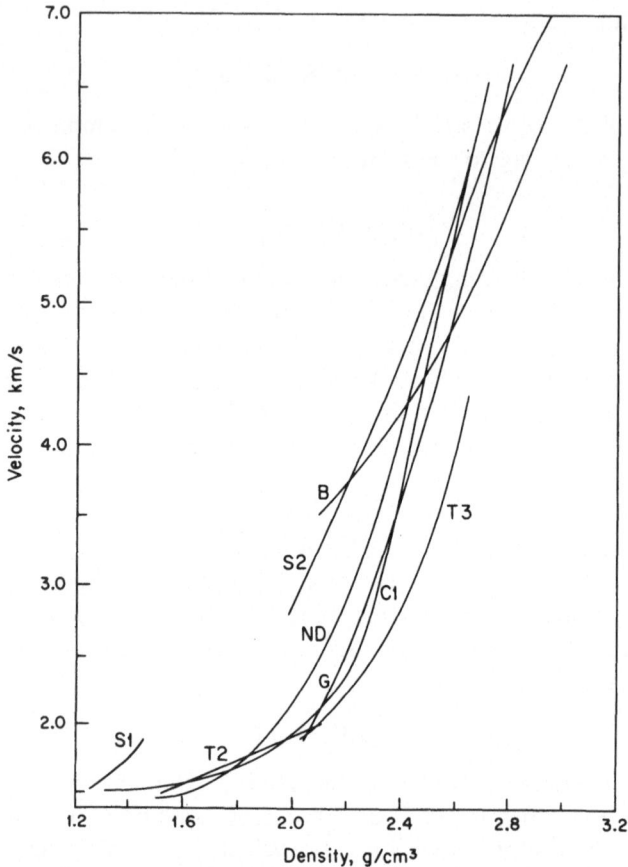

FIG. 4. A summary of compressional wave velocity versus density in various sediment and rock types of the sea floor, plus general curves from the literature. The letters beside the curves denote the sediment or rock types: Sl—diatomaceous sediments (0–500 m); S2—siliceous rocks; T2—turbidites (0–500 m); T3—mudstones, shales; C1—chalk, limestone; B—basalt. The general curves of Ludwig *et al.* (1970) (labelled ND) and Gardner *et al.* (1974) (labelled G) are included for comparison. See Hamilton (1978) for references, discussions and regression equations. (Modified from Hamilton, 1980.)

It would appear that the particle density ρ_2 is not a controlling factor in determining wet bulk density in porous rocks.

Because compressional velocity is related to the total bulk composition of a rock, it is thought that porosity plays a key role in velocity determination. Moreover, the extreme variability observed in velocities

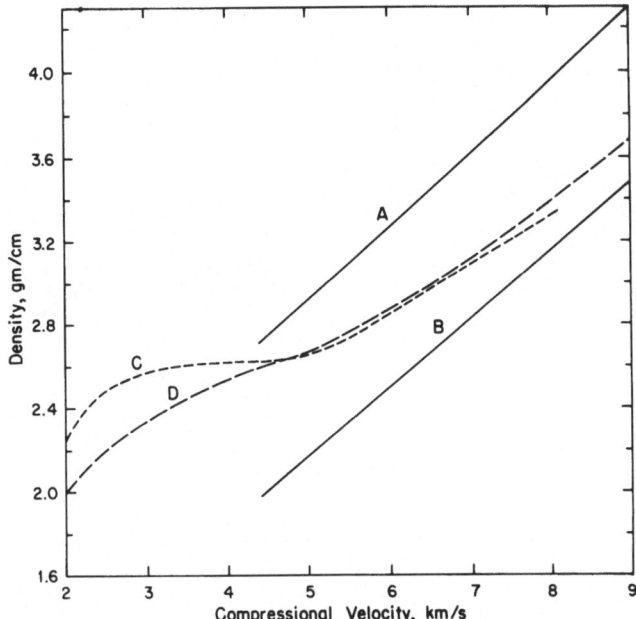

FIG. 5. Empirical velocity dependence on density and mean atomic weight. Solid lines are from Birch's (1961) laboratory results at 10 000 bars: A and B are for mean atomic weights of 26 and 21, respectively. Dashed lines are observed data from C Woollard (159) and D Nafe and Drake (1957) (as quoted in Talwani, M., Sutton, G. H. and Worizel, J. L., *J. Geophys. Res.*, **69** (1959) 1545–56 (Modified from Press, 1966.)

at low pressures and temperatures (Table 1) also reflects the wide range of porosities observed in rocks at low pressures. Plotting velocities as a function of porosities (Fig. 7), a wide range of porosities are indeed observed. Nafe and Drake (1963) compared the field and laboratory measurements in Fig. 7 with the theoretically predicted values of several authors (Fig. 7).

For example, Wood (1941) related the compressional velocity V of material in suspension to the fractional porosity $= \phi$, $\phi_2 = (1 - \phi)$, compressibility β and density ρ, where subscripts 2 represents the solid (suspended) material, 1 the interstitial water and b the bulk (combined) material:

$$V_b = [(\phi \beta_2 + \phi_2 \beta_1)(\phi \rho_2 + \phi_2 \rho_1)]^{-1/2} \qquad (7)$$

where

$$\phi \rho_2 + \phi_2 \rho_1 = \rho_b$$

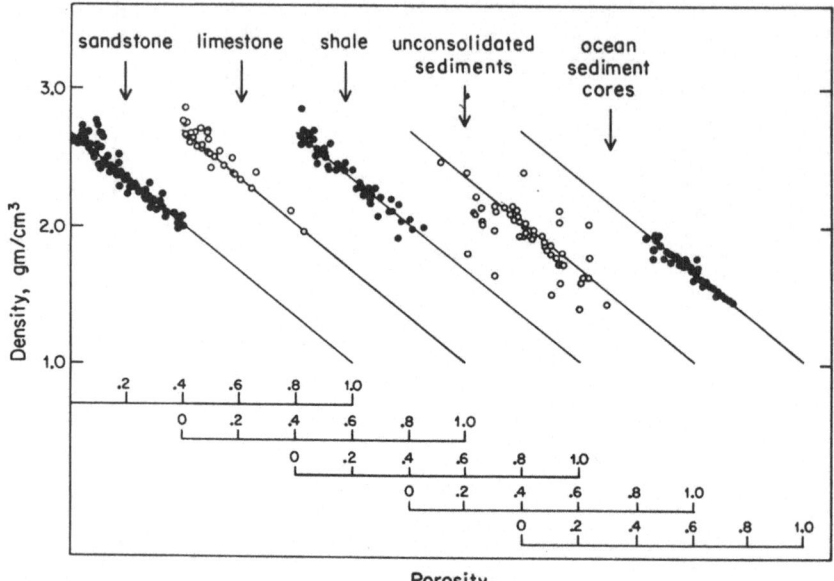

FIG. 6. Density as a function of porosity. Data for each type of material is compared with the same line $\rho = 1\cdot00\phi + 2\cdot68\phi_2$ where $\phi =$ porosity and $\phi_2 = 1 - \phi$. (Modified from Nafe and Drake, 1957.)

Wyllie et al. (1956) developed a similar relationship for consolidated sediments:

$$1/V_b = \phi/V_1 + (1 - \phi)/V_2 \tag{8}$$

From Fig. 7 we see that the Wood equation appears to be an approximate lower limit of velocity for a given porosity. If the velocity is higher than predicted by the Wood equation, the increase is attributed to an increase in the elastic constants β (bulk modulus or incompressibility) and μ (shear modulus or rigidity). Geologically this would represent an increased compaction or cementation. The basic equation relating compressional velocity and elastic constants is

$$V_p = \sqrt{(k + \tfrac{4}{3}\mu)/\rho_b} \tag{9}$$

The Wyllie et al. equation is considered to be a fair representation of the main trend of velocities for consolidated sediments, although Boyce (1976) points out that it represents an upper limit to the velocity–porosity relationship. The closest theoretical–empirical equation to pre-

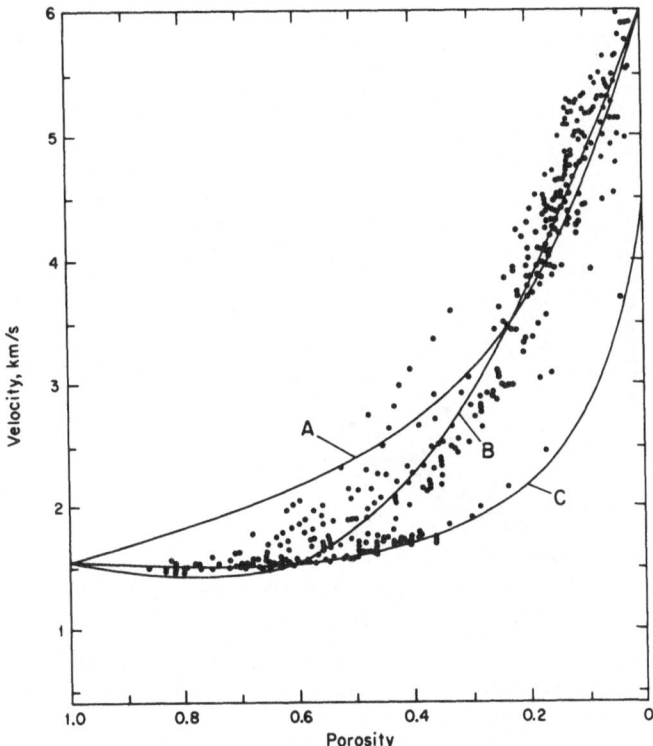

FIG. 7. Observations of compressional velocity as a function of porosity. Data are compared with (A) the Wyllie *et al.* (1956) equation, (B) an empirical equation of Nafe and Drake (1963) and (C) the Wood (1941) equation. (Modified from Nafe and Drake, 1963.)

dict the velocity–porosity relationship in marine sediments was developed by Nafe and Drake (1957; 1963):

$$V_b = \phi(V_1^2[1 + \rho_1/\rho_2]\phi_2) + (\rho_2/\rho_b)\phi_2^N V_2^2 \tag{10}$$

where setting N equal to 4 and 6 generates two functions that act as upper and lower bounds for most of the observed points in Fig. 7.

The problem of acoustic anisotropy must also be considered if porosity (and eventually velocity) are to be estimated from a CVL or resistivity log (Boyce, 1980). A scatter diagram of vertical versus horizontal velocities (Fig. 8) demonstrates the anisotropic nature of sediments and sedimentary rocks.

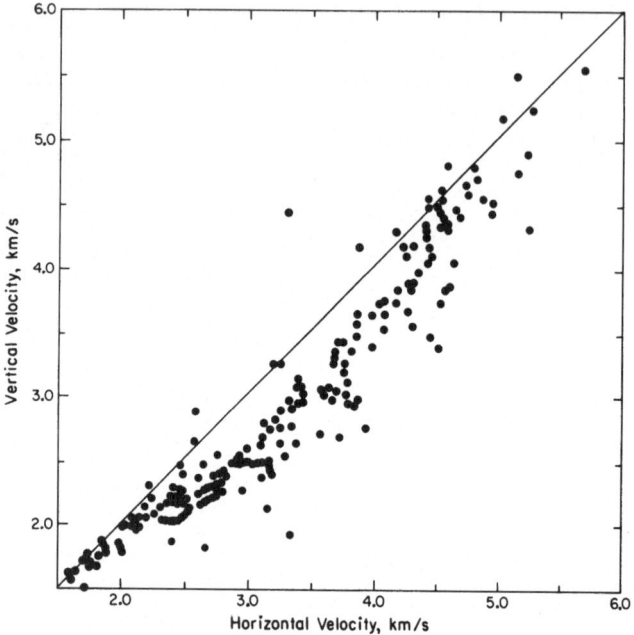

FIG. 8. Laboratory vertical velocity versus laboratory horizontal velocity for oceanic sediments. (Modified from Boyce, 1980.)

3. ELECTRICAL RESISTIVITY

Resistivity, the reciprocal of conductivity, is defined as the resistance r of a material to current flow. For a cube of length L and cross-section A, the resistivity R is given by

$$R = r A / L \tag{11}$$

where R is conventionally measured in ohm-m. Conduction in rocks is through the interstitial water in pore spaces; conduction through mineral grains occurs only if highly conductive materials are present (e.g. magnetite, graphite or pyrrhotite), or if there is no effective porosity. Hence, porosity and fluid content are the key parameters related to resistivity. (For additional background the reader is advised to review the role of well log measurements in Chapter 5).

3.1. Resistivity Versus Lithology, Age and Temperature

There is an apparent relationship between resistivity and lithology and age of rock, since these two factors are related to porosity and fluid content (Keller and Frischknecht, 1966). Although there are many exceptions, the values in Table 2 demonstrate this dependence on geologic age. Figure 9 groups sedimentary rocks of similar characteristics to further examine the role of age. Note that these resistivity measurements were made at radio frequencies and these are generally lower than DC measurements.

TABLE 2

GENERALISED RESISTIVITY RANGES FOR ROCKS OF DIFFERENT LITHOLOGY AND AGE (FROM KELLER AND FRISCHKNECHT, 1966)

Age	Marine sedimentary rocks	Terrestrial sedimentary rocks	Extrusive rocks (basalt, rhyolite)	Intrusive rocks (granite, gabbro)	Chemical precipitates (limestone, salt)
Quaternary and Tertiary age	1–10	15–50	10–200	500–2000	50–5000
Mesozoic	5–20	25–100	20–500	500–2000	100–10000
Carboniferous Paleozoic	10–40	50–300	50–1000	1000–5000	200–100000
Early Paleozoic	40–200	100–500	100–2000	1000–5000	10000–100000
Precambrian	100–2000	300–5000	200–5000	5000–20000	10000–100000

At temperatures well over 100°C, rocks are essentially water-free and resistivity is governed by mineral content according to the relationship

$$R = a \exp(b/T) \tag{12}$$

where a and b are constants and T is the absolute temperature. For fluid-filled rocks (Keller, 1966) the resistivity for temperatures T between the freezing and boiling points of water is given by

$$R = R_{20} \exp(-0.22(T-20)) \tag{13}$$

where R_{20} is the resistivity at 20°C.

3.2. Resistivity Versus Porosity and Formation Factor

In fluid-filled rocks the bulk resistivity R is related to porosity ϕ via an empirical equation developed by Archie (1942):

$$R = a R_w \phi^{-m} \tag{14}$$

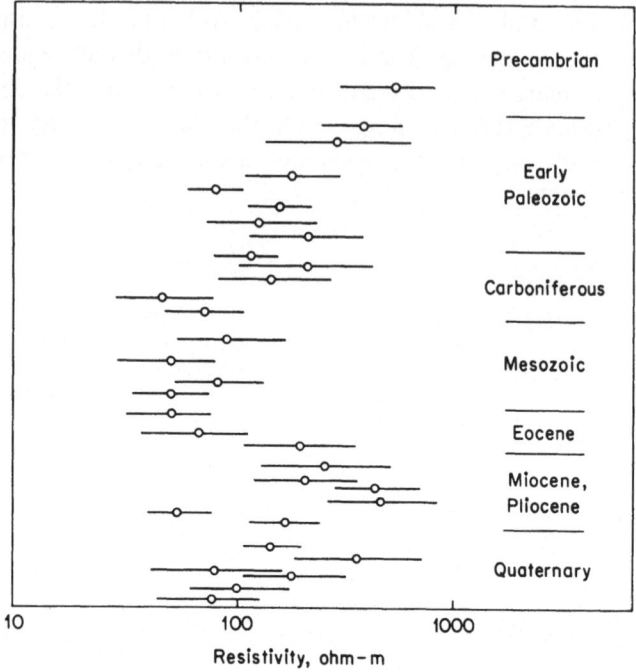

FIG. 9. Average resistivity for groups of sedimentary rocks with similar characteristics. Each circle indicates an average resistivity determined from 35–250 measurements about radio stations. The bar through each circle indicates the range within which 95% of the values for that group of data fall. (Modified from Keller and Frischknecht, 1966.)

where a and m are empirical constants chosen to fit the observations and R_w is the resistivity of the fluid, a is approximately 1 and m is approximately 2 for well cemented and sorted rocks (hence m is sometimes called the cementation factor). Figure 10 presents Archie's law for a typical suite of constants.

Log analysts often find it convenient to use the formation factor $F = R/R_w$ and represent Archie's law as

$$F = \phi^{-m} \tag{15a}$$

Although Archie derived this equation for consolidated sandstones without clay material, it appears to hold for porosities as low as 0·001 and as high as 0·25 with a wide range of grain size, mineralogy and texture (Fig. 11).

The 'true' formation factor F is related to a high resistivity matrix and

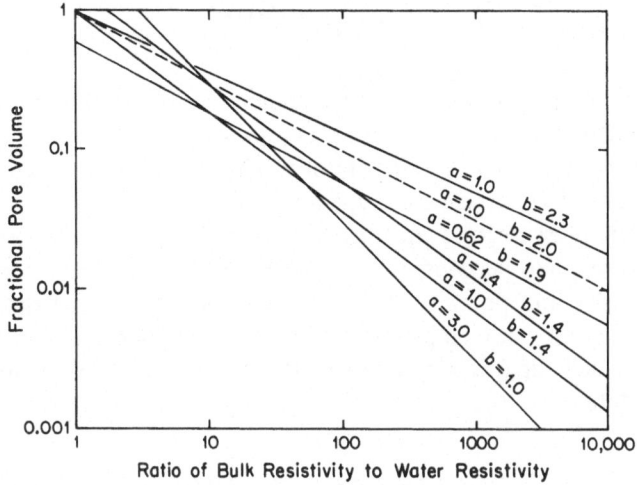

FIG. 10. Graphical presentation of the various forms of Archie's law: $R = aR_w\phi^{-b}$, where a and b are positive constants. The dashed curve is a simple inverse square relationship between resistivity and porosity which is usually a good approximation to the more exact expressions. (Modified from Keller and Frischknecht, 1966.)

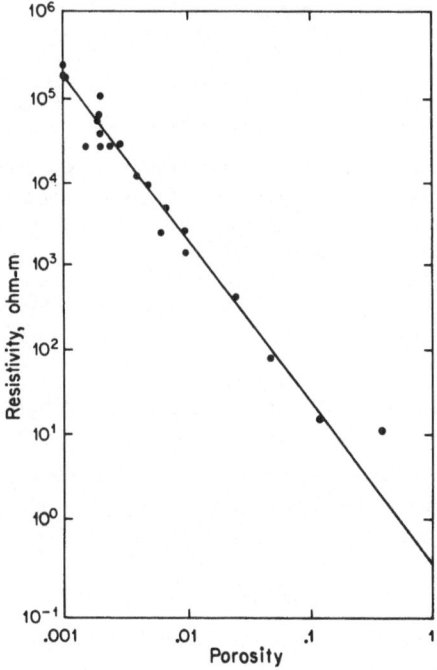

FIG. 11. Resistivity versus porosity of 22 rocks saturated with salt solution, under a confining pressure of 4 kbar. The slope of the line is -1.9. (Modified from Greenberg and Brace, 1969.)

low resistivity interstitial fluid. If the mineral conductivity approximates the conductivity of the water, other factors may influence the 'apparent' formation factor, such as distribution, shape and size of conducting grains, presence of certain clay-type minerals with ion exchange capacity, overburden pressure and anisotropy (resistivity parallel to bedding is usually lower than perpendicular to bedding) (Boyce, 1980). A plot of F versus density and porosity (Fig. 12) shows well log data from oceanic sediments along with empirical–theoretical functions published in the literature. Such figures are useful in the calculation of porosities from well log data.

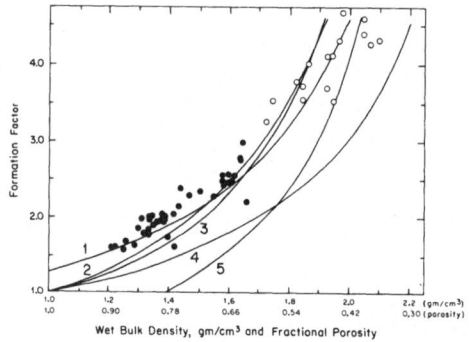

FIG. 12. Electrical 'formation factor' (horizontal) versus density and porosity. The solid data points are from laboratory studies of modern marine sediments. The open circles are Schlumberger well log resistivity and density data. (The porosity is calculated from the wet bulk density by assuming a grain density of $2 \cdot 7 \, \text{g/cm}^3$.) The curves are from the published literature as follows; (1) Boyce (1968); (2) Kermabon et al. (1969); (3) Archie (1942); (4) Maxwell (1904); and (5) Winsauer et al. (1952). (Modified from Boyce, 1980.)

Log analysts, through experience, have evolved slightly different forms of Archie's law (eqn 15a) that meet the specialised needs of the oil industry (Merkel, 1981). For example, log analysts often are concerned with the relation of water saturation S_w versus hydrocarbon saturation (100% minus the water saturation). Archie's law is thus modified to

$$S_w = (F R_w / R)^{1/2} \tag{15b}$$

Field experience has also led to specialisation of Archie's law for granular (sandstone) rocks;

$$F = 0 \cdot 62 / \phi^{2 \cdot 15} \tag{15c}$$

and for carbonate rocks:

$$F = 1/\phi^2 \tag{15d}$$

Specialised resistivity logs, such as micro-logs, mini-logs and micro-laterologs, have been developed specifically to measure the resistivity near the bore wall (the flushed zone). The ratio of this flushed zone resistivity to the mud resistivity yields the formation factor.

4. RESISTIVITY VERSUS VELOCITY

4.1. Theory

We have established a common dependence of velocity and resistivity on porosity. The advent of digitisation of well logs and transcription on to magnetic tape permits rapid computer analysis of geophysical logs. It is now possible to use conventional well logs to obtain plots of such petrophysical quantities as lithology, porosity, permeability, water saturation and hydrocarbon indices (Work et al., 1974). In this chapter, the generation of a pseudovelocity log (transit times that simulate a CVL log) from easily obtained electrical resistivity logs is emphasised.

The most readily available resistivity log is the short normal (16 in) log; unfortunately the measurement yields only the apparent resistivity R_a (a combination of mud filtrate, mudcake and rock). Derivation of the true resistivity from such data is difficult (Guyod and Pranglin, 1961) and it is simpler to use an apparent formation factor F_a:

$$F_a = R_a/R_{mf} = \phi^{-m} \tag{16}$$

where R_{mf} is the resistivity of the mud filtrate.

Faust (1953) adapted his 1951 equation to include the apparent resistivity R_a:

$$V = D Z R_a^{1/6} \tag{17}$$

where $D = 1948$ is an empirical constant applicable to most, but not all, geologic sections. Acheson (1963) empirically demonstrated that velocity was related to depth through a set of constants that varied with lithology and basin position. A more flexible equation was suggested by Delaplanche et al. (1963):

$$V = \alpha Z^\beta R_a^\theta \tag{18}$$

where α, β and θ are empirical constants adjusted for each area. Their results indicated that $\theta = 0 \cdot 3$ was the best choice with β varying from $0 \cdot 064$ to $0 \cdot 27$. Such wide variations in β demonstrates that Z is not a proper predictor for velocity in many areas.

Rewriting eqn (17) in terms of a pseudotransit time TT' ($\mu s/ft$) gives

$$1/V = TT' = 1/D \, (Z \, R_a)^{-1/6} \tag{19}$$

Equation (19) equates transit time to resistivity which is, ultimately, one of the goals of this chapter. It will be shown in Section 4.2 that inclusion of the depth factor Z into the equation is not sufficient or necessary to account for variations in lithology. Kim (1964) used porosity as the linking parameter between transit time (velocity) and resistivity.

Rewriting eqn (16), the porosity is given by

$$\phi = (R_a / R_{mf})^{-1/C} \tag{20}$$

Wyllie's equation for the velocity of an elastic wave through a porous medium (eqn 8) is modified for the case of well log data by equating fluid to the mud filtrate:

$$TT' = T T_{mf} + (1 - \phi)TT_s \tag{21}$$

where TT_s is the transit time through the solid and TT_{mf} is through the mud filtrate.

Solving eqn (21) for porosity,

$$\phi = (TT' - TT_s)/(TT_{mf} - TT_s) \tag{22}$$

Setting eqn (20) equal to eqn (22) and solving for TT', we obtain

$$TT' = TT_s + (K)R_a^{-1/c} \tag{23}$$

where $K = (TT_{mf} - TT_s)R_{mf}^{1/C}$. Equation (23) is Kim's 'scale function' relating transit time to resistivity.

Rudman et al. (1975) generalised Kim's scale function to the form

$$TT' = A + (B)R_a^{-1/C} \tag{24}$$

where A, B and C are determined empirically.

4.2. Applications

4.2.1. Faust's Equation
Faust's (1953) equation (eqn 19) equates transit time to resistivity through depth Z. Although it was originally derived only for permeable sections, it was applied by Rudman et al. (1975) to a well in southwestern

Indiana that penetrated 1790 ft of a mixture of sandstone, shale and limestone (Fig. 13). Ages ranged from Pennsylvanian to Mississippian. Comparison of the pseudovelocity log versus the CVL log shows that they appear similar in form, although the magnitudes are somewhat different.

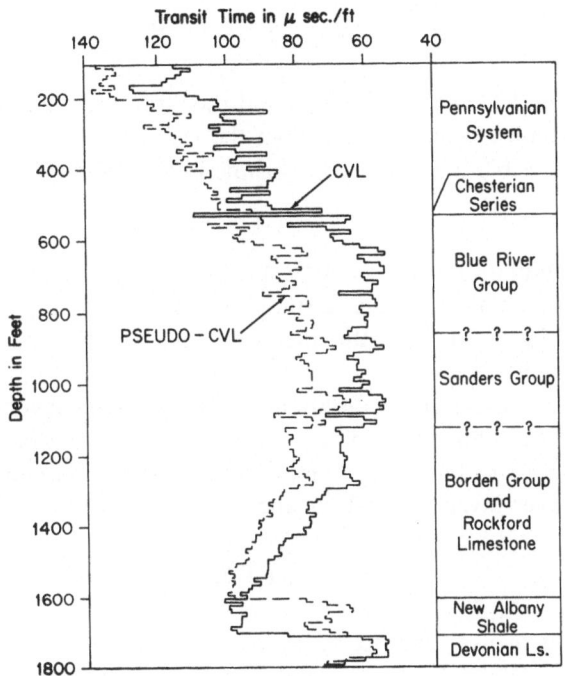

FIG. 13. Plot of transit times from Swaby well. True transit times (solid line) are from CVL; pseudo times (dashed) are generated from resistivity logs using the Faust equation (eqn (19)). (From Rudman *et al.*, 1975.)

Magnitudes of the transit times TT' for the pseudovelocity log are 10–25 μs/ft higher than the true values. Beneath the Sanders Group, comparison improves with depth, suggesting that the depth function $(Z^{1/6})$ is not as significant as the effect of porosity (lithology?) changes within the overlying Borden Group. Another major difference is related to the New Albany shale where pseudovelocities are at least 30 μs/ft greater than the true velocities. Studies of this black, carbonaceous shale show that it has unusually high resistivity without having a correspondingly high velocity. Of course, there are no parameters included in Faust's equation to accommodate this kind of anomalous condition.

4.2.2. Evaluation of Accuracy

The preceding section evaluated Faust's equation by visually comparing the true and pseudovelocity curves. Such qualitative evaluations are useful, but a quantitative measure of similarity is desirable. One possible measure of similarity s is

$$s = (1/N) \sum_{i=1}^{N} (TT_i - TT_i')^2 \qquad (25)$$

where TT_i and TT_i' are the observed and pseudotransit times observed at equal intervals for N measurements in the well.

Another possible measure is the seismic reflection coefficient RC_i:

$$RC_i = (V_i - V_{i+1})/(V_i + V_{i+1}) \qquad (26)$$

A comparison of all the RC values is a measure of overall form; it is also used in constructing synthetic seismograms and, indirectly, would be a measure of the validity of the pseudovelocity log for such synthetic seismogram generation.

The measure of accuracy followed in this chapter involves a computation of the total travel time T from a selected datum to a given depth (or geologic horizon). Computation involves summing interval times ΔTT_i over given interval distances Δx_i:

$$T = \sum_{i=1}^{N} \Delta TT_i \Delta x_i \qquad (27)$$

where ΔTT_i is in μs/ft and Δx_i is in ft.

4.2.3. Generation of the Scale Function

Computation of pseudotransit times from Kim's scale function (eqn 24) presumes knowledge of the constants A, B and C. Figure 14 is useful in illustrating some of the step-by-step procedures. The first goal is obtaining a plot of the scale function curve.

1. A CVL and a short normal resistivity log are selected from the same well.
2. Relatively non-oscillating portions of the logs are 'blocked-out', for example 2500–2575 ft in Fig. 14.
3. The average transit time and resistivity are obtained for the same 'block' of footage, for example 31 ohm-m and 69 μs/ft.
4. The values (31, 69) are plotted as one point of the scale function curve.

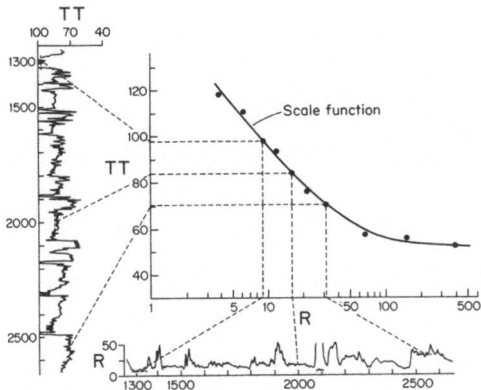

FIG. 14. Sketch illustrating generation of scale function curve. Resistivity R, in ohm-m, and transit time TT, in μs/ft from specific depth interval, form a pair of values for each data point. (Modified from Kim, 1964.)

5. Pairs of points from various depths are plotted and a scale function curve is sketched.

At this point it is possible to use the scale function curve and manually generate pseudotransit times from given resistivity values. For computer applications, however, it is necessary to evaluate the three constants A, B and C for the given scale function curve [see Appendix for derivation of eqns 33–35].

6. An initial (minimum) value of R_1 and an arbitrary constant Q are chosen so that R_1, R_2 and R_3 span the entire range of resistivity values (eqn 29, see Appendix). For example, for $R_1 = 10$ and $Q = 7$, a span of values for the Midwest well of Indiana (Fig. 15) is obtained:

$$R_1 = 10; \ R_2 = Q\,R_1 = 70; \ R_3 = Q\,R_2 = 490$$

7. TT_1', TT_2' and TT_3' are read from the curve for three specified R values and eqns 33–35 are then solved for A, B and C.

Often one set of constants do not adequately fit the scale function data points. The plot is then divided into three parts (Fig. 16) and separate constants evaluated for each part. For the Midwest well the intervals were 10–90, 90–200 and 200–450 ohm-m. Although the divisions were chosen on the basis of slope changes in the scale function, the divisions approximate the resistivity ranges of shale, sandstone and limestone.

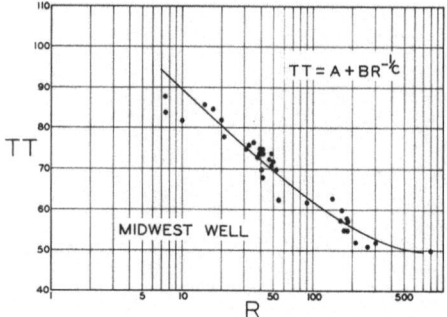

FIG. 15. Plot of data points and scale function curve for Midwest well. Resistivity R is in ohm-m and transit time TT is in μs/ft. Curve was fitted to data points by eye. (From Rudman, 1975.)

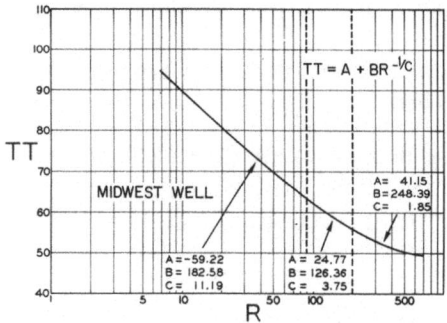

FIG. 16. Scale function curve for Midwest well showing three sets of constants (A, B, C) for three segments: $R = 10$–90; $R = 90$–200; $R = 200$–450. Resistivity R is in ohm-m and transit time TT is in μs/ft. (From Rudman, 1975.)

4.2.4. Pseudotransit Time Results

Scale function constants computed for the Midwest well (Fig. 16) were used to computer generate a pseudovelocity log (TT' in Fig. 17). Visual comparison of the pseudovelocity log with the true log (TT in Fig. 17) shows close correlation. Table 3 compares the travel times of the true and pseudovelocity logs (eqn 27) and shows a maximum error of 5%, acceptable for most purposes.

Scale functions were computed for wells drilled in two distinct geologic provinces of Indiana: the Cincinnati Arch and the Illinois Basin. The same set of constants were shown to be applicable in both provinces

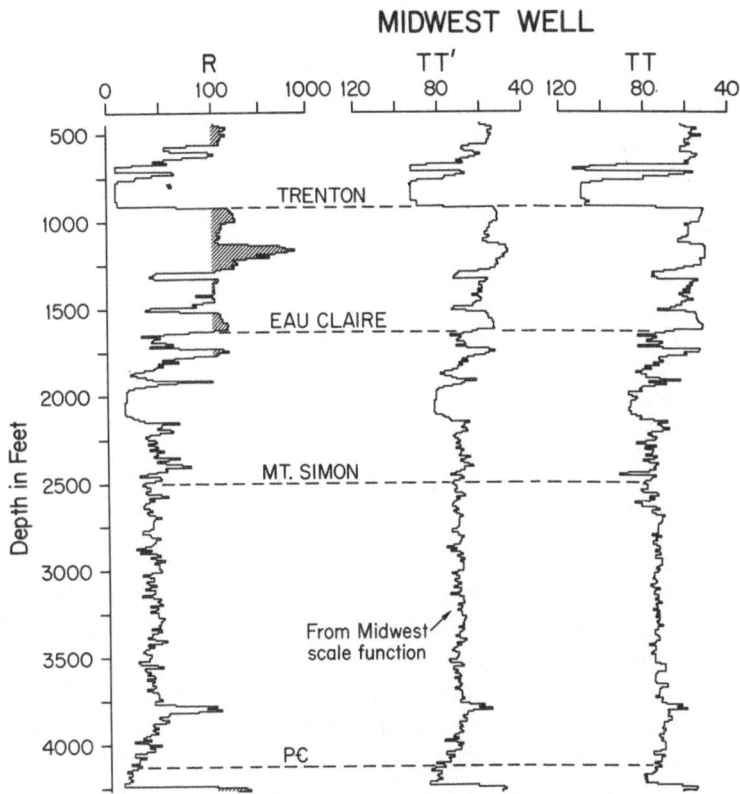

FIG. 17. Plot of Midwest well logs: R, short normal resistivity; TT', pseudo-CVL using scale function constants from Midwest well; TT, original CVL. Resistivity R is in ohm-m and transit time TT is in $\mu s/ft$. (From Rudman, 1975.)

(Rudman *et al.*, 1976). Studies of 23 wells have yielded some general rules for successful application of the method:

1. Logs drilled with mud resistivities greater than 5·0 ohm-m or less than 2·0 ohm-m should be avoided.
2. Pseudovelocity logs are most successful if applied to sections below the fresh water–salt water contact.
3. Certain stratigraphic units have anomalous velocities or resistivities and should be handled separately in the computations.
4. Scale functions of several wells should be averaged for general application in an area.

TABLE 3

TABULATION OF PSEUDOTRANSIT TIMES FOR THE
MIDWEST WELL (SEE FIG. 17). (MODIFIED FROM RUDMAN
et al. 1975)

Depth (ft)	CVL time (ms)	Midwest	
		Time	Error (%)
575	8	8	0
916	37	35	+5
1 333	61	59	+3
1 633	78	77	+1
2 498	144	139	+4
4 247	270	267	+1
4 283	272	269	+1

5. CONCLUSIONS

As a consequence of increased availability of digitised logs, geologists are continuing to generate additional data from ordinary logs. In this chapter the interrelationship of velocity and resistivity is seen to depend strongly on porosity. This theoretical relationship serves as a basis for empirical generation of pseudotransit times from ordinary resistivity logs. Application of scale functions for generation of logs has been successful in Indiana (Rudman *et al.*, 1975) and Australia (Nowak, 1978).

REFERENCES

ACHESON, C. H. (1963). Time–depth and velocity–depth relations in western Canada, *Geophysics*, **28**, 894–909.

AL-CHALABI, M. (1979). Velocity determination from seismic reflection data, in *Developments in Geophysical Exploration Methods—1*, Ed. A. A. Fitch. Applied Science Publishers, London, pp. 1–68.

ARCHIE, G. E. (1942). The electrical resistivity log as an aid in determining some reservoir characteristics, *AIME Trans. (Petroleum Development and Technology)*, **146**, 54–62.

BIRCH, F. (1961). The velocity of compressional waves in rocks to 10 kilobars, Part 2, *J. Geophys. Res.*, **66**, 2199–224.

BIRCH, F. (1964). Density and composition of mantle and core, *J. Geophys. Res.*, **69**, 4377–88.

BOYCE, R. E. (1968). Electrical resistivity of modern marine sediments from the Bering Sea, *J. Geophys. Res.*, **73**, 4759.

BOYCE, R. E. (1976). Sound velocity–density parameters of sediments and rock from DSDP drill sites 315–318 on the Line Island Chain, Manihiki Plateau, and Tuamotu Ridge in the Pacific Ocean, *Initial Reports of the Deep Sea Drilling Project*, **33**, 695–728.

BOYCE, R. E. (1980). Determination of the relationships of electrical resistivity, sound velocity, and density/porosity of sediments and rock by laboratory techniques and well logs from deep sea drilling project sites 415 and 416 off the coast of Morocco, *Initial Reports of the Deep Sea Drilling Project*, **50**, 305–18.

DAVIES, D. (1965). Dispersed Stoneley waves on the ocean bottom, *Bull. Seism. Soc. Am.*, **55**, 903–18.

DELAPLANCHE, J., HAGEMANN, R. F. and BOLLARD, P. G. C. (1963). An example of the use of synthetic seismograms, *Geophysics*, **28**, 842–54.

FAUST, L. Y. (1951). Seismic velocity as a function of depth and geologic time, *Geophysics*, **16**, 192–206.

FAUST, L. Y. (1953). A velocity function including lithologic variations, *Geophysics*, **18**, 271–88.

GARDNER, G. H. F., GARDNER, L. W. and GREGORY, A. R. (1974). Formation velocity and density—the diagnostic basics for stratigraphic traps, *Geophysics*, **39**, 770–80.

GASSMAN, F. (1951). Elastic waves through a packing of spheres, *Geophysics*, **16**, 673–85.

GREENBERG, R. J. and BRACE, W. F. (1969). Archie's law for rocks modelled by simple networks, *J. Geophys. Res.*, **74**, 2099–102.

GUYOD, H. and PRANGLIN, A. J. (1961). Now get true resistivities from conventional electric logs, *Oil and Gas J.*, **59** (24), 113–18.

HAMILTON, E. L. (1978). Sound velocity–density relations in sea floor sediments and rocks, *J. Acoustic Soc. Am.*, **63**, 366–77.

HAMILTON, E. L. (1980). Geoacoustic modelling of the sea floor, *J. Acoustic Soc. Am.*, **68**, 1313–40.

KELLER, G. V. (1966). Electrical properties of rocks and minerals, in *Handbook of Physical Constants*, Ed. S. P. Clark, GSA Mem. 97, p. 553–77.

KELLER, G. V. and FRISCHKNECHT, F. C. (1966). *Electrical Methods in Geophysical Prospecting*. Pergamon Press, Oxford, 519 p.

KENNETT, P. (1979). Well geophone surveys and the calibration of acoustic velocity logs, in *Developments in Geophysical Exploration Methods—1*, Ed. A. A. Fitch. Applied Science Publishers, London, pp. 93–114.

KERMABON, A., GEHIN, C. and BLAVIER, P. (1969). A deep-sea electrical resistivity probe for measuring porosity and density of unconsolidated sediments, *Geophysics*, **34**, 554.

KIM, D. Y. (1964). Synthetic velocity log. Paper presented at 33rd Annual Intern. SEG Meeting, New Orleans.

LUDWIG, W. E., NAFE, J. E. and DRAKE, C. L. (1970). Seismic refraction, in *The Sea*, Vol. 4, Ed. A. E. Maxwell. Wiley-Interscience, New York, pp. 53–84.

MAXWELL, J. C. (1904). *Electricity and Magnetism*, Vol. 1, 3rd edn. Clarendon Press, Oxford.

MERKEL, R. H. (1981). Well log formation evaluation, in Continuing Education Course Note Series No. 14, publication of the AAPG·Education Dept., 82 p.

NAFE, J. E. and DRAKE, C. L. (1957). Variations with depth in shallow and deep water marine sediments of porosity, density and the velocities of compressional and shear waves, *Geophysics*, **22**, 523–52.

NAFE, J. E. and DRAKE, C. L. (1963). Physical properties of marine sediments, in *The Sea*, Vol. 3, Ed. N. M. Hill. Interscience, New York, pp. 794–815.

NOWAK, I. R. (1978). Pseudovelocity applications in the Carnarvon Basin, *Ann. Rept Geol. Survey W. Australia*, pp. 75–9.

PRESS, F. (1966). Seismic velocities, in *Handbook of Physical Constants*, Ed. S. P. Clark, GSA Mem. 97, p. 195–218.

RUDMAN, A. J., WHALEY, J. F., BLAKELY, R. F. and BIGGS, M. E. (1975). Transformation of resistivity to pseudovelocity logs, *AAPG Bull.*, **59**, 1151–65.

RUDMAN, A. J., WHALEY, J. F., BLAKELY, R. F. and BIGGS, M. E. (1976). Geologic note: Transformation of resistivity to pseudovelocity logs, *AAPG Bull.*, **60**, 879–82.

WEATHERBY, B. B. and FAUST, L. Y. (1935). Influence of geological factors on longitudinal seismic velocities, *AAPG Bull.*, **19**, 1–18.

WINSAUER, W. O. *et al.* (1952). Resistivity of brine-saturated sands in relation to pore geometry, *Am. Assoc. Petrol. Geol. Bull.*, **36**, 253.

WOOD, A. B. (1941). *A Textbook of Sound*. Macmillan, New York.

WOOLLARD, G. P. (1959). Crustal structure from gravity and seismic measurements, *J. Geophys. Res.*, **64**, 1521–44.

WORK, P. L. and MEADOW, H. M. (1974). Digitize well logs in Morrow sand exploration, *Oil and Gas J.*, **72** (7), 61–3.

WYLLIE, M. R. J., GREGORY, H. R. and GARDNER, L. W. (1956). Elastic waves in heterogeneous and porous media, *Geophysics*, **21**, 41–70.

APPENDIX: CALCULATION OF SCALE FUNCTION CONSTANTS

Calculation of the scale function constants (A, B and C of eqn 24) involves solving three non-linear equations of the form

$$TT_i = A + (B) R_i^{-1/C}, \quad i = 1, 2, 3 \tag{28}$$

Computations are eased if the values of R_i are interrelated through an arbitrary constant Q, where

$$R_2 = Q R_1 \text{ and } R_3 = Q R_2 \tag{29}$$

(Q is chosen such that R_i spans the entire range of resistivity values on the log). Substituting eqn (29) into eqn (28) and combining equations gives

$$TT_1' - TT_2' = (B) R_1^{-1/C}(1 - Q^{-1/C}) \tag{30}$$

$$TT_2' - TT_3' = (B) R_2^{-1/C}(1 - Q^{-1/C}) \tag{31}$$

Dividing eqn (30) by eqn (31), taking the log of both sides and using eqn (29) gives

$$\log\,[(TT_1' - TT_2')/(TT_2' - TT_3')] = -(1/C)\,\log\,(1/Q) = +(1/C)\,\log\,Q$$
$$(32)$$

Solving eqn (32) for C:

$$C = \log\,Q/\log\,[(TT_1' - TT_2')/(TT_2' - TT_3')] \tag{33}$$

Using eqn (33) and solving eqn (30) for B:

$$B = (TT_1' - TT_2')R_1^{1/C} \tag{34}$$

and from eqn (28) (with $i = 1$), solving for A gives

$$A = TT_1' - (B)R_1^{-1/C} \tag{35}$$

Chapter 3

FOCUSED RESISTIVITY LOGS

AMALENDU ROY

National Geophysical Research Institute, Hyderabad, India

SUMMARY

Focused resistivity logging tools, such as Laterolog 7, have been used extensively, because they record large deflections—as compared with those for the unfocused normal and lateral sondes—against highly resistive thin formations. These large deflections, however, do not signify any inherent superiority of these tools, but are the result of using a set of fictitious and genetically unrelated currents in the apparent resistivity formulae— currents that do not exist in the ground at the time of measurement, do not ensure the necessary condition for null or near-null of potential, and do not generate the measured potential. When the real currents are used, the apparent superiority disappears altogether. By a suitable manipulation of the apparent resistivity formulae, even the normal or the lateral device can be made to yield large deflections.

It is also demonstrated that (i) the unfocused two-electrode normal sonde has a far larger radius of investigation than the 'focused' seven-electrode Laterolog 7 of equal spacing; (ii) the response of any sonde, focused or unfocused, can be synthesised exactly from that of a normal device of suitable spacings; (iii) Laterolog interpretation charts, published by different companies or by the same company at different times, exhibit very wide variations; and (iv) the extraordinarily high refinement of present-day resistivity log interpretation is inconsistent with the drastic simplifications, idealisations and other ambient factors governing the complex logging problems.

1. INTRODUCTION

The visual and intuitive appeal of focusing has always intrigued the exploration geophysicist, both in well logging and in ground surface prospecting. Focused resistivity logging tools, such as Laterologs 3 and 7, appeared in the petroleum industry around 1949 and were hailed as very significant improvements over the conventional unfocused normal and lateral devices. Believed to be the direct result of focusing the current from a central electrode into the target formation in the form of a thin horizontal beam, these improvements are (Doll, 1951; Anon., 1958, 1969, 1972; Pirson, 1963; Moran and Chemali, 1979; and others)

(1) Much larger radius of investigation as compared with those of normal and lateral sondes;
(2) Sharper definition of formation boundaries;
(3) Primarily, measured apparent resistivity very much closer to true value, especially when the beds are thin and highly resistive relative to the mud column and the adjacent formations.

Consequently, these two and several other related logging devices have been and are being used extensively all over the world for over 30 years.

Following the usual logging notation, AM will represent the normal (or potential) sonde where A and M are respectively the current-emitting and potential-measuring electrodes. AMN or AO will stand for the lateral (or gradient) device where A is the current electrode and the potential gradient is measured between the two closely spaced electrodes M and N with O as their midpoint. $A_1M_1'M_1A_oM_2M_2'A_2$ or $A_1O_1A_oO_2A_2$ will refer to Laterolog 7 (LL7), where A_1 and A_2 are the two 'auxiliary' power electrodes shorted to each other, A_o is the centre or 'measure' current source, M_1–M_2 and M_1'–M_2' are two pairs of shorted potential probes that monitor and ensure null or near-null of potential, and O_1 and O_2 are the midpoints of $M_1'M_1$ and M_2M_2' (see Fig. 1).

Figure 2 reproduces from Doll (1951) and Anon. (1958, 1969, 1972) two illustrations, based on laboratory measurements, on the comparative performance of normal and LL7 sondes against highly resistive targets. While the normal readings are much too small, the LL7 records apparent resistivities which, at the central locations, are very much larger and are within 20% of the true values. This remarkable enhancement is ascribed by the above authors to the neatly focused current pattern from the LL7 as compared schematically in Fig. 7(a) with the unfocused emanation from the normal (and the lateral) device. The focused current rays from

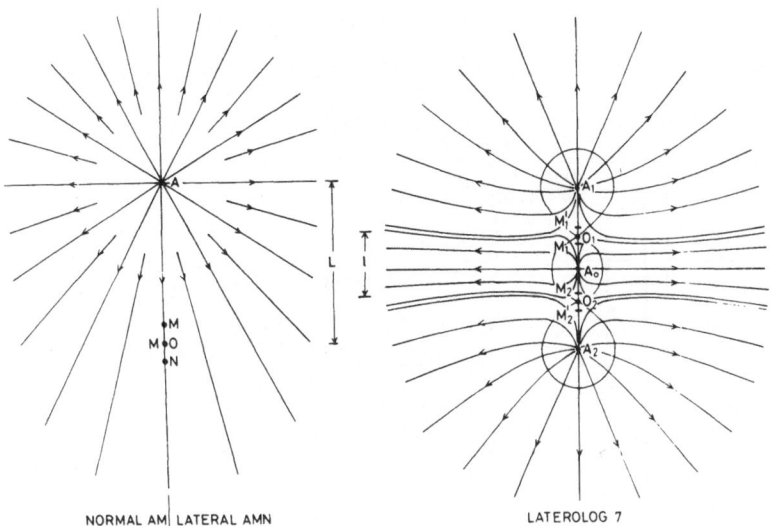

NORMAL AM | LATERAL AMN   LATEROLOG 7

FIG. 1. Electrode nomenclature and arrangement in normal, lateral and LL7 sondes with current lines for homogeneous medium. Note the triple loop in LL7 formed by the equipotential surface passing through O_1 and O_2. $O_1 O_2 = l = 0.4 A_1 A_2 = 0.4L$.

the LL7 are more nearly perpendicular to the borehole wall and have therefore smaller lengths in the mud column than the unfocused oblique current filaments of a normal sonde. Also, the focused beam flows effectively within the target formation alone as long as its thickness is smaller than that of the formation. In comparison, the nearly radial lines from the power electrode in a normal device would have to cross the formation boundaries rather early and travel a substantial part of their effective distance through the adjacent formations. These two features, it was argued, would ensure that the measured signal in the focused device is largely free from the effects of the mud column and the shoulder beds and is therefore characteristic of the target formation alone. It will be shown that this is not the case.

It is only recently (Roy, 1975; Roy and Apparao, 1976; Roy, 1977; also Roy and Dhar, 1971; Roy and Apparao, 1971) that doubts have been expressed on the special merits of so-called focusing. Also, Moran and Chemali (1979) now concede that 'prediction of sonde responses on the basis of current patterns can in some cases be misleading', as pointed out earlier (Roy, 1975).

For several practical reasons, focused arrays for resistivity profiling on

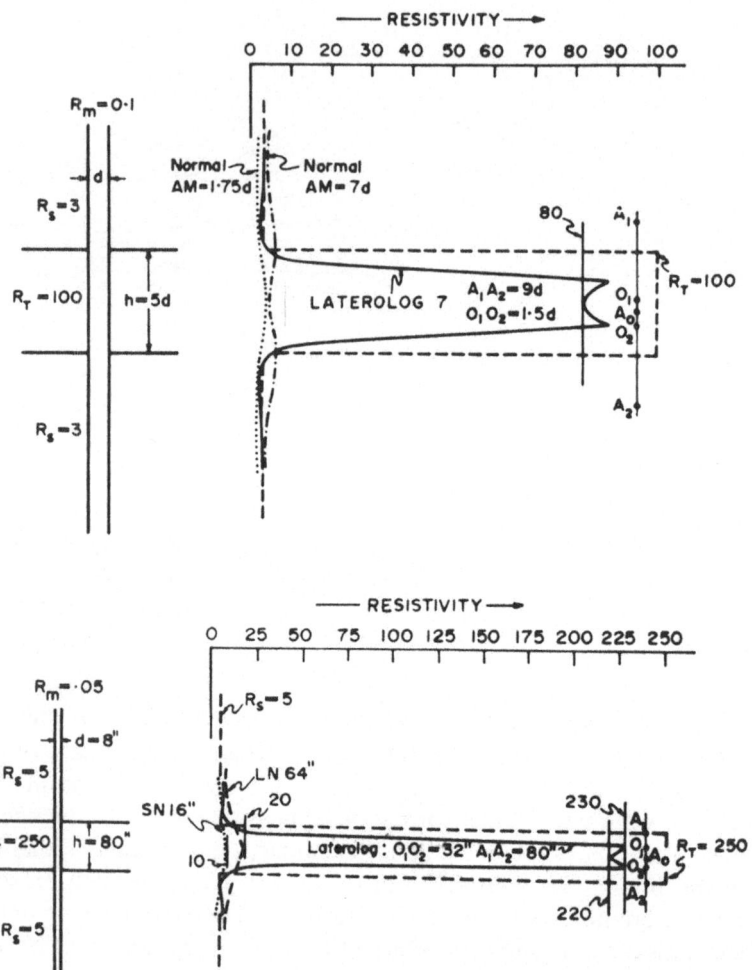

FIG. 2. Laboratory results from Doll (1951) and Anon (1958, 1969, 1972) on relative merits of LL7 and normal sondes against thin highly resistive formations. R_m, R_s and R_T are resistivities respectively of mud, shoulder beds and target formation; d is borehole diameter; and SN and LN are short and long normals. (Permission for reproduction granted by Society of Petroleum Engineers(AIME) and Schlumberger Well Services, USA.)

the ground surface have not developed very far, although attempts have been made. First, the axial symmetry of the borehole is non-existent on the ground surface and point electrodes are no longer adequate for maximum possible focusing. Secondly, extended electrodes as in

Laterolog 3 (LL3) or a large number of electrodes as in LL7 are impractical on the ground surface for many reasons, one of which is the difficulty associated with making solid–solid contacts between each electrode and the ground. Thirdly, the surface array moves in discrete jumps and continuous automatic adjustments of currents and voltages are therefore no longer possible.

2. RADIUS OF INVESTIGATION OF LOGGING DEVICES

According to Anon. (1950), 50% of the signal measured by the normal device AM in an infinite homogeneous medium arises out of the material within a thick *spherical* shell centred at A and having radii AM and 2AM. Similarly, 90% of the signal originates from the thick *spherical* annulus of radii AM and 10AM. The radius of investigation of a normal device can thus be taken as 2AM, provided one agrees with Anon. (1950) to define it as the external radius of the volume of ground that contributes half of the total measured potential.

The above concept on radius of investigation, however, is not entirely satisfactory. First, a sphere, especially one not centred at the midpoint of AM, is incompatible with well logging geometry. Resistivity interfaces in logging are circular *cylinders* coaxial with the borehole and orthogonal planes representing horizontal bedding. A cylindrical coordinate system is obviously the only one that is consistent with this geometry, and the radius of investigation in such a case should refer to the radius of a *cylindrical* surface, not spherical, coaxial with the drill hole and the sonde. Secondly, the definition is not applicable to other sondes. For the lateral device AMN or AO, where the signal is a potential gradient at O, it is not possible to mark out separate zones that contribute half, or any other fraction, of the total measured signal. The radius of investigation for this sonde is consequently assumed, for no specific reason, to be equal to its spacing AO. The concept does not also work for four-electrode arrays—such as Wenner and Schlumberger—should one want to use them. For the LL7 which also has multiple sources and non-spherical equipotential surfaces, the signal is again a potential, but Doll (1951, in discussion and reply) speaks of a radius of investigation in terms of the potential drop from the borehole wall. These treatments do not reconcile mutually and do not provide a common basis for comparing the radii of investigation of the different resistivity logging tools.

Yet, such an analysis is possible—an analysis that is logical, easily visualised and equally applicable to all sondes and arrays in well logging

as well as in ground surface prospecting. This follows from the following
formula (Roy, 1978):

$$\Phi_P = \frac{1}{4\pi} \int_\tau \nabla\Phi \cdot \nabla(1/r) d\tau \tag{1}$$

where Φ_P is the measured or calculated direct current potential at any
point P in an infinite medium τ of any description, Φ is the potential
distribution in that medium due to any system of point or finite-sized
current electrodes, r is the distance between P and any volume element
$d\tau$, the gradients are evaluated at the element, and the integration covers
the entire infinite space τ. Thus, each space element $d\tau$ behaves as an
electric dipole of moment $(1/4\pi)\nabla\phi d\tau$ and contributes $(1/4\pi)$
$\nabla\phi \cdot \nabla(1/r)d\tau$ units of potential to the total potential Φ_P at P.

Integration of these elementary contributions over a cylindrical surface
coaxial with the borehole and of radius ρ yields a curve that shows the
variation with ρ of the signal contributed by a thin cylindrical shell of
ground to the total measured signal. For a homogeneous ground to
which radii of investigation are usually referred, this curve, the radial
investigation characteristic (RIC), begins with a zero value at $\rho=0$,
passes through a small negative zone, and then reaches maximum before
decaying to zero again at large radii (Fig. 3). This behaviour of the RIC
allows one to define, quite unambiguously, the radius of investigation of
any logging sonde as that radius at which its RIC attains the maximum
value. In other words, the radius of that cylindrical shell of ground which
contributes the largest share to the recorded potential (or potential
gradient, or potential difference) is to be taken as the radius of investigation for the particular sonde. One may as well integrate the RIC
with respect to ρ, obtain a cumulative RIC, and determine the radius of
investigation on the basis of 50% contribution or any other fraction.

According to this analysis, the radii of investigation of the three
devices in an infinite homogeneous medium are (Roy and Dhar, 1971):

Radius of investigation by

	Max contribution	50% contribution	
Normal	0·60L	1·42L	$L=AM$
Lateral	0·40L	0·58L	$L=AO$
Laterolog 7	0·16L	0·48L	$L=A_1A_2$

where L is the array length or spacing and equals the distance between

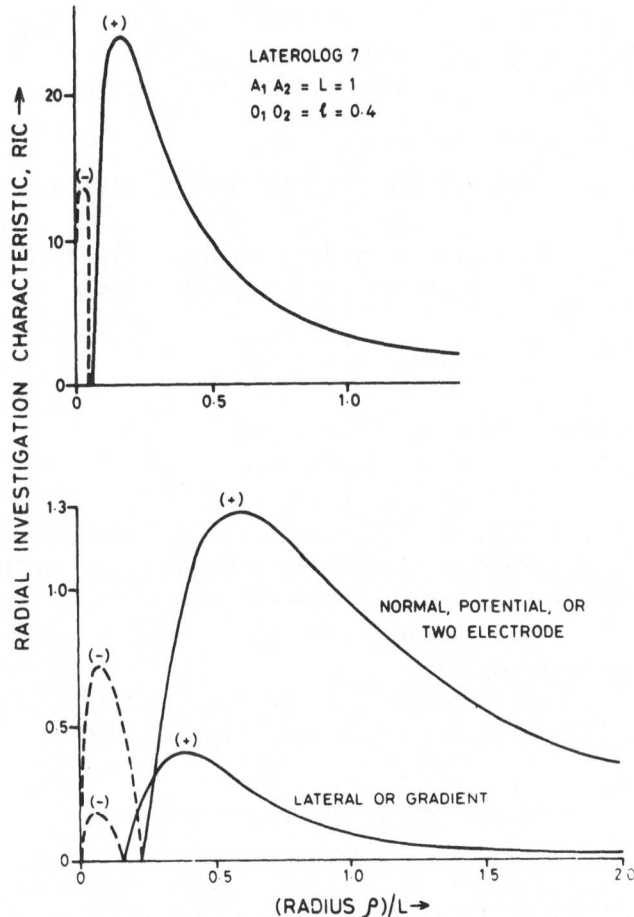

FIG. 3. RIC for normal, lateral and LL7 devices. (Reproduced from Roy (1974) with permission from Society of Exploration Geophysicists, USA.)

the two outermost active electrodes of any sonde or array. Thus, for a common value of the spacing L, the radius of investigation of the simplest unfocused two-electrode or normal device is far larger than that of the complex 'focused' seven-electrode LL7. The last paragraph of Section 4 gives another independent proof of the same result. This result is entirely consistent with the writer's experience on the ground surface with resistivity sounding on horizontal structures as well as profiling across laterally bounded conducting targets. In all cases of theoretical computations, laboratory model experiments and test field surveys, the two-electrode array defined the sounding curve and optimally detected the conducting target with far smaller spacings than any other array. See, for instance, Apparao and Roy (1973) for field examples.

3. CONCEPT OF APPARENT RESISTIVITY

For an infinite homogeneous medium, its true resistivity R_T can be determined from LL7 measurements through the formula

$$R_T = \frac{4\pi L\Phi(O)}{\dfrac{I}{0.7} + \dfrac{I_o}{0.2} + \dfrac{I}{0.3}} \tag{2}$$

where $L = A_1 A_2$ is the array length, $O_1 O_2 = 0.4L$, I is the *measured* current from each of the outer power electrodes A_1 and A_2, I_o is the *measured* current from the central electrode A_o, $I/I_o = 2.76$ for null of potential at O_1 and O_2, and $\Phi(O) = \Phi(O_1) = \Phi(O_2)$ is the *measured* potential at O_1 or O_2. Note especially that all the *four* quantities—three currents and a potential—*are actually measured under null condition and then used in eqn (2)* to obtain the true resistivity.

In inhomogeneous ground, the apparent resistivity R_{LL7} for the Laterolog 7 should therefore be calculated from the formula

$$R_{LL7} = \frac{4\pi L\bar{\Phi}(O)}{\dfrac{\bar{I}}{0.7} + \dfrac{I_o}{0.2} + \dfrac{\bar{I}}{0.3}} \tag{3}$$

where $\bar{I} = (\bar{I}_1 + \bar{I}_2)/2$, $\bar{I}_1$ and $\bar{I}_2$ are the new generally unequal currents from the two outer power electrodes that re-establish null or near-null at O_1 and O_2 under inhomogeneous surroundings, $\bar{\Phi}(O)$ is the new signal or potential, and I_o is maintained constant although this is not necessary. That is, one carries out *exactly the same operations and measurements* as for homogeneous ground, substitutes the new potential and the new currents in eqn (2) or (3), and obtains not R_T but the apparent resistivity R_{LL7}. It is necessary to emphasise that the potential and the currents in eqn (3) are the ones that actually exist in the ground at the time of measurement and the potential $\bar{\Phi}(O)$ is normalised with respect to the currents that generate this potential as well as restore the null condition prescribed for LL7 operation.

In LL7 logging, however, eqn (3) has never been used. Instead, the

apparent resistivity for LL7 in inhomogeneous ground is computed from
the relation (Moran, 1976)

$$R_{LL7} = \left[\frac{4\pi L}{\dfrac{I/I_o}{0\cdot7} + \dfrac{1}{0\cdot2} + \dfrac{I/I_o}{0\cdot3}}\right]_{\substack{\text{homogeneous}\\\text{ground}}} \times \frac{\Phi(O)}{I_o} \tag{4}$$

$$= 1\cdot41\frac{\Phi(O)}{I_o} \tag{4a}$$

where $L = 2\cdot032$ m and $I/I_o = 2\cdot76$ *as for homogeneous ground regardless of
its actual value under inhomogeneous environment.* Since I_o is held constant
throughout, the currents used for normalisation in relation (4) or (4a)
continue to be those that relate only to the homogeneous case. Except
for I_o, these currents are fictitious and non-existent in the ground at the
time of measurement, are genetically unrelated to the observed signal
$\Phi(O)$, and do not bring about the necessary null at O_1 and O_2. Since null
or near-null at these points is an essential precondition for LL7 measurements, the correct formula for R_{LL7} must combine the observed signal
$\Phi(O)$ with only those currents that satisfy this condition and give rise to
this signal. Besides I_o, these currents are $\bar{I}_2$ and $\bar{I}_2$ which are very
different in general from $I = 2\cdot76I_o$. As it stands, formula (4) or (4a) is a
relationship between unrelated quantities.

It is not very meaningful for formula (4a) to have retained an explicit
identity for I_o. This current is maintained at a fixed value throughout
and could very well have been absorbed within the numerical multiplier.
There is no precondition that the multiplier must have a dimension of
length and cannot be $1\cdot41/I_o$ in place of $1\cdot41$. Retention of this identity,
however, promotes the misconception that it is only the 'measure'
current I_o that is important, the 'auxiliary' currents from A_1 and A_2 are
not. The real situation is exactly the reverse and nearly the entire signal
$\Phi(O)$ originates from the so-called 'auxiliary' currents. Indeed, there is no
reason why the central power electrode A_o should be considered func-
tionally different from A_1 and A_2, and why it should be called the
'main' or 'exploring' or 'measuring' electrode while the other two are
described as 'auxiliary', even 'screening'.

Against a thin highly resistive formation for which the laterologs were
specially designed, the currents $\bar{I}_1$ and $\bar{I}_2$ from A_1 and A_2 are very much
larger for a given I_o than I emanating from each of the same electrodes
under homogeneous conditions (see Fig. 8, Section 5), and $\Phi(O)$ is
correspondingly higher. Because of this, the denominator in formula (4)

is much smaller in magnitude than it should be. This of course boosts the quotient to an artificially high value and results in the large deflections of LL7 in Fig. 2. These large numbers do not represent apparent resistivity and are unrelated to the intrinsic merit or otherwise of the particular tool.

Figures 4 and 5 present the results of repeating the laboratory

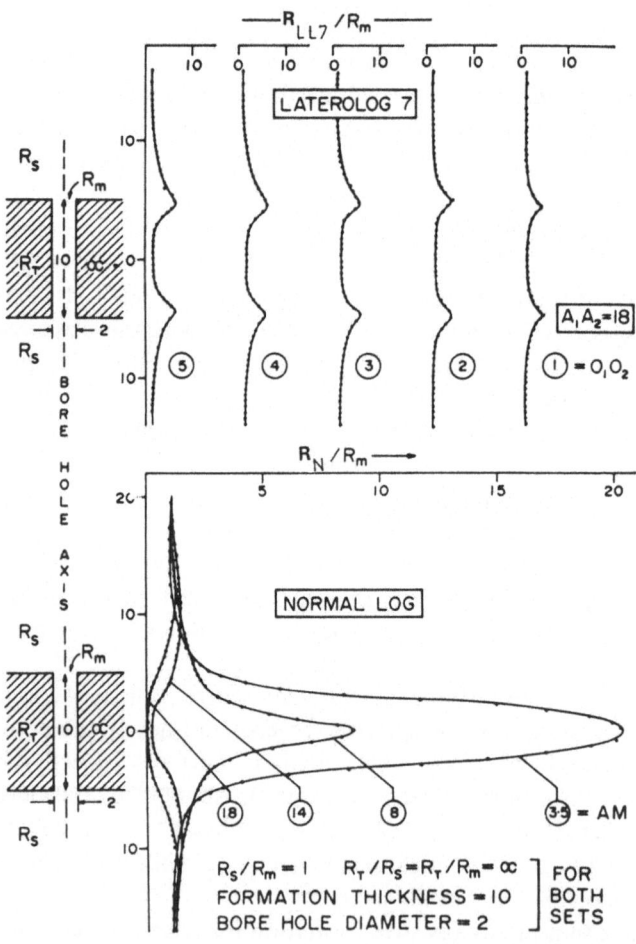

FIG. 4. Results of repeating the laboratory experiments of Fig. 2 (top) using eqn (3) for computing apparent resistivity R_{LL7} for LL7. R_N=apparent resistivity for normal device. (Reproduced from Roy and Apparao (1976) with permission from European Association of Exploration Geophysicists, The Netherlands.)

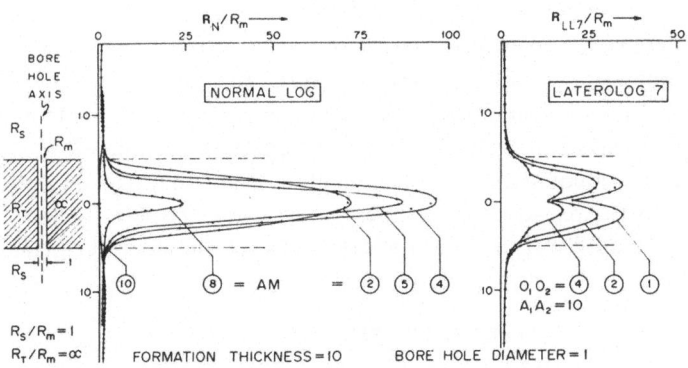

FIG. 5. Results of repeating the laboratory experiments of Fig. 2 (bottom) using eqn (3) for computing apparent resistivity R_{LL7} for LL7. (Reproduced from Roy and Apparao (1976) with permission from European Association of Exploration Geophysicists, The Netherlands.)

experiments of Fig. 2 and computing the LL7 response from the correct apparent resistivity expression, eqn (3). Compare Fig. 2(top) with the LL7 curve for $O_1O_2 = 3$ and the normal logs for $AM = 3\cdot5$ and 14 in Fig. 4, and Fig. 2(bottom) with the LL7 response for $O_1O_2 = 4$ and the normal records for $AM = 2$ and 8 in Fig. 5. As expected, the apparent superiority of the LL7 responses in Fig. 2 has altogether disappeared. Indeed, the situation makes a complete turnabout and the performance of the normal sondes emerges as distinctly the better of the two. For details of these repeat experiments, see Roy and Apparao (1976).

If a large number must somehow be extracted, one does not have to design a complex array like LL7 or LL3. The formula for the conventional normal sonde AM can be manipulated to achieve the same result with much less ado. In a homogeneous medium, the true resistivity R_T in the case of a normal device is given by

$$R_T = 4\pi L \frac{\Phi(M)}{I} \tag{5}$$

where $L = AM$ and $\Phi(M)$ is the potential at M. For inhomogeneous conditions, one can then write

$$R_n = 4\pi L \left[\frac{\bar{\Phi}(M)/\bar{I}}{\Phi(M)/I} \right]^n \times \frac{\bar{\Phi}(M)}{\bar{I}} \tag{6}$$

where n is any positive number, and the absence or presence of the overhead bar continues to refer respectively to measurements under homogeneous and heterogeneous conditions. For the homogeneous case, eqn (6) reduces to eqn (5) and R_n gives the correct value of true resistivity. In an inhomogeneous environment, R_n can be made as large (resistive target) or as small (conductive target) as one pleases by simply choosing an adequately large value of n. The quantity $\Phi(M)/I$ corresponds to the resistivity base line against thick shoulder beds.

Curves 1 and 2 in Fig. 6 are replots of conventional short normal and

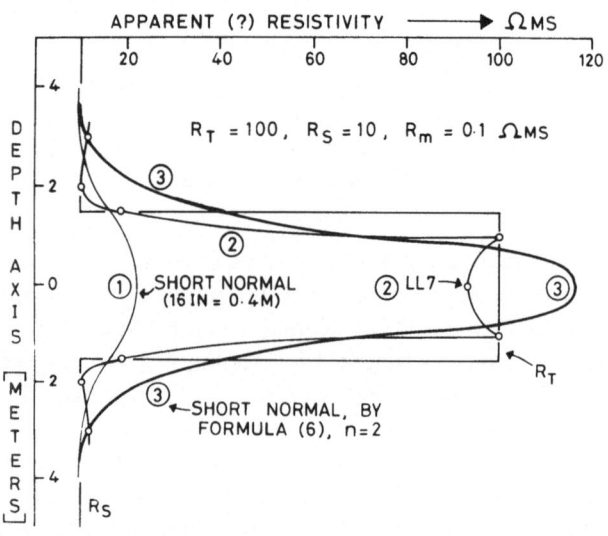

FIG. 6. Curve (1): conventional short normal log; curve (2): conventional LL7 response; curve (3): enhanced short normal, R_n as computed from eqn (6) for $n=2$. Data for curves (1) and (2) taken from Moran and Chemali (1979, Figure 11 (top), p. 922).

LL7 data taken from Moran and Chemali (1979) for $AM = 0.4\,\text{m}$ (16 in), $A_1A_2 = 2\,\text{m}$ (80 in), bed thickness $= 3\,\text{m}$, hole diameter $= 0.2\,\text{m}$, $R_T = 100\,\Omega\text{m}$, $R_s =$ shoulder bed resistivity $= 10\,\Omega\text{m}$, and $R_m =$ mud resistivity $= 0.1\,\Omega\text{m}$. Curve 3 is what becomes of the short normal response when recomputed by eqn (6) for $n=2$. It will be seen that there is not much to choose between curves 2 and 3; both produce large deflections by virtue of artificial and incorrect formulae. Operationally, however, production of curve 3 would be considerably simpler and therefore less expensive.

It is worth emphasising that the objective of the laterologs is not merely to record a deflection that is as large as it can be made. An apparent resistivity that is too high is as undesirable as one that is too low, in the sense that both need substantial corrections. Recall that, for resistive beds thicker than its spacing, the lateral or gradient device has a maximum response that is much higher than the true resistivity and is considered unacceptable without correction. Recall also that, for the LL7 itself, the apparent resistivity is not read at the maximum deflection points. This results in the mild enigma that, at the central symmetrical location where the LL7 reading is taken, the tool does not record the highest apparent resistivity although focusing is at its best in this position. The highest values appear (Figs. 2 and 6) at two off-centre points where the current lines are heavily distorted (Fig. 7b) and remain unutilised even though they are nearer to the true resistivity.

The initial objective of the laterologs was to measure an apparent resistivity that equals the true value, to the extent that the former can be used directly without correction. Such an objective is unattainable because, in a strongly heterogeneous ground involving several resistivities, no tool can be expected to respond selectively to only one and remain unaffected by the others. This is why laterolog apparent resistivities do need to be corrected for the mud, side beds and invasion. With a suitable scheme for similar corrections, eqn (6) for the normal sonde can perform equally well.

3.1. Laterolog 3 or Guard Electrode

The calculation of apparent resistivity for LL3 or Guard Electrode is conceptually similar to that of LL7. For homogeneous ground, the basic LL3 formula is (Owen and Greer, 1951)

$$R_T = \frac{4\pi k}{\ln[(c+k)/a]} \times \frac{\Phi}{I_T} \tag{7}$$

where Φ is the common potential of the three sections of the Guard Electrode, I_T is the total current leaving the entire electrode, $2c$ and $2a$ are its length and diameter respectively, and $k^2 = c^2 - a^2$. Equation (7) can be rewritten as

$$R_T = \frac{4\pi k}{\ln[(c+k)/a]} \times \frac{I_o}{I_T} \times \frac{\Phi}{I_o} \tag{8}$$

$$= \frac{4\pi b}{\ln[(c+k)/a]} \times \frac{\Phi}{I_o}$$

where I_0 is the current from the central ring of length $2b$, and $I_0/I_T = b/k$ *in a homogeneous medium*. Although only I_0 is explicitly visible in eqn (8), one must note that Φ is really being normalised by the total current $I_T = (k/b)I_0$, as seen in the basic eqn (7).

When the medium is inhomogeneous, the apparent resistivity R_{LL3} is computed from a formula of the type

$$R_{LL3} = \frac{4\pi b}{\ln[(c+k)/a]} \times \frac{\bar{\Phi}}{\bar{I}_0} \qquad (9)$$

where the overhead bar has the same meaning as before. The total implicit current in eqn (9) is $(k/b)\bar{I}_0$. As seen earlier in the case of LL7, this does not equal the actual ground current and would not comply with the prescribed condition of maintaining the three sections of the Guard Electrode at the common potential $\bar{\Phi}$. The correct expression for the LL3 apparent resistivity should thus be

$$R_{LL3} = \frac{4\pi k}{\ln[(c+k)/a]} \times \frac{\bar{\Phi}}{\bar{I}_T} \qquad (10)$$

where $\bar{I}_T$ is the total current that flows into the inhomogeneous ground from the sonde as whole.

4. SUPERPOSITION PRINCIPLE

Because of the operation of the principle of superposition in linear fields, the response of any complex array of discrete point electrodes—focused or unfocused—can be synthesised exactly from those of a normal device with suitable spacings. In this section, this equivalence will be demonstrated (Roy, 1980) for the lateral and the LL7 devices.

4.1. Lateral Device
In an infinite homogeneous medium of true resistivity R_T, the potential Φ at a vertical distance z from a point current source of strength I is

$$\Phi = \frac{R_T I}{4\pi} \times \frac{1}{z} \qquad (11)$$

Under inhomogeneous conditions, therefore, the apparent resistivities R_N and R_L for the normal and lateral devices can be determined from the formulae

$$\bar{\Phi} \equiv \bar{\Phi}_M = \frac{R_N \bar{I}}{4\pi} \times \frac{1}{z} \tag{12}$$

$$-\frac{\partial \bar{\Phi}}{\partial z} = \frac{\bar{\Phi}_{MN}}{MN} = \frac{R_L \bar{I}}{4\pi} \times \frac{1}{z^2} \tag{13}$$

where $z = AM$ in eqn (12), and $\bar{\Phi}_{MN} = \bar{\Phi}_M - \bar{\Phi}_N$ and $z = AO$ in eqn (13). While eqn (13) was derived directly from eqn (11), one can also take a z-derivative of eqn (12) and write the following alternative formula for $\partial \bar{\Phi}/\partial z$:

$$\frac{\partial \bar{\Phi}}{\partial z} = -\frac{R_N \bar{I}}{4\pi} \times \frac{1}{z^2} + \frac{\bar{I}}{4\pi z} \times \frac{\partial R_N}{\partial z} \tag{14}$$

Combining eqns (13) and (14), one obtains

$$R_L = R_N - z\frac{\partial R_N}{\partial z} \tag{15}$$

$$= \frac{R_N(AM) + R_N(AN)}{2} - AO\frac{R_N(AN) - R_N(AM)}{MN} \tag{16}$$

where $z = AM = AO$, $\partial R_N/\partial z$ is the slope of the normal departure curve at $z = AM = AO$, and $R_N(AM)$ and $R_N(AN)$ are the normal apparent resistivities for spacings AM and AN. Equation (15) or (16) can be used to compute the response of the lateral sonde from that of the normal. Published departure curves (Anon., 1955) confirm that (i) $R_L = R_N$ at the maximum or minimum points of the normal departure curves where $\partial R_N/\partial z$ is zero; and (ii) R_L is greater or less than R_N according to whether the spacing is longer or shorter than that for the extremum, as required by eqn (15).

Formulae (15) and (16) are valid for any ground geometry. Note also that, in this particular case, one can take the reverse step of computing the normal response from that of the lateral from the formula

$$R_N = AM \int_{\substack{z=AM \\ =AO}}^{\infty} \frac{R_L}{z^2} dz \tag{17}$$

which follows from eqns (12) and (13). In general, however, the process is not reversible.

4.2. Laterolog 7 Device

The problem of direct current potential in media containing cylindrical and orthogonal plane interfaces has so far eluded a direct theoretical solution, although it is well defined in terms of a differential equation and source and boundary conditions. Laterolog 7 has its own peculiar difficulties, in addition to the above, which makes it ill-defined and therefore theoretically intractable in general. First, the currents leaving or entering the shorted electrodes M_1, M_1', M_2 and M_2' are unknowns that cannot be taken into account. Secondly, the potentials at O_1 and O_2 are generally unequal—otherwise, the shorting becomes meaningless—and their measurement correspondingly ambiguous. Thirdly, simultaneous minima or null of potential at the predetermined positions O_1 and O_2 cannot in general be established unless $\bar{I}_1$ and $\bar{I}_2$ are separately adjustable. Since the power electrodes A_1 and A_2 are shorted in an LL7, such individual adjustment is not possible and the device therefore seeks only an approximate null, resulting again in a situation that is analytically vague.

The difficulties characteristic of the LL7 disappear under two relevant geometries: (i) infinitely thick invaded or uninvaded formation without top and bottom, so that only coaxial cylindrical boundaries exist; and (ii) centrally or symmetrically positioned sonde against a finitely thick bed sandwiched between identical upper and lower shoulders. In both these cases, $\bar{I}_1 = \bar{I}_2 = \bar{I}$ for simultaneous null at O_1 and O_2, $\Phi(O_1) = \Phi(O_2)$, and the shorting leads become ineffective. Enforcement of null at O_1 and O_2, under these conditions, requires that

$$\frac{R_L' I_o}{4\pi} \frac{1}{(0.2L)^2} + \frac{R_L'''\, \bar{I}}{4\pi} \frac{1}{(0.7L)^2} = \frac{R_L''\bar{I}}{4\pi} \frac{1}{(0.3L)^2}$$

or

$$\frac{\bar{I}}{I_o} = \frac{25R_L'}{11.11R_L'' - 2.04R_L'''} = \bar{\beta}, \text{ say,} \tag{18}$$

where R_L', R_L'' and R_L''' are the lateral apparent resistivities for $AO = 0.2L$, $0.3L$ and $0.7L$ respectively. One can of course use eqn (15) or (16) to obtain an expression for $\bar{\beta}$ either in terms of three normal apparent resistivities and their gradients at $AM = 0.2L$, $0.3L$ and $0.7L$ or in terms of six normal apparent resisitivities with $AN = 0.2L \pm M_1 M_1'/2$, $0.3L \pm M_1 M_1'/2$ and $0.7L \pm M_1 M_1'/2$. Note that, for homogeneous ground, $R_L' = R_L'' = R_L'''$ in eqn (18) and $\bar{\beta} \to \beta = 2.76$.

For the apparent resistivity R_{LL7} of a Laterolog 7, one has either

$$R_{LL7}(N) = \frac{4\pi L}{5 + 4 \cdot 76\bar{\beta}} \frac{\bar{\phi}(O)}{I_o} \quad \text{(NGRI version, from eqn 3)} \quad (19)$$

or

$$R_{LL7}(S) = \frac{4\pi L}{18 \cdot 125} \frac{\bar{\Phi}(O)}{I_o} \quad \text{(Schlumberger version, from eqn 4)} \quad (20)$$

where currents $\bar{\beta}I_o$, I_o and $\bar{\beta}I_o$ emanate from the electrodes A_1, A_o and A_2, and $\bar{\Phi}(O)$ is the potential measured at O_1 or O_2. But, by the principle of superposition, one simultaneously has

$$\bar{\Phi}(O) = \bar{\Phi}'(O) + \bar{\Phi}''(O) + \bar{\Phi}'''(O) \tag{21}$$

where

$$\bar{\Phi}'(O) = \text{potential at } O_1 \text{ or } O_2 \text{ due to current } I_o \text{ from } A_o$$

$$= \frac{R'_N I_o}{4\pi} \frac{1}{0 \cdot 2L};$$

$$\bar{\Phi}''(O) = \text{potential at } O_1 \text{ due to current } \bar{\beta}I_o \text{ from } A_1$$

$$= \text{potential at } O_2 \text{ due to current } \bar{\beta}I_o \text{ from } A_2$$

$$= \frac{R''_N \bar{\beta}I_o}{4\pi} \frac{1}{0 \cdot 3L};$$

$$\bar{\Phi}'''(O) = \text{potential at } O_1 \text{ due to current } \bar{\beta}I_o \text{ from } A_2$$

$$= \text{potential at } O_2 \text{ due to current } \bar{\beta}I_o \text{ from } A_1$$

$$= \frac{R'''_N \bar{\beta}I_o}{4\pi} \frac{1}{0 \cdot 7L};$$

and R'_N, R''_N and R'''_N are the normal apparent resistivities for $AM = 0 \cdot 2L$, $0 \cdot 3L$ and $0 \cdot 7L$ respectively. Substitution for $\bar{\Phi}(O)$ from eqn (21) into eqns (19) and (20) yields

$$R_{LL7}(N) = \frac{5R'_N + \bar{\beta}(3 \cdot 33R''_N + 1 \cdot 43R'''_N)}{5 + 4 \cdot 76\bar{\beta}} \tag{22}$$

or

$$R_{LL7}(S) = \frac{5R'_N + \bar{\beta}(3 \cdot 33R''_N + 1 \cdot 43R'''_N)}{18 \cdot 125} \tag{23}$$

as the case may be. Equations (18), (22) and (23) allow one to compute the LL7 apparent resistivities from those of three normal and three lateral sondes or from those of six normal sondes alone. For a bed of infinite thickness, these six apparent resistivities are independent of the vertical positions of the six constituent sondes along the borehole axis. With the insertion of horizontal boundaries, however, they must refer to the specific vertical locations relative to the boundaries.

It will be clear from the above discussion that the two-electrode array or the normal sonde is the basic unit from which *all* other arrays or sondes can be built up, without exception. The response of *any* sonde is nothing more than a weighted average of the *unfocused* responses from several normals of specified lengths (and relative locations). This rather obvious result of the principle of superposition is not always accepted without a certain amount of scepticism and reluctance, although one must assume that the principle itself is well known. Much of the popular appeal of focusing stems from a lingering but unfounded feeling that a focused current pattern, because of its streamlined elegance, has some special properties or attributes not possessed by unfocused configurations.

For a complex array of length L between its two extreme active electrodes, the longest of the constituent dipoles can have a length at most equal to L, although it is usually smaller. It follows immediately that, for a given spacing, no resistivity sonde or array can have a radius or depth of investigation larger than that of a normal sonde or two-electrode configuration of the same spacing.

5. FURTHER OBSERVATIONS ON LATEROLOG 7

Figure 7(a) reproduces a qualitative diagram from Doll (1951) and Anon. (1958) for the 'comparative distribution of current lines for normal device (left) and laterolog (right) opposite a thin resistive bed'. Acutally, however, only the left half of this diagram corresponds to the caption; the right half does not. The right half pertains really to a homogeneous environment, not to a thin highly resistive formation with $R_T \gg R_s$ and $R_m \doteq R_s$ for which the laterologs were specially devised. Indeed, such flawless parallelism is not attainable even in a homogeneous medium (compare with Fig. 1). Figure 7(b) from Guyod (1966) illustrates quantitatively how heavily defocused the central shaded current lines from LL7 can be for a far off-centre position. Guyod (1966) concludes that '... Laterolog tools are not true focusing devices, although they are fre-

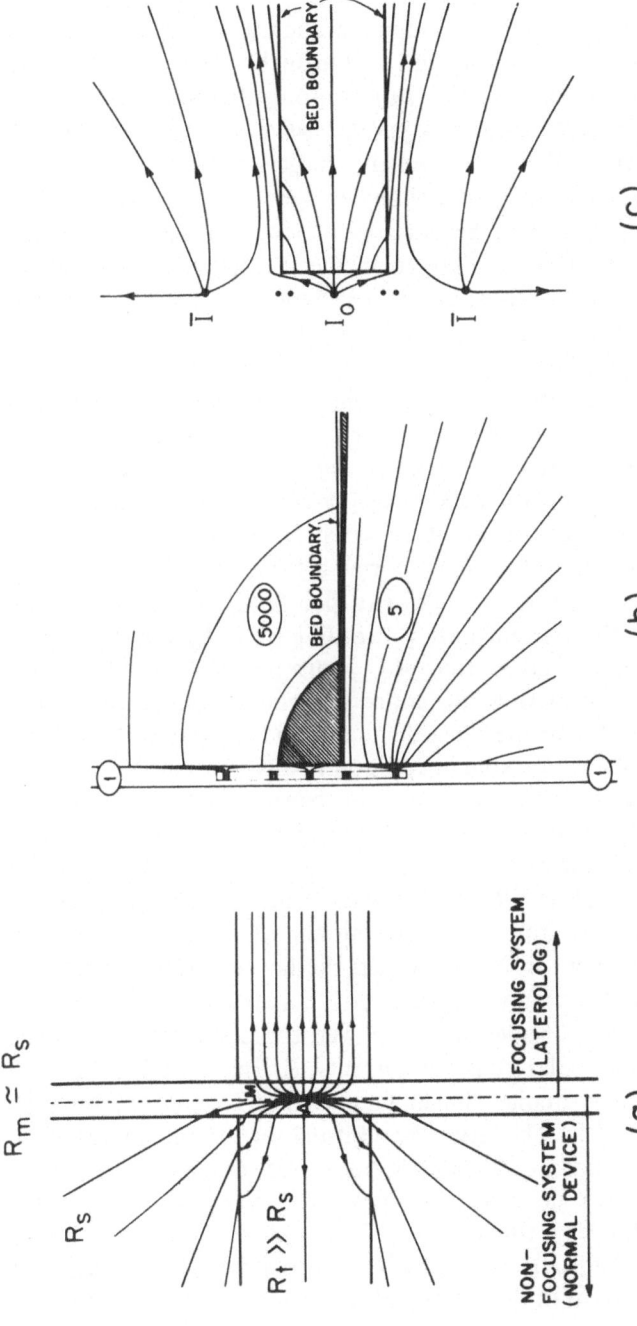

FIG. 7. (a) Schematic current lines from normal and LL7 devices against a thin highly resistive bed, reproduced from Doll (1951) and Anon. (1958) with permission from Society of Petroleum Engineers (AIME) and Schlumberger Well Services, USA; (b) quantitative LL7 current lines when positioned asymmetrically near lower edge of highly resistive formation, reproduced from Guyod (1966) with permission from Society of Professional Well Log Analysts, USA: (c) more realistic but still qualitative LL7 current pattern for same conditions as in (a). (Reproduced from Roy and Apparao (1978) with permission from European Association of Exploration Geophysicists, The Netherlands.)

quently referred to as such'. Figure 7(c) from Roy and Apparao (1978) is a more realistic, but still qualitative, representation of the current lines for an on-centre position of LL7 against a thin highly resistive target and should replace the right half of Fig. 7(a). Not only do the I_o current lines deviate strongly from the ideal of a parallel beam, a large fraction of I_o can altogether bypass the highly resistive target and flow in the much more conducting saline mud and shoulder beds. With increase in formation resistivity, defocusing becomes more severe and a progressively larger fraction of I_o avoids the target. In the theoretical limit when the formation is infinitely resistive, no current penetrates the target and I_o sidetracks the formation entirely. See Fig. 11 in Guyod (1964) for an illustration of this kind.

Even if the current pattern from the central electrode *were* focused in the manner shown in the right half of Fig. 7(a), no purpose is served because the signal is not measured *within* the focused region. The potential is measured on the sonde axis, at $O_1 O_2$, where the deviation from parallelism happens to be at its worst. The equipotential line or surface passing through O_1 and O_2 forms a triple loop around A_0, A_1 and A_2, and intersects itself at those two points (see Fig. 1, for homogeneous ground). The current lines proceeding from A_0 and A_1 toward O_1, and those from A_0 and A_2 toward O_2, turn nearly 90° at or near O_1 and O_2. Indeed, considering the entire neighbourhood of the sonde, locations O_1 and O_2 are the most disturbed and most unsuitable from the standpoint of focusing. Note in addition that the current lines, even if they are parallel in a vertical section as in the right half of Fig. 7(a), do not form a parallel system in three dimensions. They diverge radially outward from the sonde axis.

Moran and Chemali (1979) have drawn attention to the fact that the current configurations from the conventional LL7 and its reciprocal mode are quite different from each other; the first is 'focused' while the second is not. An idea of how *very* different these two configurations are can be acquired by considering the power sources in the reciprocal system: (i) two monopolar constant currents, $I_o'/2$ each, at M_1 and M_2; and (ii) two *strong* dipolar 'bucking' currents, $\bar{I}_1'$ and $\bar{I}_2'$, between $M_1 \rightarrow M_1'$ and $M_2 \rightarrow M_2'$. The latter are adjusted for zero potentials at A_1 and A_2; the potential at A_0 is then measured as signal. In homogeneous ground, such an adjustment makes the moment of each of the two current dipoles equal to $0.2625(A_1 A_2)I_o'$ so that, with $M_1 M_1' = M_2 M_2' = 2$ in(say), $I_1' = I_2' = 10 I_o'$ approximately. Although the overall current patterns in the direct and the reciprocal modes of LL7 are thus so vastly dissimilar, their responses are identical under all conditions—underscoring once again the misplaced emphasis on focusing. Moran and Chemali consider the two patterns as 'equally focused since they yield the same R_a ($\equiv R_{LL7}$) in all cases'. They are also of course equally unfocused by the same token.

The same process that 'focuses' the current from A_0 into the target also focuses the currents from A_1 and A_2 into the adjacent formations at the same time. The potential measured at O_1 or O_2—or at any other point— is generated not by the current from A_0 alone but also by those from A_1 and A_2. Although not shown in Fig. 7(a), currents $\bar{I}_1$ and $\bar{I}_2$ flow entirely through the shoulder beds and the mud column, and account for *most* of the measured signal. The statements that 'With a constant I_0 current, the potential varies directly with formation resistivity' (Anon., 1972, p. 20) and that 'As the current to the I_0 electrode is constant, the potential of the M electrodes is a measure of the resistivity of the rock opposite the center of the device' (Lynch, 1962, p. 150) are, in the author's opinion, unfounded.

In a homogeneous medium, the current I from each of the outer electrodes is about 2·76 times larger than I_0, for nulls at O_1 and O_2. A simple computation shows that, even in homogeneous ground, the contribution to $\Phi(O)$ by these two outer currents is about 2·63 times that from the central 'measure' current I_0. In the presence of a highly resistive formation of finite thickness, $\bar{I}/I_0$ is very much higher than 2·76, attaining values up to several hundreds or more, as illustrated in Fig. 8. The ratio of the contributions to $\bar{\Phi}(O)$ by ($\bar{I}_1 + \bar{I}_2$) and I_0 increases correspondingly to values much beyond 2. Thus, although the 'focused' sheet of current from A_0 in Fig. 7(a) travels almost entirely through the highly resistive target, the effect on measurement from the currents residing in the adjacent formations is considerably larger. In fact, after adjustment for null, I_0 in such cases can be switched off without materially altering the $\Phi(O)$ value. Moran and Chemali (1979, p. 921) also recognise that the two outer currents contribute 'the major part of the measure signal'; but they ignore these currents for apparent resistivity computation.

The quantity $\bar{\Phi}(O)/I_0$ in formulae (4) and (4a) can be interpreted as the electrical resistance of the ground material contained within the outermost envelopes of the I_0 current as it flows to infinity from the three-looped equipotential surface passing through O_1 and O_2 (Anon., 1958, p. 67). Since the region bounded in this manner includes the target formation, this resistance has been considered as specially significant and directly controlled by the formation resistivity. Notice, however, that

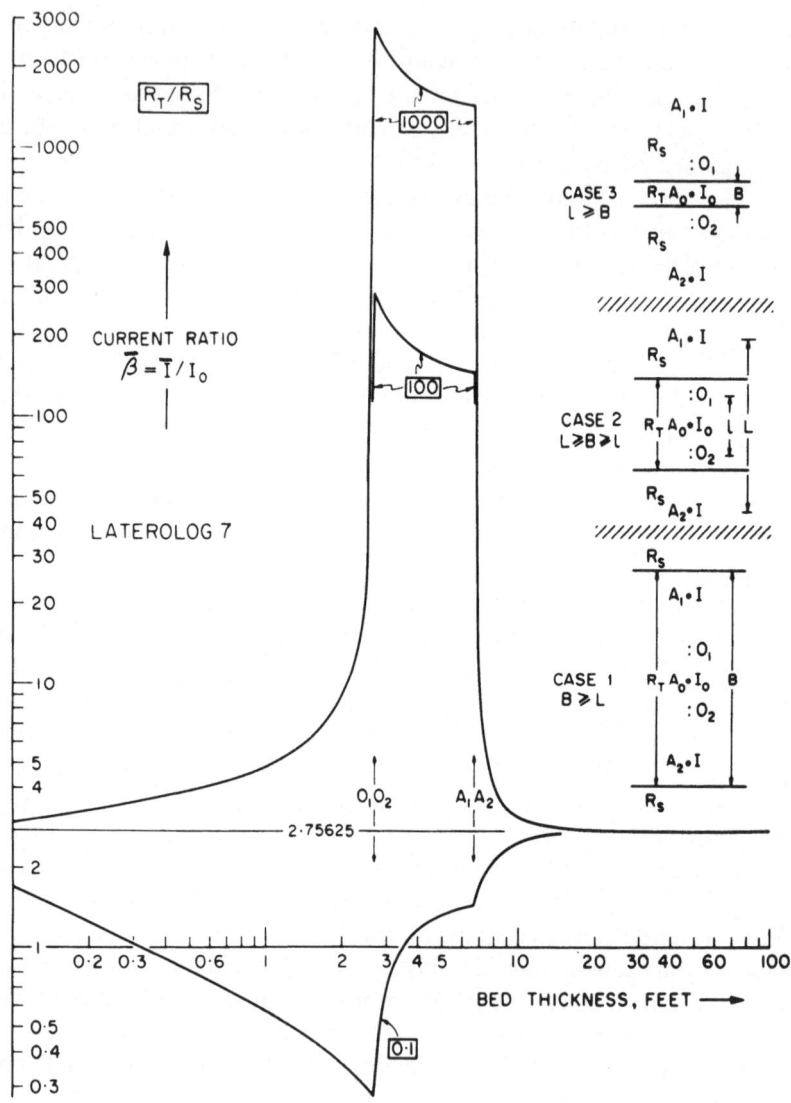

FIG. 8. Variation of $\bar{\beta} = \bar{I}/\bar{I}_0$ with bed thickness B for centrally located LL7. No borehole, $R_T/R_s = 1000$, 100 and 0·1. (Reproduced from Roy (1980) with permission from European Association of Exploration Geophysicists, The Netherlands.)

$\bar{\Phi}(O)/I_o$ is also the resistance, between the same equipotential surface and infinity, of the ground confined between any two current surfaces *emanating from A_1 or A_2* and enclosing a current equal to I_o. Since I_o is far smaller than $\bar{I}_1$ or $\bar{I}_2$, there would be many such regions contained strictly within the shoulder beds and the mud column. No special meaning can thus be attached to $\bar{\Phi}(O)/I_o$ relating it preferentially to the formation resistivity.

It must be re-emphasised that the theoretical analysis of a so-called 'focused' array is in no way different from that of an unfocused array. There is no special way of incorporating 'focusing' into the theory of a device, and no privileges can thus accrue because of 'focusing'. For all arrays, focused or non-focused, one has to consider each current electrode independently and then superpose the individual effects. Considered in this manner, the current lines from each power electrode in an LL7 cross the mud column and the formation boundaries exactly as in the left half of Fig. 7(a).

6. INTERPRETATION CHARTS

In studying the available charts for quantitative log interpretation, one comes up against two persistent difficulties. First, the publications rarely contain adequate information on how they were constructed and thus inhibit independent evaluation of their applicability, accuracy and limitations. Secondly, versions of the same chart published by different sources—or even the same source at various times—exhibit significant, even radical, variations. One such example is displayed in Figs. 9 and 10.

Figure 9 is a superposition of two borehole correction charts for LL3, published by Anon. in 1969 and 1972, for infinitely thick uninvaded beds. The coordinates of the initial point on the extreme left are (1, 0·98) in 1968 and (1·25, 0·78) in 1972—a variation of about 25% for the abscissa and 20% for the ordinate. The discrepancy between the two sets at other points is less but still considerable. Under homogeneous conditions, the tool does not read the true resistivity of the medium and thus violates the definition of apparent resistivity. What the sonde registers in this environment remains undefined in the 1972 version.

The Halliburton Oil Well Cementing Company (1952) and Welex Incorporated (Bulletin A127) published Guard Electrode response curves for uninvaded and invaded formations of infinite thickness, with tool dimensions similar to those for Schlumberger LL3. The Halliburton tool

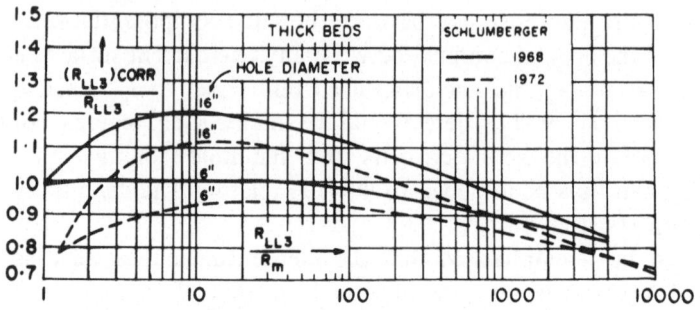

FIG. 9. Superposition of borehole correction charts for LL3 (Guard Electrode) published by Anon. in 1969 and 1972. Infinitely thick bed without invasion. $R_{LL3} \equiv R_{LL3}(S)$, and (R_{LL3})CORR means '$R_{LL3}(S)$ corrected'. (Reproduced from Roy (1980) with permission from European Association of Exploration Geophysicists, The Netherlands.)

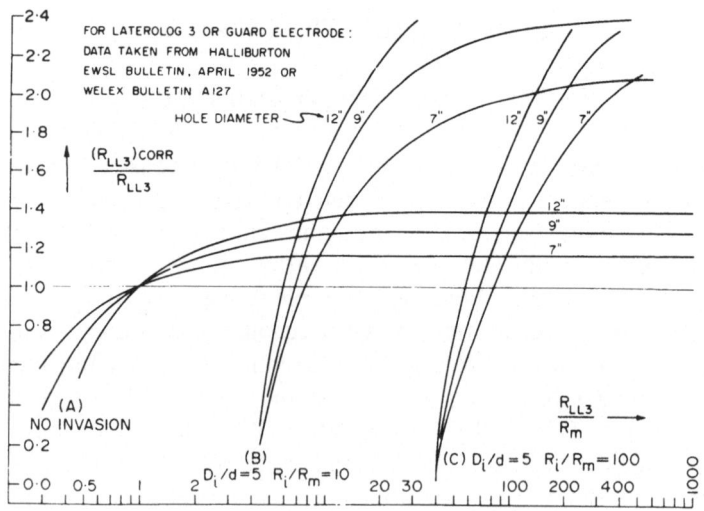

FIG. 10. Borehole correction chart for Guard Electrode (LL3) plotted from data published by Halliburton Oil Well Cementing Company (1952) and Welex Bulletin A127. Infinitely thick uninvaded and invaded beds. R_i and D_i are resistivity and diameter of invaded zone. Compare (A) with Fig. 9. (Reproduced from Roy (1980) with permission from European Association of Exploration Geophysicists, The Netherlands.)

has an overall length of 10 ft, a diameter of $3\frac{7}{8}$ in, and a measuring electrode length of 3 or 4 in. The Schlumberger device has two 5 ft guards, an unspecified electrode diameter, and a measuring electrode 'about 12 inches' long. These differences in linear dimensions are not important, since the response is independent of the length of the central measuring ring as long as it remains small relative to the total tool length. The Halliburton–Welex data have been recombined and plotted in Fig. 10 in the same format as in Fig. 9. The disagreement between the two diagrams is startling and goes beyond what the slightly differing tool dimensions could cause. Figure 10 behaves correctly in a homogeneous environment and reads consistently lower or higher than the true resistivity according to whether the formation is more or less resistive than the mud.

Very similar discrepancies occur for LL7 also in the data published by Anon. in 1952, 1969 and 1972 (see Roy, 1980, for details of these references).

Because of such disagreements among the published editions of log interpretation charts, new computations were made (Roy, 1980) for some of them, only to arrive at yet other versions. Figure 11 reproduces the shoulder bed correction charts for LL3 and LL7 from Anon. (1972). In the upper chart for LL7, some sample points have also been plotted from an earlier version (Anon., 1969) for $R_a/R_s \equiv R_{LL7}/R_s = 500$, 50 and 0·05. In the lower chart for LL3, the sample points are taken from a similar diagram published by Go International (1972).

Anon. (1969, 1972) requires that the borehole effect must be removed or corrected for before the charts in Fig. 11 are used. Also, the charts themselves are free of borehole parameters. For the new computation, therefore, it was assumed that these charts deal with a finitely thick bed sandwiched, without a borehole, between two semi-infinite shoulder beds of equal resistivity R_s. The location of the sonde is central and symmetrical, so that the electrode delivering the current I_0 lies midway between and is orthogonal to the formation boundaries (see right margin of Fig. 8).

The reconstructed shoulder bed correction charts for LL3 and LL7 are displayed in Fig. 12. In these recomputations, the conventional Schlumberger-type apparent resistivity eqns (4), (9) and (20) are used. The difference between Figs. 11 and 12 is therefore not related to the argument over the use of the real magnitudes of the two outer currents.

Although a vague similarity is discernible between the two LL7 charts in Figs. 11 and 12, the disagreement is still very substantial: (i) the LL7

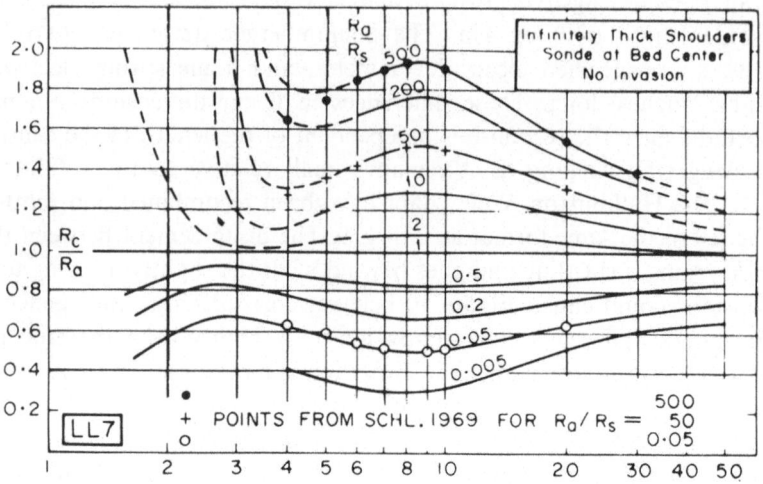

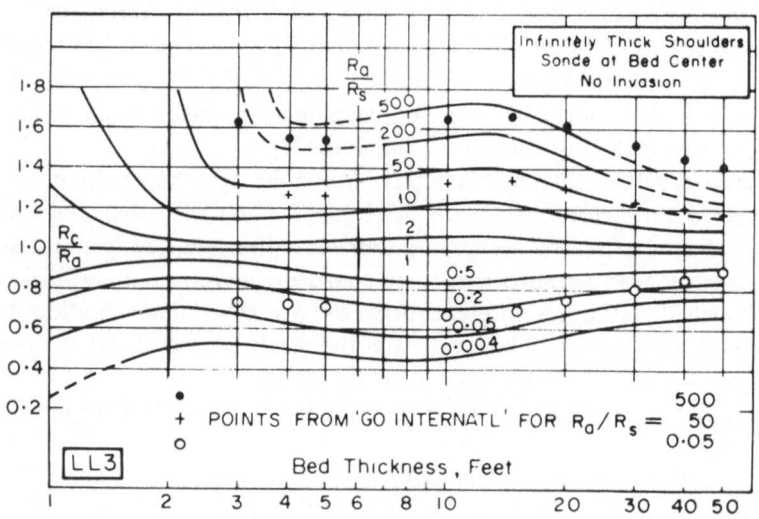

FIG. 11. Shoulder bed correction charts for LL7 (top) and LL3 (bottom). $R_a \equiv R_{LL7}(S)$ or $R_{LL3}(S)$, and $R_c \equiv R_{LL7}(S)$ corrected or $R_{LL3}(S)$ corrected, as the case may be, ● + ○ are points from similar charts by Anon. (1969) and Go International (1972). (Reproduced from Anon. (1972) with permission from Schlumberger Well Services, USA.)

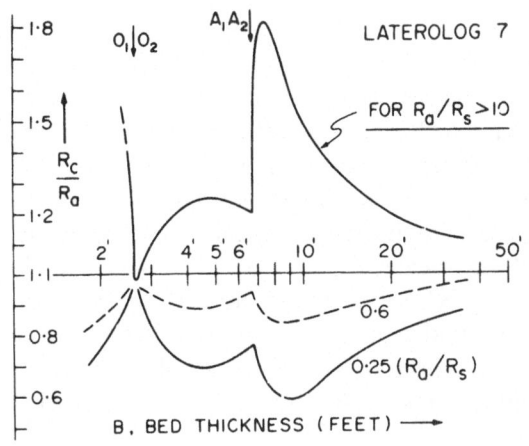

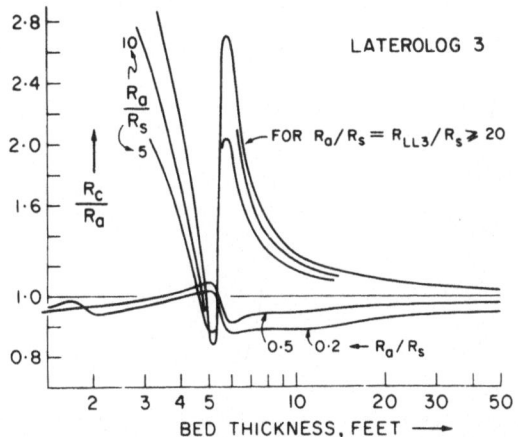

FIG. 12. Reconstructed shoulder bed correction charts for LL7 (top) and LL3 (bottom) using conventional apparent resistivity eqns (4) and (9). $R_a \equiv R_{LL7}(S)$ or $R_{LL3}(S)$, and $R_c \equiv R_{LL7}(S)$ corrected or $R_{LL3}(S)$ corrected, as the case may be. Compare with Fig. 11. (Reproduced from Roy (1980) with permission from European Association of Exploration Geophysicists, The Netherlands.)

chart in Fig. 11 does not exhibit any discontinuities at bed thickness $= O_1 O_2$ and $A_1 A_2$. These breaks occur as the discrete electrodes A_1, O_1, O_2 and A_2 cross the formation boundaries and should in principle exist as in Fig. 12—the borehole effect having been already removed earlier. The mud column does not exist and does not therefore smother the corners; (ii) the correction factor at bed thickness $= O_1 O_2$ is slightly smaller than unity in Fig. 12 for all values of $R_a/R_s \equiv R_{LL7}(S)/R_s$;

(iii) most importantly, for all values of R_a/R_s greater than about 10, Fig. 12 has only one correction factor curve, giving adequate accuracy. The curves separate and spread out only for low values of R_a/R_s.

The LL3 chart in Fig. 12 has little resemblance to that in Fig. 11. It has no discontinuities because each guard is a continuous metallic body and its transition across a formation boundary is gradual. Again, to a sufficient degree of approximation, one curve serves for all values of $R_a/R_s \equiv R_{LL3}(S)/R_s$ greater than about 20.

Figure 13 is the NGRI version of the shoulder bed correction chart for LL7 where the actual ground currents are taken into account, as in eqns (3), (19) and (22). The tool, according to this view, is totally unsuitable for beds thinner than A_1A_2 (see Fig. 4) and works well only for

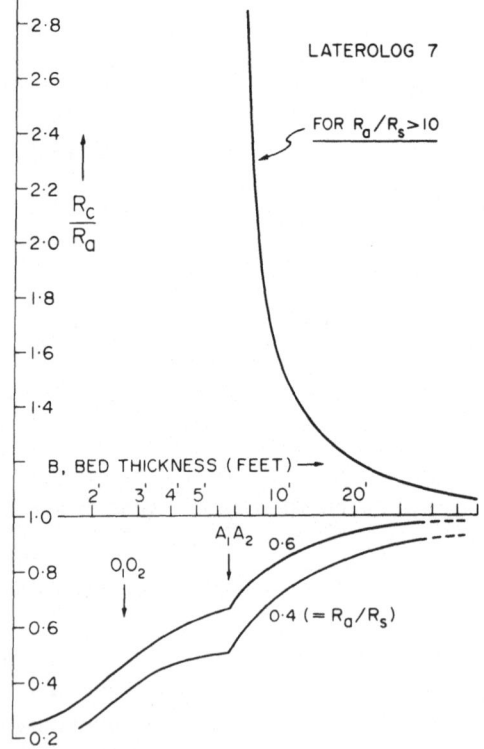

FIG. 13. Shoulder bed correction chart for LL7 using eqn (3) for computing apparent resistivity. $R_a \equiv R_{LL7}(N)$ and $R_c \equiv R_{LL7}(S)$ corrected. (Reproduced from Roy (1980) with permission from European Association of Exploration Geophysicists, The Netherlands.)

formations thicker than 10 ft, the sonde length being 6 ft 8 in. The only favourable feature that persists is that one curve is still adequate as long as $R_a/R_s \equiv R_{LL7}(N)/R_s$ is greater than 10.

7. CONCLUDING REMARKS

Looked at from any angle, the advantages resulting from focusing turn out to be illusory. The large deflections in LL3 and LL7 are the outcome of impermissible manipulations with the apparent resistivity formulae. As seen in eqn (6) and Fig. 6, such manipulations are possible with the normal and lateral devices as well, and results very similar to those yielded by focused devices can be or could have been obtained from the old and conventional non-focused sondes without the added complications and expenses of the former.

It is a welcome sign that Moran and Chemali (1979) agree to two of the basic points at issue, namely (i) the current pattern of an array, by itself, may not be necessarily and immediately indicative of its performance, and (ii) the measured signal comes primarily from the two 'auxiliary' or 'bucking' currents. This is the first essential step toward a correct appreciation of the focused resistivity logging tools. Besides LL3 and LL7, these tools include Laterolog 8, Dual Laterolog (LL9 ?), microlaterolog, spherically focused log (SFL), micro-SFL, proximity log and perhaps others. Actually, the basic issues involved are extraordinarily simple and do not go beyond first principles. Their eventual acceptance is inevitable, although the background of worldwide use of these devices over the past decades will be the cause for reluctance and delay. Equally inescapable is the requirement that the formulae for apparent resistivity must use the real currents that satisfy the necessary conditions of measurement and generate the measured signal. These necessary conditions are the establishment of null at O_1 and O_2 in an LL7 and the equalisation of potential over the three sections in an LL3.

Use of diagrams and concepts valid only for homogeneous ground to justify the operation of these devices in strongly heterogeneous media has been a persistent feature. Even in their 1979 paper, Moran and Chemali include the all-too-familiar spider-web diagram for the Guard Electrode in a homogeneous medium, which would have little resemblance to the current configurations in actual situations involving formations of high resistivity contrasts. Oversimplified also is the frequent analogy drawn between the guard-ring condenser and LL3. One deals with a homogeneous sample, the other does not. The guard-ring capacitor would be less than satisfactory for measuring the dielectric constants in a highly inhomogeneous specimen.

A ground surface geophysicist can but wonder at the variety and abundance of log interpretation charts, offering made-easy solutions to all kinds of problems that continue to baffle him theoretically and/or experimentally. In this respect, the log interpreters appear to be substantially better off than their ground surface counterparts, even though geophysical well logging began originally as an outgrowth or extension of the surface geophysical methods. More organisations have been working on ground surface geophysics for a longer time, with apparently less luck, to solve parallel problems with techniques founded on the same basic principles of physics. For instance, after over 60 years of surface resistivity profiling across vein-type ore deposits, the surface geophysicist cannot yet obtain simple charts or nomograms to translate his field data quantitatively into plan location, depth, thickness and conductivity. For the corresponding problem in logging, interpretation charts were already available in the late 1940s before the days of digital computation, although the operating conditions, the geometry and other parameters are generally more difficult in logging than in ground surveys. One wonders why surface geophysics appears to have fallen so far behind its own offspring.

One obvious reason for the proliferation of log interpretation charts is the drastic simplifications made of highly complex logging problems— simplifications so immoderate as to have been unacceptable to a surface geophysicist. For instance, the theory of induction log and its interpretation neglected the skin effect for a long time and used a geometric factor concept that, again, is valid for a homogeneous medium only. As late as 1972, induction logs were being 'automatically corrected' for the skin effect—a primary and a secondary correction—as if this phenomenon is an extraneous disturbance, not an integral part of the induction process itself. For another instance, in determining the true resistivity of a formation from LL3 and LL7, the present practice seems to be to correct the central reading *separately and successively* for borehole mud, side bed and invasion—reducing the real problem into a sequentially additive set of near trivialities. This is far removed from the actual situation where these three effects are mutually interlinked and inseparable. In inductive and resistive prospecting on the ground surface, such excessive approximations have been rarely used, if at all.

A not-so-obvious reason may be that the professional log interpreter, by and large, accepts the log interpretation charts at their face values all too readily and depends very greatly on what the relatively few logging contractors provide, as part of business. While logging operations remain the worldwide prerogative of a few specialised companies, ground geophysical surveys are carried out by very many organisations, groups and even individuals who rely little on black-box interpretation packages. Yet another latent reason may be that the quantitative predictions of a surface geophysicist looking for, say, a vein of ore are more readily and directly verifiable, and he therefore needs to exercise greater caution generally.

While excessively strong approximations are not uncommon and the charts themselves are as variation-prone as illustrated in Section 6, interpretations are sometimes carried through to an extraordinary degree of detail and refinement. For example, consider the problem of determining the mudcake thickness. According to chart Rxo–1 on page 42 of *Log Interpretation Charts* (Anon., 1972) and the text on page 34 of *Log Interpretation, Volume I—Principles* (Anon., 1972), mudcake thickness can be ascertained to an accuracy of $\frac{1}{16}$ in (1·6 mm) or better. This can apparently be done either from the caliper or the microcaliper log simply by subtracting the measured borehole diameter from the drill-bit size (when the former is the smaller of the two) or from chart Rxo–1 by using two microlog resistivities. In view of the very high accuracy involved, the first alternative is not compatible with the fact that holes are larger than the drill-bit size by unknown amounts depending on the type of formation, speed of drilling, mud characteristics, etc. For measuring a linear dimension of the order of 1·6 mm, indirectly as in the second alternative, at a depth of several kilometres, under a pressure of several hundred atmospheres and at a temperature of a few hundred degrees Celsius, such factors as unevenness of the borehole wall, electrode dimensions, accuracy of electrode spacing, temperature expansions in the tool, possible presence of sea-shells, pebbles or minor irregularities, leakage in microlog pads, etc., can become overridingly important. And all along, one must keep hoping that the tool itself, which pushes and slides against the borehole wall, does not disturb the mudcake. In fact, such accuracy would be hard to achieve even under the controlled conditions of an idealised laboratory model. With a scaling factor of 1/25, a 10 in drill hole would have a diameter of 1 cm in the model and a mudcake of actual thickness $\frac{1}{16}$ in would be but a smear, 0·06 mm thick, on the wall of the model bore.

Log interpretation in shaly sands makes a good second example.

Anon.(1972) speaks of laminar, structural, dispersed, bedded and average shales, and of shale being differentiated into its clay and silt components. Such ultrafine details coexist with formulae which are either empirical or involve extreme simplifications. The empirical relations are necessarily approximate as evident from the sizeable scatter of field points, and contain constants and exponents that vary over wide ranges. The exponent x in the R_{sh}–R_{clay}–R_{silt} relation, for instance, is usually taken as 2 although it is known to vary between 1·4 and 2·4—an uncertainty of 20–30% in the value of this exponent alone. The idealisations necessary for a theoretical derivation, quite apart from the simplifications already made, invariably depart from real ground conditions. No boundary has the exact shape assigned to it, nor is any formation really isotropic and homogeneous, nor does the current flow only along or across the bedding planes, nor are the sonde and the borehole coaxial, and so on. Formulae so assumed or derived require values of such quantities as true formation resistivity, formation water resistivity, porosity or formation factor, etc.—all of which are approximate at best.

An *a priori* evaluation of the overall effect on log interpretation of these numerous sources of uncertainty and error is not possible. A reasonable estimate can, however, be made after the event by statistically comparing the predictions over the past decades with the subsequently determined ground truths. In surface geophysical exploration for ores where simplifications are less severe in general, a quantitative interpretation is considered exceptionally good and rare if the predictions turn out to be within 10% of the actual values found by drilling. A 20% deviation is good interpretation, while 30% is still acceptable. In geophysical well logging, the overall reliability of predictions can hardly be any better. It follows that hyperfine details, not consistent with the inherent approximations and sources of error, serve no useful purpose. Rather, they can mislead and create a false sense of high reliability. At all times, the log interpreter like any other must be fully aware of the assumptions and limitations associated with each step of the interpretation process.

Curiously enough, exactly the same sentiment was expressed decades ago within the logging fraternity itself. 'Much of recent literature imply that our knowledge is sufficient to permit accurate and precise reservoir evaluation in all but a small percentage of the reservoirs encountered in electrical logging. This is a questionable implication fraught with danger, ... There are producing reservoirs that show minute resistivity contrasts at the oil–water contact. And there are also reservoirs with considerable contrast that produce only saltwater. All experts become adroit at rationalising these experiences, but that rationalisation awaits the results of production tests. ... Undoubtedly much of the inaccuracy stems from incomplete knowledge and facile assumption. ... These limitations must be ever present in the mind of the astute analyst' (Bulletin A101, Welex Inc.). The caution sounded 40 years ago by the father of quantitative log interpretation is as valid today as it was then. 'It should be remembered that the equations given are not precise and represent only approximate relationships' (Archie, 1942).

ACKNOWLEDGEMENT

This chapter is largely a summary of five papers (Roy, 1975, 1977, 1980; Roy and Apparao, 1976; Roy and Dhar, 1971). It is written entirely in a personal capacity, not on behalf of UNESCO, where the writer is currently employed.

REFERENCES

ANON. (1950). *Interpretation Handbook for Resistivity Logs*, Document No. 4, Schlumberger Well Surveying Corporation, USA.

ANON. (1955). *Resistivity Departure Curves (Beds of Infinite Thickness)*, Document No. 7, Schlumberger Well Surveying Corporation, USA.

ANON. (1958). *Introduction to Schlumberger Well Logging*, Document No. 8, Schlumberger Well Surveying Corporation, USA.

ANON. (1969) and (1972). (i) *Log Interpretation Principles*, and (ii) *Log Interpretation Charts*, Schlumberger Limited, USA.

APPARAO, A. and ROY, A. (1973). Field results for direct-current resistivity profiling with two-electrode array, *Geoexploration*, **11**, 21–44.

ARCHIE, G. E. (1942). The electrical resistivity log as an aid in determining some reservoir characteristics, *Petroleum Technol.*, **5**, TP 1422, 54–62.

DOLL, H. G. (1951). The laterolog: a new resistivity logging method with electrodes using an automatic focussing system, *J. Petroleum Technol.*, **3**, TP 3198, 305–16.

GO INTERNATIONAL (1972). *Log Interpretation Reference Data Handbook*, Gearhart-Owen Industries Inc., USA.

GUYOD, H. (1964). Factors affecting the responses of laterolog type logging systems (LL3 and LL7), *J. Petroleum Technol.*, **16** 211–19.

GUYOD, H. (1966). Examples of current distribution about laterolog sondes, *Log Analyst*, **7**, 27–33.

HALLIBURTON OIL WELL CEMENTING COMPANY (1952). *Guard Electrode*

Logging, Electrical Well Services Laboratory Bulletin, April 1952, Duncan, Oklahoma, USA.

LYNCH, E. J. (1962). *Formation Evaluation*. Harper and Row Publishers, New York, Evanston and London, 422 p.

MORAN, J. H. (1976). Comments on 'New results in resistivity well logging (*Geophysical Prospecting*, **23**, 426–448)', *Geophys. Prospecting*, **24**, 401–2.

MORAN, J. H. and CHEMALI, R. E. (1979). More on the laterolog device, *Geophys. Prospecting*, **27**, 902–30.

OWEN, J. E. and GREER, W. J. (1951). The Guard Electrode logging system, *J. Petroleum Technol.*, **3**, TP 3222, 347–56.

PIRSON, S. J. (1963). *Handbook of Well Log Analysis for Oil and Gas Formation Evaluation*. Prentice-Hall Inc. USA, 326 p.

ROY, A. (1974). Correction to 'Radius of investigation in DC resistivity well logging (*Geophysics*, **36**, 754–760)', *Geophysics*, **39**, 566.

ROY, A. (1975). New results in resistivity well logging, *Geophys. Prospecting*, **23**, 426–48.

ROY, A. (1977). The concept of apparent resistivity in Laterolog 7, *Geophys. Prospecting*, **25**, 730–7.

ROY, A. (1978). A theorem for direct current regimes and some of its consequences, *Geophys. Prospecting*, **26**, 442–63.

ROY, A. (1980). On some resistivity log interpretation charts, *Geophys. Prospecting*, **28**, 453–91.

ROY, A. and APPARAO, A. (1971). Depth of investigation in direct current methods, *Geophysics*, **36**, 943–59.

ROY, A. and APPARAO, A. (1976). Laboratory results in resistivity logging, *Geophys. Prospecting*, **24**, 123–40.

ROY, A. and APPARAO, A. (1978). Reply to comments on 'New results in resistivity well logging' and 'Laboratory results in resistivity logging', *Geophys. Prospecting*, **26**, 481–4.

ROY, A. and DHAR, R. L. (1971). Radius of investigation in dc resistivity logging, *Geophysics*, **36**, 754–60. Errata in *Geophysics*, **39**, 566 (1974).

WELEX INCORPORATED. (i) Bulletin A101, *Interpretation Charts for Electric Logs and Contact Logs*, and (ii) Bulletin A127, *Application of Radiation-Guard Surveys to Carbonate Reservoirs*, Fort Worth, Texas, USA. Dates of publication not given.

Chapter 4

GAMMA–RAY LOGGING
AND INTERPRETATION

P. G. KILLEEN

Geological Survey of Canada, Ottawa, Canada

SUMMARY

Gamma-ray logging has evolved from a qualitative estimation of the variation of natural radioactivity with depth in a borehole, using a Geiger-Müller detector, to a sophisticated quantitative logging technique which yields information on the identification and amount of the radioelements present, as well as data on the likely geologic history of mobilisation of these radioelements in the rock. To extract this additional information, a keener appreciation of the natural radioactive decay series, the isotopes involved, their half-lives, the energies of their gamma-rays and the interaction of these gamma-rays with the rock through which they travel is necessary. This has led to the development of improved data acquisition systems (digital logging systems), new methods of data processing (application of digital time series analysis—inverse filtering), the introduction of gamma-ray spectral logging, and improvements in variety and type of detectors to enhance the logging measurements and facilitate the new data processing techniques. Many of the new developments relate primarily to uranium exploration problems but they also have ramifications in other geologic applications of gamma-ray logging.

1. INTRODUCTION

The most important new developments in gamma-ray logging include advances in the hardware and data acquisition techniques, as well as advances in the methods of data interpretation. The former are the result of the development of new and improved gamma-ray detectors, the introduction of digital data acquisition to logging systems, and more recently the addition of the minicomputer or microcomputer to logging systems. The latter stems from the application of inverse filtering (or deconvolution) to the gamma-ray logs which has resulted in probably one of the most significant advances in the development of improved methods of quantitative uranium, thorium and potassium determinations. This technique has enabled a logging system to produce quantitative uranium determinations in 'almost' real time at the site of a uranium exploration borehole.

Many of these innovative developments have been designed to solve one or more of the problems which earlier techniques could not. For example, the gamma-ray spectral log distinguishes the three naturally occurring radioelements, thorium, uranium and potassium, while the gross count gamma-ray log assumes all radioactivity detected is from one source, e.g. the uranium decay series alone. Another example is the high resolution gamma-ray detector which makes it possible to produce an accurate log and interpretation even if there is radioactive disequilibrium. Both of the previously mentioned systems assume there is radioactive equilibrium.

To appreciate a review of developments in gamma-ray logging and interpretation, we must first, at least briefly, review the physical basis of gamma-ray logging and point out those areas in which there have been new developments.

2. TECHNOLOGICAL BASIS FOR GAMMA-RAY LOGGING

Gamma-rays are detected and recorded (or logged) in boreholes drilled for exploration or development. These logs can yield information on the characteristics of the rock, its geological history, the presence of radioelements, and their concentrations.

Besides the obvious application to uranium exploration, gamma-ray logging has been applied to exploration and development for oil and gas, coal, potash, site evaluation for potential radioactive waste management vaults, lithologic/stratigraphic identification problems and other geological applications wherein the gamma-rays provide a useful characteristic signature.

2.1. Gamma-Radiation in Nature

The gamma-rays of interest in the present application originate from radionuclides found in nature, primarily potassium and the uranium and thorium decay series. Although potassium is fairly abundant (about 2·6%) in the earth's crust, the radioactive isotope K-40 represents only a small fraction of this (0·0118%). However, the radioactivity from potassium often dominates the total detected gamma-ray count rate, since uranium and thorium and their radioactive daughter isotopes are usually present in abundances on the parts-per-million level. Thus, in oil and gas exploration, for example, the characteristic gamma-radiation from a shale, resulting from its potassium content, makes it easily distinguishable from a low potassium rock such as limestone, dolomite or sandstone. The gamma-radiation contributed by uranium and thorium becomes important in cases where minerals containing these elements are present. Uranium and thorium may either represent interference by producing a gamma-ray count which is interpreted as due to potassium, or they may represent additional lithologic information which can be used to identify certain rocks by their characteristic uranium and thorium content. In potash mining applications, the gamma-ray log can be used to compute quantitative potassium assays from the measurements. In coal mining applications, the coal seams can be accurately delineated by their lack of radioactivity as shown on a gamma-ray log. Of course, there are exceptions to all of these and broad interpretations using generalised rules must be viewed with suspicion. In the case of spectral logging the individual radioelements K, U and Th can be measured, reducing the ambiguities and expanding the information base. Variations in ratios of U/Th, U/K or Th/K may be diagnostic of certain geologic processes. Applications and interpretation of gamma-ray spectrometry to uranium exploration including discussion of the use of radioelement ratios in geology has been reviewed by Killeen (1979).

In uranium exploration, where the quantities of that radioelement are anomalously high, the object of gamma-ray logging is to make quantitative determinations of the grade and thickness of the uraniferous zones intersected by the borehole. It is often assumed that potassium and thorium do not contribute significantly to the total detected gamma-ray count rate. This may not be true in certain low grade uranium deposits, and is certainly not true in the conglomeratic uranium ore of the Elliot Lake uranium mining area or the pegmatite-type ore of the Bancroft uranium mining area, both in Ontario, Canada. In these areas, thorium concentrations often exceed the uranium concentrations. In cases where more than one radioelement are present, gamma-ray spectral logging must be used to separate the contributions from each of the radioelements to the total count rate.

Figure 1 shows the radioactive decay series for uranium-238 (the most abundant U isotope in natural uranium), the decay series for uranium-235 (an isotope of U representing only about 0·72% of natural uranium) and the thorium-232 decay series. Since the relative isotopic abundance

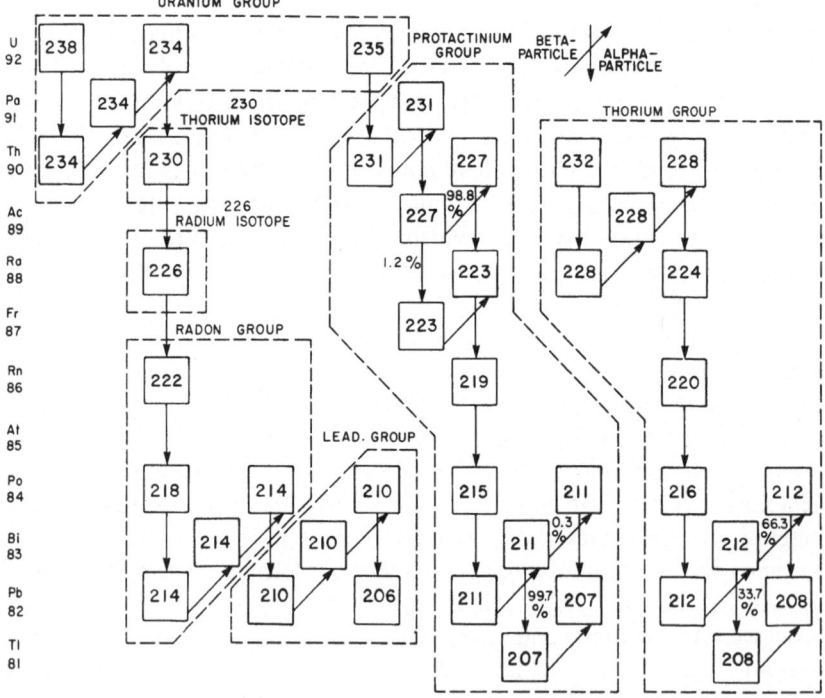

FIG. 1. The radioactive decay series for uranium-238, uranium-235 and thorium-232 (after Rosholt, 1959). In gamma-ray logging, gamma-rays emitted by the various daughter products in these decay series are detected, counted and displayed as a function of depth in boreholes. Information about these daughters, their relative location in the decay scheme, their half-lives and their gamma-ray energies and relative intensities is required for a complete interpretation of a modern gamma-ray log.

of U-235 is so low, the U-235 series is usually insignificant in consideration of gamma-radiation. Within the U-238 decay series, a number of daughters actually produce most of the detectable gamma-radiation; U-238 does not emit gamma-rays. Thus, determination by gamma-counting is an indirect method. Gamma-rays from the daughter bismuth-214 are the main component of the radiation, and these are then related to the quantity of parent uranium at the top of the decay series. Similarly for thorium, gamma-rays emitted by the daughter isotope thallium-208 are counted and related to the quantity of the parent.

Gamma-ray energies are characteristic of the emitting nuclide as shown in Fig. 2. This figure shows the principal gamma-rays in the energy spectrum for uranium-238 and thorium-232 and their daughters in radioactive equilibrium. Although only these gamma-rays need be considered for gamma-ray or gamma-ray spectral logging with a scintillation detector such as sodium iodide, much more detailed information is required for high resolution spectral logging (see Section 5.)

If only one of the three radioelements is present, it may be measured by gross count gamma-logging or total count logging (generally using a low energy threshold of around 0·4 MeV). If more than one is present, a gamma-ray spectral logging system can measure the energies of the gamma-rays, and determine the proportions due to K, U or Th (actually K-40, Bi-214, or Tl-208, since the energies usually measured and counted are from these daughters).

2.2. Radioactive Equilibrium

Most of the gamma-radiation emitted by nuclides in the uranium decay series is not actually from uranium, but from daughters in the series. The measured count rate can be related to the amount of parent by assuming there is a direct relation between the amount of daughter and parent. This assumption is valid when the radioactive decay series is in a state of secular equilibrium.

A radioactive decay series such as that of U-238 is in secular equilibrium when the number of atoms of each daughter being produced in the series is equal to the number of atoms of that daughter being lost by radioactive decay. When this condition exists it is possible to determine the amount of the parent of the decay series by measuring the radioactivity from any daughter element.

It is important to know whether the assumption of secular equilibrium, required for analysis by gamma-ray logging techniques, is valid for the geological material being analysed for its uranium content. If one or

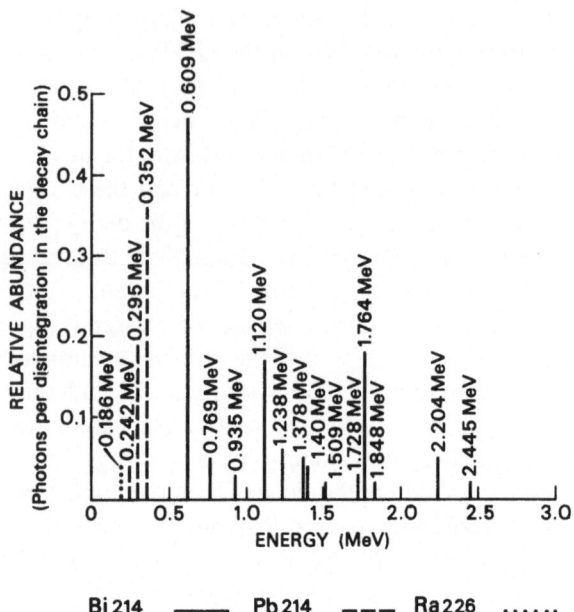

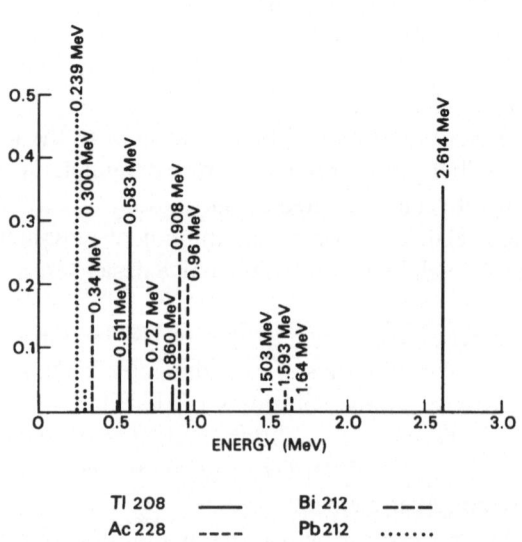

FIG. 2. The principal gamma-rays emitted by the uranium-238 decay series (top) and by the thorium-232 decay series (bottom).

more of the daughter products is being lost by any process other than radioactive decay, or if the parent was recently deposited, the assumption of secular equilibrium is not satisfied. Since each daughter product is an element with its own characteristic physical and chemical properties it may behave differently within a given environment. For example, in the U-238 decay series, radon-222 is a gas with a 3·85 day half-life. Also, the solubilities of radium, uranium and thorium differ, and preferential leaching of elements may occur.

Radioactive disequilibrium is commonly found in the roll front or sandstone-type uranium deposits of Colorado and Wyoming in the United States. The ore occurs along a front, usually C-shaped in cross-section. The front is formed by ions moving in solution in the sandstone which deposit along an oxidation—reduction interface. The roll front is bounded on the concave side by oxidised altered rock and on the convex side by relatively reduced rock. The disequilibrium occurs because uranium is relatively more mobile within the sandstone than the daughter products. After a period of time this leads to a distribution of radioelements wherein the slow-moving daughter products (e.g. Bi-214) form a region in the sandstone of daughter excess (or parent deficiency state), with strong gamma-ray activity. At some nearby location the more mobile parent uranium forms a (relatively) weakly radioactive uraniferous zone.

The degree of disequilibrium will vary with circumstances, and a number of factors are involved. These include the mineralogy of the radioelements and the mineralogy of their surroundings which may create a reducing or oxidising environment, for example the presence or absence of sulphides or carbonates. Uranium can exist in two valence states, either hexavalent or quadrivalent. In the hexavalent state it is very mobile in nature, while otherwise it is not. In an oxidising environment uranium will revert to the hexavalent state. Thus the degree of disequilibrium is strongly dependent upon climate, topography and surface hydrology. The detection of disequilibrium is also dependent on the sample volume studied. A small drill core specimen is much more likely to show extreme disequilibrium than will a large bulk sample such as in gamma-ray logging, i.e. in some cases the parents and daughters may have moved apart on the scale of a hand specimen, but not on the scale of a cubic metre of rock. The degree of disequilibrium is not easily established with direct field measurements, although it can be determined in a number of ways in the laboratory, either by comparing chemical analyses with estimates based on decay product radioactivity, or by measuring the radioactivity of different decay products.

Disequilibrium in a uranium deposit or occurrence does not necessarily rule out gamma-ray logging as an evaluation tool. Satisfactory
results may still be obtainable by spot-checking the state of disequilibrium of the deposit by laboratory chemical analyses, and applying
appropriate corrections to the gamma-ray logs. Gamma-ray logging has
been shown to be an effective quantitative evaluation tool in the
sandstone-type uranium deposits in the United States, even though these
deposits are known to be in radioactive disequilibrium to various
degrees.

2.3. Interaction of Gamma-Rays with Rocks
In borehole logging two types of interactions between gamma-radiation
and matter are important. These interactions are Compton scattering
and the photoelectric effect. In Compton scattering, the photon imparts
some of its energy to an electron and is scattered at an angle to its
original direction. This process predominates for moderate energies in a
wide range of materials. In the photoelectric effect, the photon gives up
all its energy in ejecting a bound electron from an atom. This process
predominates at low energy, especially in matter with a high atomic
number Z. In the case of rocks which represent a complex medium the
atomic number is the 'effective' or 'equivalent' atomic number Z_{eq}. The
equivalent atomic number Z_{eq} of a complex medium is defined as the
atomic number of the element having the same ratio of Compton
attenuation to photoelectric attenuation (for a given photon energy) as
the complex medium.

A third interaction, pair production, may occur with gamma-rays
having energies higher than $1 \cdot 022$ MeV, and even then not in significant
proportion except for large values of Z_{eq} of the rock. The proportion in
which these three interactions occur depends on gamma-ray energy and
Z_{eq} as shown in Fig. 3. It can be seen from the figure that for the range of
atomic numbers in common rocks (approximately $Z = 12$ to $Z = 18$)
about 95% of the gamma-ray interactions with the rock are Compton
scattering for the range of gamma-ray energies (see Fig. 2) emitted by the
natural radioelements and their daughters. If, however, the uranium
content of the rock causes a significant increase in the Z_{eq} of the rock,
then at low energies the percentage of interactions due to Compton
scattering decreases and the percentage of photoelectric interaction
increases. Total absorption by the photoelectric effect decreases the
gamma-ray count rate, and results in a non-linear relation between true
ore grade and ore grade computed on the basis of gamma-ray count
rates. This so-called 'Z-effect' becomes an important source of error

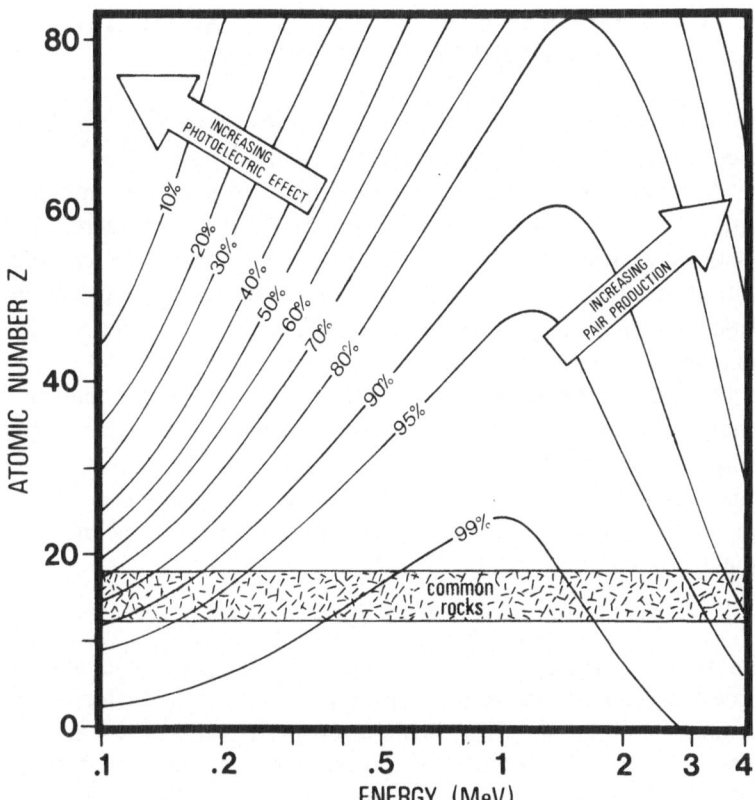

FIG. 3. The percentage of Compton scattering interactions in materials of different atomic number Z for different gamma-ray energies (after Conaway and Killeen, 1980). The equivalent atomic number of common rocks is indicated.

above 0·4% uranium content in the rock. It can be avoided by counting only gamma-rays of energies which are high enough as to not be affected by the change in Z_{eq}. This may be accomplished with the use of a shielded (i.e. filtered) detector or an electronically set threshold which only permits high energy gamma-rays to be counted.

The ratio of the number of high energy to low energy gamma-rays in an energy spectrum is constant unless the Z-effect causes a decrease in the number of low energy gamma-rays. Consequently, a gamma–gamma-density tool, modified to produce a gamma–gamma-spectral tool, could be used to provide information on the effective atomic number of the rock by observing this variation in photoelectric effect. This is the basis of a relatively new lithological identification log offered by some service companies.

3. GAMMA-RAY LOGGING AND QUANTITATIVE
URANIUM MEASUREMENTS

Many of the advances in gamma-ray logging have been made with respect to the problem of evaluation of uranium deposits. First, however, some discussion of the basic premise behind all standard methods of quantitative interpretation of gamma-ray logs will be considered. It will be assumed in the following discussion that all count rates considered have been deadtime corrected. Scott (1980) discussed the most accurate method of determining deadtime, and pointed out the inaccuracies in some commonly used methods.

It has been determined that the area beneath a gamma-ray log anomaly is directly proportional to the quantity of radioactive material causing the anomaly (Scott *et al.*, 1961). Note that the amplitude of the anomaly is not generally proportional to the uranium grade. The area, however, is proportional to the product of the grade and thickness of the radioactive layer, i.e.

$$GT = KA \qquad (1)$$

where G is the average ore grade (ppm eU or %eU) over a zone of thickness T; K is the constant of proportionality or K-factor; A is the area under the curve. Thus with a properly calibrated gamma-ray logging tool, the GT product of any uraniferous zone could be computed. The thickness has generally been considered to be the full-width-half-maximum (FWHM) of the anomaly (not always true—see Section 3.5.4), and hence an average grade could be estimated.

3.1. The Iterative Technique Applied to Gamma-Ray Logs

For a complex anomaly, the principle of superposition was used. This means that a complex anomaly can be considered to be the sum of the individual anomalies produced by a number of adjacent thin ore zones. This principle is demonstrated in Fig. 4 where the anomaly from each of four 5 cm thick zones is summed to produce the anomaly corresponding to a 20 cm thick zone. Using an iterative technique any anomaly can be broken down into a series of adjacent 5 cm thick zones of different grades. The calibration of the gamma-ray tool to obtain the shape of the basic 5 cm thick anomaly must be done in model holes such as those in Ottawa, Canada (Killeen, 1978; Killeen and Conaway, 1978) or in Grand Junction, USA (Mathews *et al.*, 1978). The iterative technique requires that the entire anomaly shape be used in the computation, and the

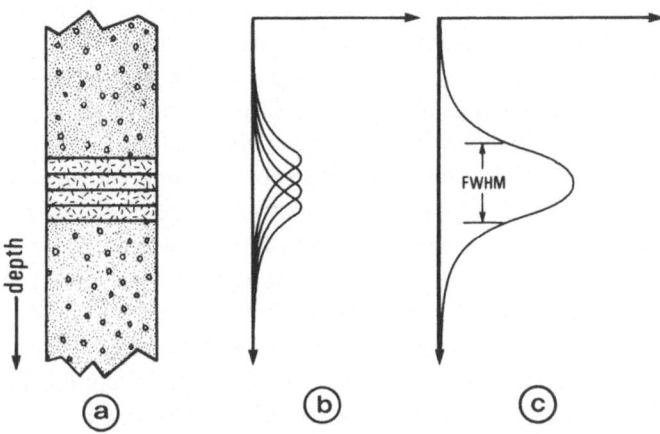

FIG. 4. Illustration of the principle of superposition wherein the four adjacent 5 cm thick radioactive zones (a) produce the four individual gamma-ray anomalies shown in (b) which, when summed, produce the measured gamma-ray log shown in (c). To interpret a gamma-ray log requires finding a method to go from the measured log (c) to the ore grade distribution (a).

number of iterations (adjustments of the grades of the thin zones and recomputing the sum of the individual anomalies) determines the accuracy of the final results.

3.2. Inverse Filtering of Gamma-Ray Logs

If the gamma-ray log is deconvolved by an inverse filtering technique instead (Conaway and Killeen, 1978a), the amplitude of the deconvolved log is everywhere proportional to grade along the log. That is

$$G = KI \qquad (2)$$

where I is the intensity (or count rate) of the gamma-ray log at any point on the log and G is the grade at that point using the same K-factor as mentioned above. It is assumed that all gamma-rays originate from the uranium decay series, and that it is in radioactive equilibrium.

The ultimate objective of gamma-ray logging is the determination of the 'true radioelement distribution' as a function of depth in the borehole. However, what is actually measured is the response of the detector system to the distribution of gamma-rays emitted by the desired 'true radioelement distribution'. The accuracy of any scheme to recover the desired information will be determined by the precision and resolution of the natural gamma-ray or gamma-ray spectral logs.

At the Geological Survey of Canada, Conaway and Killeen (1978a) considered the application of digital time series analysis to the problem, equating the gamma-ray log to the output of the logging system after passing through a series of low pass filters. Each filter had special characteristics associated with parameters which affect the output, such as detector length, digital sampling interval (or in an analogue system the time constant), and the gamma-ray distribution near an infinitesimally thin layer of the 'true radioelement'. This latter was coined the 'geologic impulse response' (GIR), that is, the response measured by an ideal point detector as it logged past the infinitesimally thin layer of radioactive material. Inverse filters were developed (to remove the effects of the above-mentioned filters) and applied to the data, deconvolving it to produce the 'true radioelement distribution'. The radioelement concentration as a function of depth produced in this way is termed the Radiometric Assay log or RA-log.

In a series of papers (commencing with Conaway and Killeen, 1978a) published by the Borehole Geophysics Research Group at the Geological Survey of Canada, the inverse filtering technique was developed and applied to gamma-ray logging. Most of the following pages regarding inverse filtering are a description of that development, and the results of their investigations into the factors which influence this method of deconvolving gamma-ray logs.

3.2.1. The Response of a Gamma-Ray Detector to a Geologic Impulse
Starting with an ideal gamma-ray logging tool with a point detector, having no counting errors, and an infinitesimally small sampling interval, results are modified to approach the real detector situation. Figure 5(a) illustrates a geologic column containing an infinitesimally thin layer of radioelement (say uranium). The 'true radioelement distribution' as a function of depth is illustrated in Fig. 5(b). The gamma-ray log produced by an ideal gamma-ray logging tool run past this thin layer is shown in Fig. 5(c). The equation describing the ideal gamma-ray log (which is the geologic impulse response) shown in Fig. 5(c) is

$$\phi(z) = \frac{\alpha}{2} \exp(-\alpha|z|) \tag{3}$$

where z is depth, ϕ is the response function, α is a constant and the infinitely thin layer is at depth $z = 0$. This equation was first used by Suppe and Khaikovich (1960), and later by Davydov (1970) and Czubek (1971, 1972) to describe the shape of the 'ideal' gamma-ray log. The

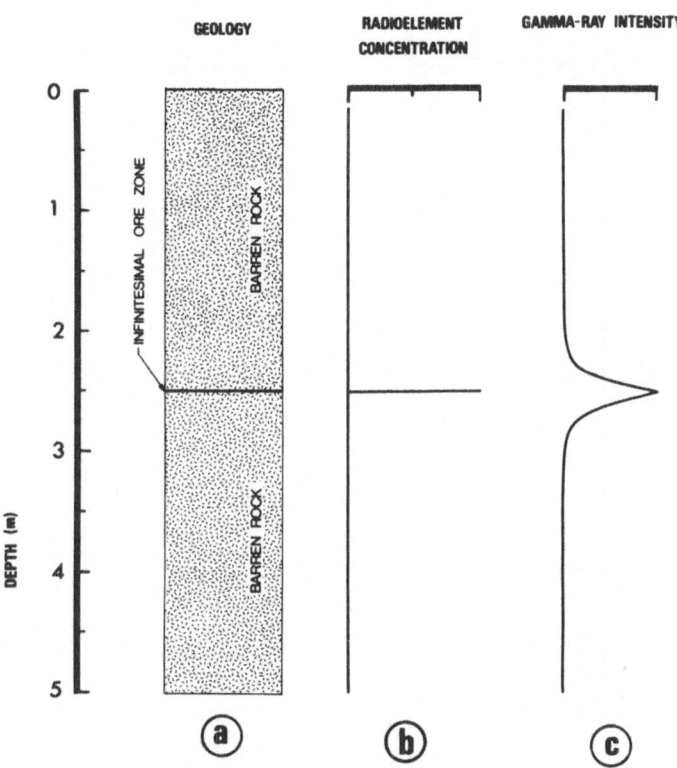

FIG. 5. (a) A 'geological log' showing an infinitesimally thin radioactive ore zone in an otherwise barren homogeneous column of rock. (b) The RA-log or true radioelement concentration as a function of depth. (c) The 'ideal' gamma-ray log produced by a point detector logging past the thin ore zone (after Conaway and Killeen, 1978a).

equation is approximate but it has been shown to be an extremely good approximation. The constant α determines the shape of the geologic impulse response and methods of determining α will be described later. First the effect of the geologic impulse response on a simulated complex distribution of radioactive material along a borehole was considered. Figure 6(a) shows a simulated RA-log (or 'true radioelement distribution') digitised at 1 cm intervals ($\Delta z = 1$ cm) and Fig. 6(b) shows the ideal gamma-ray log corresponding to that distribution. This was produced by convolving (see Hsu, 1970; Kanasewich, 1973; Robinson, 1967) the RA-log with the geologic impulse response. This log was produced

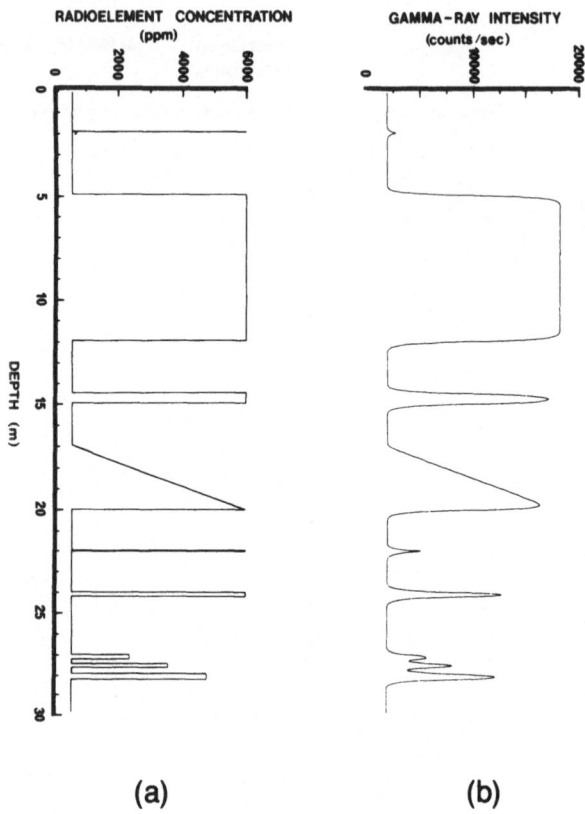

(a)   (b)

FIG. 6. (a) A complex RA-log or true radioelement distribution digitised at 1 cm intervals. (b) The ideal gamma-ray log from this complex distribution produced by an ideal point detector logging at 3 m/min, a sampling interval $\Delta t = 0.2$ s and a K-factor of 0·33 ppm/cps (after Conaway and Killeen, 1978a).

using $\alpha = 0.11$, logging velocity $v = 3$ m/min (5 cm/s), sampling time $\Delta t = 0.2$ s, sampling distance $\Delta z = 1$ cm, and a system sensitivity or K-factor of 0·33 ppm/cps. The degrading effect in the shape of the log is evident, showing especially the large departure of the gamma-ray log from the true radioelement distribution in thin zones, and the lack of separation (i.e. poor resolution) of closely spaced beds.

Since the RA-log is the desired end-product, and the gamma-ray log is what is measured, the latter must be deconvolved using an inverse filter or deconvolution operator. The advantage of using eqn (3) to describe the geologic impulse response is that the inverse operator is relatively easy to derive. The derivation of the digital inverse operator was described by Conaway and Killeen (1978a) and is illustrated in Fig. 7(a). This operator can be applied to the gamma-ray log using discrete convolution to remove the effects of the geologic impulse response evident in Fig. 6(b), recovering the RA-log of Fig. 6(a).

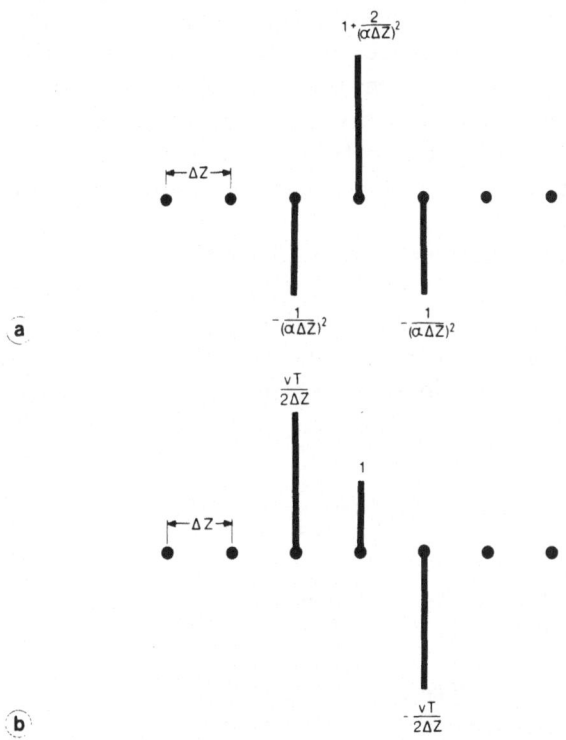

FIG. 7. (a) The digital inverse operator derived by Conaway and Killeen (1978a) for deconvolving an ideal gamma-ray log, removing the effect of the geologic impulse response. (b) The digital inverse operator for removing the effect of the ratemeter time constant T in an analogue logging system.

The coefficients for this digital inverse operator are

$$\left(-\frac{1}{(\alpha \Delta z)^2}, \; 1 + \frac{2}{(\alpha \Delta z)^2}, \; -\frac{1}{(\alpha \Delta z)^2} \right) \tag{4}$$

For a digital logging system using a sampling time Δt (which may not be 1 s), the sampling interval will be

$$\Delta z = v \Delta t \tag{5}$$

and incorporating the sensitivity or K-factor of the logging system into the inverse operator, eqn (4) becomes

$$\left(-\frac{K}{\Delta t(\alpha\Delta z)^2},\ \frac{K}{\Delta t}+\frac{2K}{\Delta t(\alpha\Delta z)^2},\ -\frac{K}{\Delta t(\alpha\Delta z)^2} \right) \tag{6}$$

The error introduced from approximations used in this derivation depends on Δz, and as Δz becomes shorter, the approximation becomes better (Conaway and Killeen, 1978a). A method of developing an exact inverse operator using z-transform techniques was discussed in detail in a later paper (Conaway, 1980a). However, when $\Delta z = 1$ cm, the results from the exact and approximate inverse operators were virtually indistinguishable.

In developing these inverse filters, the application to analogue logging systems, which are still widely used, was also discussed. In this case the degrading filter whose effects had to be removed was that of the ratemeter impulse response. Again using simulated logs and complex radioelement distributions the use of inverse filtering to remove the ratemeter distortion was illustrated. The three-term operator which was derived is

$$\left(\frac{vT}{2\Delta z},\ 1,\ -\frac{vT}{2\Delta z} \right) \tag{7}$$

where T is the ratemeter time constant in seconds. This digital inverse operator is depicted in Fig. 7(b).

3.2.2. Extending the Theory to Real Detectors

Introducing the finite length detector. The effect of a finite detector length L on the ideal gamma-ray log was considered (Conaway and Killeen, 1978a) to be the same as convolving the ideal (point detector) gamma-ray log with a 'boxcar' function or rectangular pulse of length L as shown in Fig. 8. The ramification of this in the frequency domain is that certain information is irretrievably lost by the effect of detector length. This is illustrated by deconvolving another simulated gamma-ray log of a complex radioelement distribution (Fig. 9a). In this case the simulated log is produced as before (by convolving the complex radioelement distribution (Fig. 6a) with the GIR) but also convolved with a boxcar (unweighted smoothing function) of length $L = 9$ cm, representing the degrading effects of a 9 cm long detector. Applying the inverse filter of eqn (6) produces Fig. 9(b). It can be seen that the original radioelement

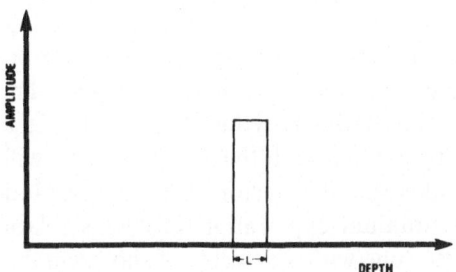

FIG. 8. The boxcar or rectangular function of length L which is applied (by convolution) to the ideal gamma-ray log if a finite detector of length L is used.

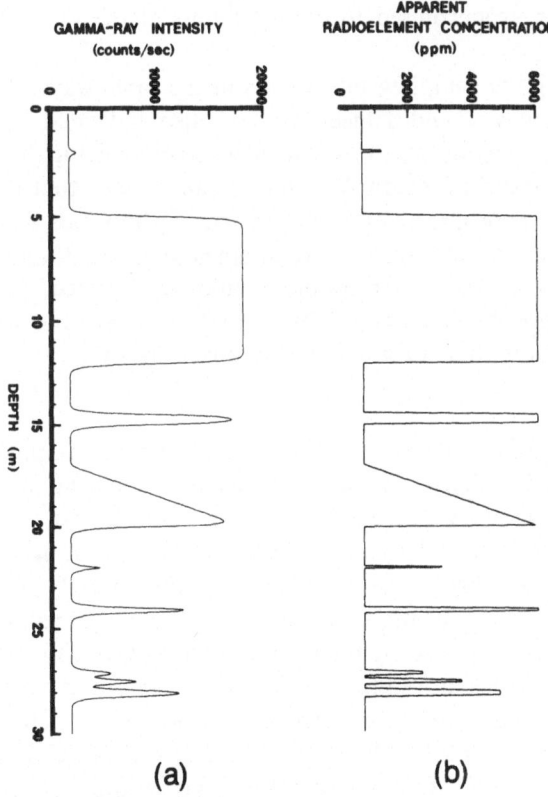

(a)

(b)

FIG. 9. (a) An ideal gamma-ray log similar to that shown in Fig. 6(b), but as 'seen' by a detector of length 9 cm. (b) The RA-log produced by deconvolving the log of (a) with the inverse filter of eqn 6. Note that instead of completely recovering the true radioelement distribution as shown in Fig. 6(a) the rounding effects of the finite detector length still remain in the RA-log.

distribution is not quite recovered. In particular, the infinitely thin layer becomes a layer of thickness L (i.e. equal to the detector length), and the anomaly amplitude becomes $1/L$ of its former value. The log of any bed of thickness less than L will be reduced in amplitude and increased in width. However, the area beneath the anomaly curve will not be changed from that of the ideal point detector. Thus the resolution of thin beds (often desired in uranium exploration problems relating to thin high grade layers) is best done with a detector of short length. However, this is usually accompanied by a requirement for longer counting times over a given depth interval Δz, and hence a slower logging velocity. Conaway and Killeen (1978a) suggested the possibility of a more complex solution to permit faster logging, using two detectors of different lengths with their counting synchronised to obtain counts at the same points along the borehole.

Considering the sampling interval Δz in the same way as the detector length L, Conaway and Killeen (1978a) also indicated that for continuous motion logging this was equivalent to convolving the gamma-ray log with a 'boxcar' of length Δz. Thus it can be seen that the larger Δz becomes, the more information will be irretrievably lost. If incremental logging (step-wise, stationary measurements) is used and the depth interval Δz is small, similar results should be expected. For optimum results it is preferable to count over small intervals and apply digital smoothing filters than to attempt smoothing by using a long counting interval.

Introducing statistical noise. Continuing the development further (Conaway and Killeen, 1978a), statistical or counting noise was introduced to the simulated gamma-ray log. Having first considered the effects in the frequency domain, it was apparent that the inverse operator would have the effect of amplifying the noise greatly at the higher frequencies. The application of smoothing filters to reduce the high frequency noise amplified by the inverse filtering was also demonstrated. Figure 10 illustrates an example with a large amount of statistical noise. Figure 10(a) shows the simulated log produced by convolving the complex radioelement distribution of Fig. 6(a) with the GIR, a boxcar of length $L=9$ cm for a 9 cm detector, and adding statistical noise ($\Delta z = 1$ cm, $v = 5$ cm/s). Figure 10(b) shows the same simulated log produced with $\Delta z = 20$ cm instead of 1 cm, illustrating an attempt to reduce noise using larger sampling intervals. Figure 10(c) is that log deconvolved and smoothed showing the RA-log or apparent radioelement

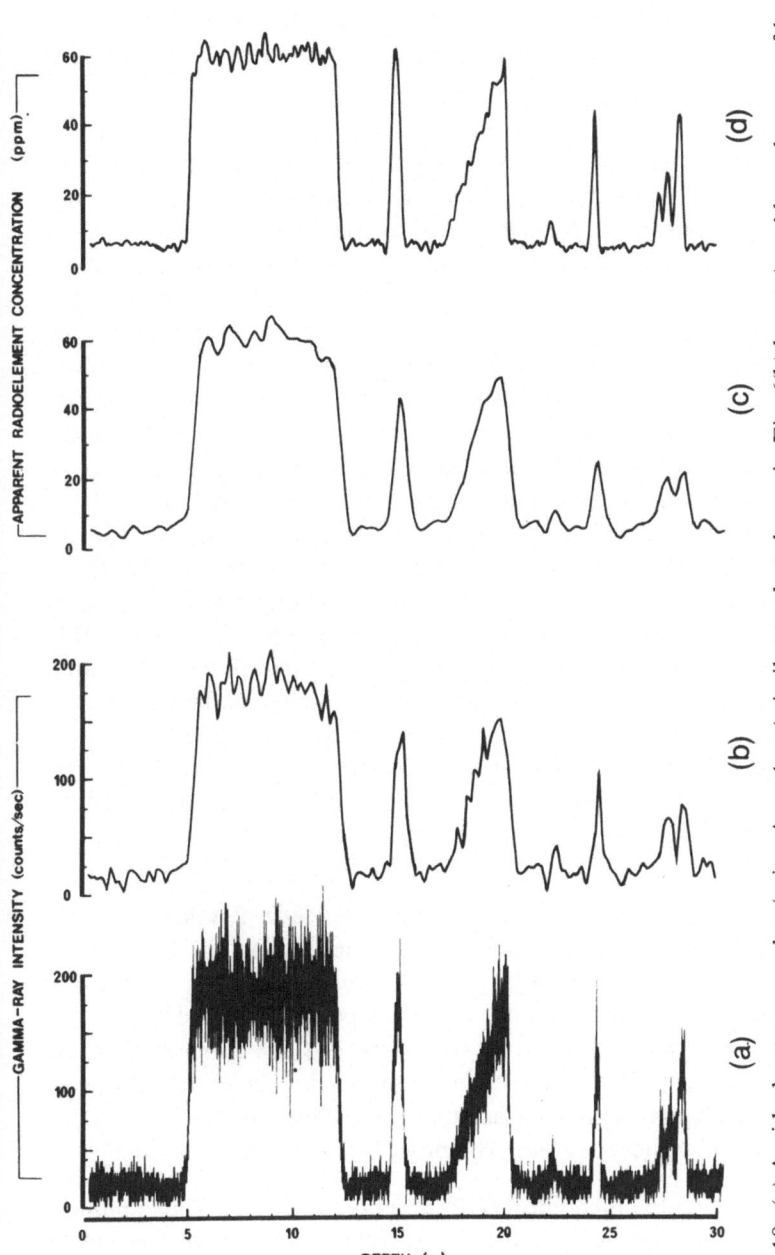

FIG. 10. (a) An ideal gamma-ray log (using $\Delta z = 1$ cm) similar to that shown in Fig. 6(b) but as 'seen' by a detector of length 9 cm and with added statistical noise. (b) The same ideal log as in (a) but produced using a sampling interval $\Delta z = 20$ cm instead of $\Delta z = 1$ cm. (c) The RA-log produced by deconvolving and smoothing the ideal gamma-ray log in (b). (d) The RA-log produced by deconvolving and smoothing the ideal gamma-ray log in (a). It is obvious that more detail has been recovered in this case and that the 20 cm sampling interval causes information to be lost.

distribution which results. Figure 10(d) is the RA-log as given by inverse filtering (deconvolving) and smoothing of the log with $\Delta z = 1$ cm. This is obviously much closer to the true radioelement distribution of Fig. 6(a).

Introducing field data. Inverse filtering of real data for the purpose of quantitative uranium determinations from gamma-ray logs was first applied by Conaway and Killeen (1978*a*) to data obtained in model boreholes. This represents a situation wherein the true radioelement distribution is accurately known as opposed to a field borehole where the true distribution is inferred from drill core, which may or may not be representative. It was apparent from the example that the deconvolved gamma-ray log produced an RA-log with much sharper boundaries on the anomaly than was evident in the raw gamma-ray log. The model holes used in the example were those of the Geological Survey of Canada, located in Ottawa, and constructed for the purpose of calibrating gamma-ray spectral logging systems. The concrete models contain ore zones 1·5 m thick sandwiched between two barren zones 1·2 m thick. The example they used to illustrate inverse filtering was from one of three thorium models (Killeen, 1978; Killeen and Conaway, 1978). Examples of inverse filtering of field gamma-ray logs will be shown later, but the model holes are important to the discussion below, on the determination of the shape factor α, to be used to compute the discrete inverse filter operator of eqn (6). A computer program using inverse filtering of gamma-ray logs to produce the RA-log was published by Conaway (1979*b*).

3.3. Determination of the Shape Constant α in Model Boreholes
The discrete inverse operator of eqn (6) is derived on the basis of the geologic impulse response (GIR) having the form of eqn (3). The factor α in the exponent will determine the actual shape of the GIR. This must be obtained experimentally for a given logging system. A method of determining α from data obtained in a model borehole containing an infinitely thick ore zone such as those in Ottawa has been described (Conaway and Killeen, 1978*a*). The procedure consists of

(a) computing the theoretical GIR for a point detector using a number of different values for α. Figure 11(a) shows a family of type curves for $\alpha = 0·08$ to 0·18 in steps of 0·02;

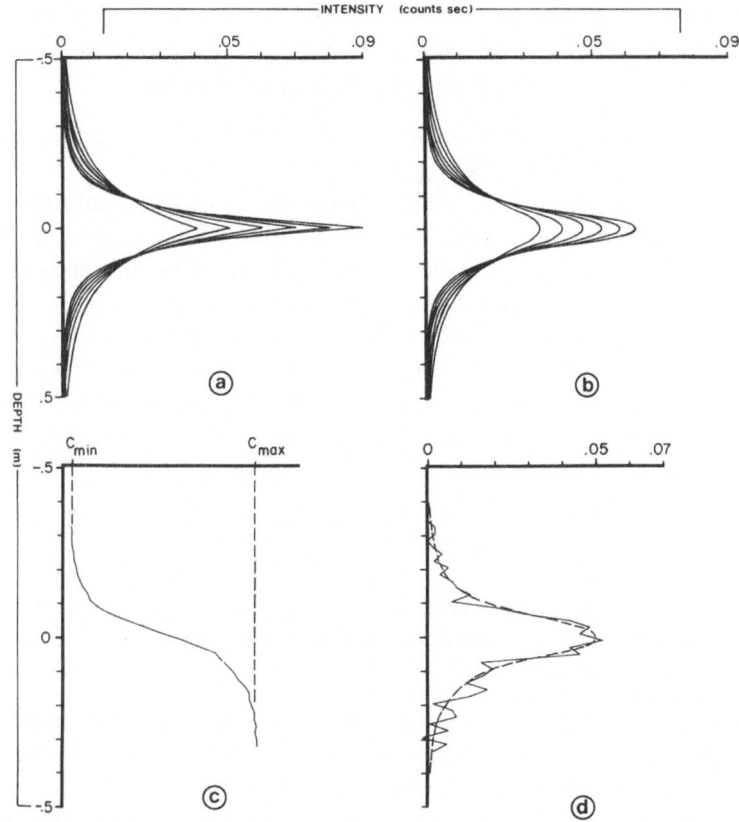

FIG. 11. (a) A family of type curves (GIRs) computed from eqn (3) for values of α from 0·08 to 0·18 in steps of 0·02. (b) The type curves of (a) modified to take detector length into account. This is done by convolving each of the curves with the boxcar function for the appropriate detector length L (in this example $L = 9$ cm). (c) The step response or measured gamma-ray log with a 9 cm detector in a model borehole showing the portion recorded when passing from a barren zone into the ore zone. (d) The impulse response derived from the step response in (c) by numerical differentiation. The best fitting type curve from (b) is overlain on this experimental data and the value of α for that curve is then considered to be the best for that detector.

(b) computing a modified set of type curves from (a) above, incorporating the detector length L. This will produce a set of type curves as shown in Fig. 11(b) for each detector length. The example of Fig. 11(b) is for a detector of length 9 cm;

(c) producing an experimental GIR from gamma-ray logs of a model
hole, and comparing this to the type curves for that particular
detector length to determine which value of α produces the type
curve that most closely fits the experimental curve.

The experimental determination of the GIR which is necessary for step
(c) is as follows (for a digital logging system):
(1) In a model borehole, log slowly from the barren zone into the ore
zone including at least 0·5 m of each zone. Use a small sampling
interval ($\Delta z = 1$ or 2 cm) and if the logging speed is sufficiently slow,
the counting statistics within these intervals will be acceptable. This
log is then the step response of the logging system as shown in Fig.
11(c). The impulse response is the derivative of the step response.
(2) The impulse response is obtained by plotting successive differences
in the step response by use of eqn (8) (from Conaway and Killeen,
1978a):

$$\phi(z_i) = \frac{C(z_{i+1}) - C(z_i)}{(\Delta z)(C_{max} - C_{min})} \qquad (8)$$

where $C(z)$ is the step response and C_{max} and C_{min} are the asymptotic maximum and minimum values of the step response curve as
shown in Fig. 11(c).

An example is plotted in Fig. 11(d) showing the experimentally
determined impulse response overlain by the best-fitting type curve,
hence determining the value of the shape factor α.

The sensitivity or K-factor may also be determined from the same
model hole data as described earlier, and incorporated into the inverse
operator of eqn (6).

It is interesting to note that the iterative technique proposed by Scott
et al. (1961) and Scott (1963) for computing uranium ore grade and
thickness from gamma-ray logs uses the step response from model holes
to obtain not the impulse response but the response to a 15 cm thick
layer of radioactive material. The response is obtained from gamma-ray
logs digitised with $\Delta z = 15$ cm. The technique assumes the entire 'true
radioelement distribution' is composed of 15 cm thick layers of radioactive material of different uranium grades. In the iterative technique, the
procedure is first to guess the grades of these 15 cm thick layers, then
compute the synthetic gamma-ray log by summing the computed response for each 15 cm thick layer, and compare this result to the actual
measured gamma-ray log. The differences are used as an error correction

to the original estimate of the grades of the 15 cm thick layers; the process is repeated and the estimates of the grades keep improving. In a later paper (Conaway and Killeen, 1978b), the equivalence of the inverse filtering technique to the iterative technique carried to an infinite number of iterations was demonstrated. The important point to be made here is that any conclusions based on consideration of the inverse filtering of gamma-ray logs very likely hold also for the iterative technique. This will become more apparent when the factors which affect the shape of the impulse response are considered. These include the borehole diameter, borehole fluid, casing, eccentricity of the probe in the hole, detector size, gamma-ray energies, pore fluid, rock density, and equivalent atomic number Z_{eq} of the rock.

3.4. Determination of the Shape Constant α in Field Boreholes

A method of determining the geologic impulse response (or response function of the gamma-ray logging system) which may be used in field boreholes as well as model boreholes has been described by Conaway (1980b). The advantage of the technique is that the field borehole contains all of the factors which are expected to influence the shape constant. Hence, the value derived in the field will include the effects of all of these factors. Only the sensitivity of K-factor must then be determined in controlled, model boreholes.

The method involves plotting the natural logarithm of the step response which has been determined in a model hole (Fig. 11c). The natural logarithm plot (Fig. 12a) shows a linear portion, the slope of which is the value of the shape constant α. In order to emphasise the linear portion, the 'background' of the barren zone was subtracted before taking the logarithms. The background is the average gamma-ray count rate measured in the barren zone, sufficiently far from the ore zone as to be unaffected by it.

Although this method eliminates the numerical differentiation, and the requirement for an infinitely thick ore zone in a model hole, it can only give a value for α outside the ore zone (Conaway, 1980b). Recall that Fig. 11(d) shows the response function determined experimentally, overlain by the computed response function which it best fits. By plotting the logarithm of the experimentally determined response (in model holes with infinitely thick ore zones), values for α both outside and inside the ore zone may be determined, as shown in Fig. 12(b). Thus if any parameter, such as density, varies from the barren zone to the ore zone, the shape constant may change and this latter double-sided plot should

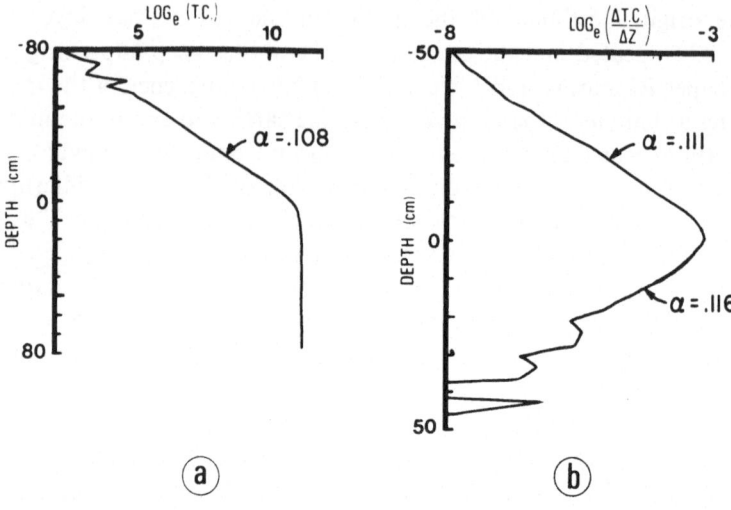

FIG. 12. (a) The natural logarithm of a similar step response to that shown in Fig. 11(c). The slope of the linear portion will be the value of the shape constant α. (b) The natural logarithm of a similar impulse response to that shown in Fig. 11(d). The slopes of the linear portions will give a value of α for both the barren zone and the ore zone. If these are appreciably different, some physical property of the ore zone (e.g. density) must be different from the barren zone (after Conaway, 1980b).

make it possible to detect differences in the shape constant α. The penalty is the added complexity of having to carry out the numerical differentiation. In that case also the ore zone must be infinitely thick, and counting statistics very good.

The above arguments were extended (Conaway, 1980b) to include the determination of the constant α in field boreholes. The borehole need only contain one distinct transition from a radioactive zone to a barren zone. The barren zone should be uniform for a distance of about 1·5 m from the ore zone. If there are several such transitions in the hole, several determinations of α will be possible.

An example of the direct determination of α in field boreholes (Conaway, 1980b) is shown in Fig. 13. Three anomalies were selected at various depths in the hole, which was located in pegmatite-type uranium occurrences of the Bancroft area in Canada. The gamma-ray logs were recorded with a 25×75 mm CsI(Na) detector in a 60 mm diameter, uncased, water-filled borehole. The natural logarithms of the gamma-ray logs shown in Fig. 13 have linear portions approximately 30 cm wide on the sides of each radioactive zone. These linear portions are usually even

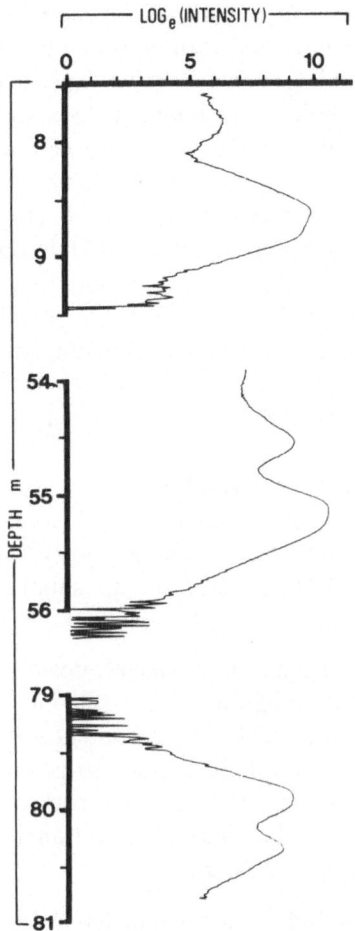

FIG. 13. The natural logarithm of gamma-ray logs through three radioactive zones in a field borehole (after Conaway, 1980*b*). The slope of the linear portion of each anomaly yields a value of the shape constant α. (The mean count rate in the adjacent 'barren' zones has been subtracted before taking the logarithms.)

shorter and are difficult to use to obtain a value of α. The mean gamma-ray count rate of the adjacent 'barren' zones (which may not be truly barren) has been subtracted from each of the anomalies before taking the logarithm, to produce the curves of Fig. 13. The subtraction of the average gamma-ray intensity of the 'barren' zone is an essential step in determining the correct value of α (Conaway, 1980*b*).

In summarising experience with field determinations of α, it was concluded that the value of α can vary considerably from anomaly to anomaly along the borehole, and variations of α by nearly a factor of two were noted in some holes (Conaway, 1980b). Since determinations by this technique in field boreholes were found to be statistically reliable, and repeatable, the variations were considered to be due to geologic causes such as variations in rock density, porosity and/or fluid content. Variations in α may ultimately prove to be an additional piece of information which can aid in interpretation of borehole logs. It should be noted that for years the iterative technique has been used effectively with a single response function for any probe in any borehole. This new knowledge of variation in the response function (or variation in α) can only improve the data processing by the fact that it can be corrected or modified for many of the situations encountered.

3.5. Variation of the Shape Constant α

The shape constant α varies according to variations in parameters associated with the detector, the borehole and the formation. All of the parameters known to affect the value of α are subdivided into three main categories:

(a) The detector: e.g. length and diameter, position in the hole (eccentricity) and probe configuration.
(b) The borehole: e.g. diameter, fluid content, casing type and thickness (or mudcake), dip angle relative to geology.
(c) The formation: e.g. density, fluid in the rock pores, effective atomic number and weight (i.e. linear attenuation coefficient of the formation) and gamma-ray energy.

The variation of α has not been investigated with respect to all parameters. Results regarding those which have are summarised below.

Variations in the physical properties of the borehole were investigated at the Geological Survey of Canada (Conaway et al., 1979; Conaway, 1980c) by using model holes of various diameters and varying the fluid content (air and water) and thickness of steel casing in the hole to measure the variations in the shape constant α.

3.5.1. The Effect of Steel Casing

Figure 14(a) shows a gamma-ray log measured in a model borehole containing a 1·5 m thick 'ore' zone between two barren zones, for no casing and for six thicknesses of steel casing (1·6, 3·2, 4·8, 6·4, 9·5 and 12·7 mm). The borehole is water-filled and 11·4 cm in diameter. All of the

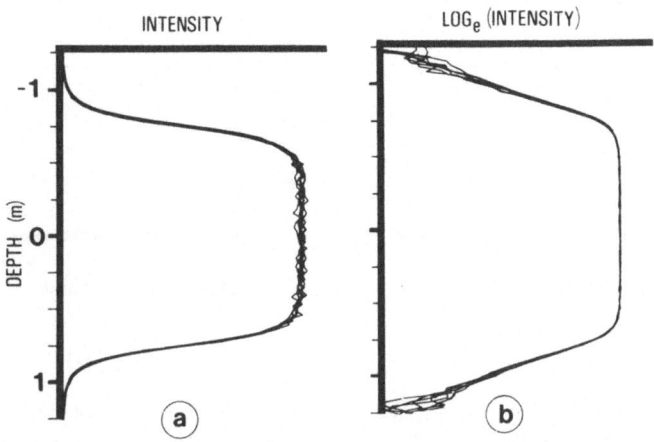

FIG. 14. (a) Gamma-ray logs through an ore zone of a model hole, using six different thicknesses of steel casing, and no casing. (The logs are normalised to equal areas). (b) The natural logarithm (background subtracted as described in the text) of the gamma-ray logs in (a). A least-squares fit of the linear portion of the flank of each log yields values of the shape constant α for each thickness of steel casing (after Conaway *et al.*, 1979).

curves have been normalised so that their areas are equal, for comparison; this has no effect on the computed value of α. The digital sampling interval is 3 cm, logging velocity is 0·3 m/min and the detector is a 25×75 mm NaI(Tl) with the lower energy threshold of the instrument set at 100 keV.

It is difficult to see from Fig. 14(a) whether the casing affects the shape of the log. Subtracting the background and plotting the natural logarithm as described above gives Fig. 14(b). Some divergence of the curves can be seen in this figure as distance from the ore zone increases. Figure 15(a) shows a plot of α as a function of casing thickness based on a least-squares fit on the linear portion of the anomaly flanks in Fig. 14(b), over the range from 6 to 36 cm outside the ore zone, on both sides. With this information, the effect of casing on the total area of the anomaly, normalised to 1 for no casing, is shown in Fig. 15(b) (Conaway *et al.*, 1979). The commonly used multiplicative casing correction factors can be obtained from the inverse of this curve, and applied to gamma-ray logs obtained with different casing thickness. The graphs shown in Fig. 15(a) and (b) are not universally applicable and depend upon instrument characteristics. Correction factors should be individually determined for each probe and logging system.

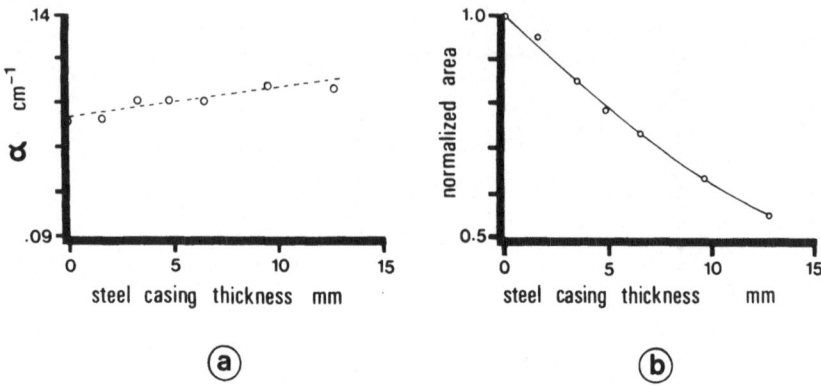

FIG. 15. (a) Variation of the shape constant α with thickness of steel casing as derived from Fig. 14(b). (b) Variation in the area under the anomaly curve with thickness of steel casing, normalised to a value of 1 for no casing (after Conaway *et al.*, 1979).

3.5.2. The Effect of Hole Diameter and Fluid Content

Data were obtained in five water-filled model boreholes of different diameters (8·9, 11·4, 17·8, 22·9 and 33 cm) through the same 1·5 m thick ore zone (Conaway *et al.*, 1979). In all cases the probe was kept in contact with the borehole wall throughout the log. The logs were normalised to give uniform area, 'background' corrected, and plotted on semi-log coordinates as before. The corresponding normalised logs for the same boreholes air-filled were also given. The differences caused by the various borehole diameters are more pronounced in the case of air-filled boreholes. Also in large diameter boreholes, the ore zone will be detected from a greater distance if the borehole is air-filled rather than water-filled due to increased radiation passing through the borehole fluid (i.e. air). For this reason there is a greater difference between the results for air and water in large diameter boreholes than in small diameters.

The values of α determined using the semi-log slope technique, based on the above data, are plotted in Fig. 16(a) (Conaway *et al.*, 1979). Figure 16(b) shows the total area under the anomaly curve for the five borehole diameters, both air- and water-filled, normalised to the 8·9 cm diameter case. In the case of an air-filled borehole there is very little change in area, whereas for the water-filled borehole the curve asymptotically approaches a constant area corresponding to that of infinite borehole diameter (i.e. logging along a flat wall, or 2π geometry). If the probe were centred, the curve for the water-filled case would approach zero instead.

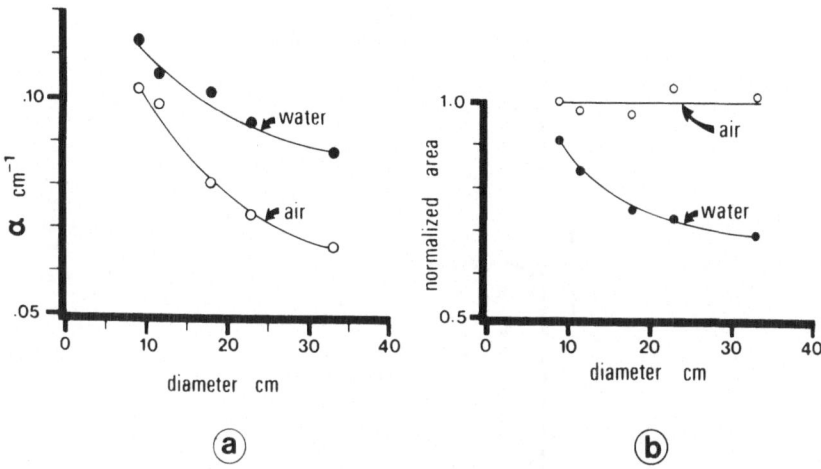

(a)   (b)

FIG. 16. (a) Variation of the shape constant α with borehole diameter for both air- and water-filled holes. (b) Variation in area under the anomaly curve with borehole diameter for air- and water-filled holes, normalised to 1 for the air-filled hole of 8·9 cm diameter (after Conaway *et al.*, 1979).

3.5.3. The Effect of Dipping Beds

If a borehole is not drilled perpendicular to the ore zone being assayed, the dip angle must be considered in the computations. The dip angle is defined as the angle between the normal to the borehole axis and the plane of the radioactive zone (Conaway, 1979a). This problem is depicted in Fig. 17. The ideal gamma-ray logs past a perpendicular and a dipping ore zone are shown in Fig. 18. The stretched log of curve 2 is the result of applying the cosine correction to the value of the shape constant which

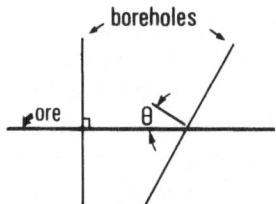

FIG. 17. Diagram of two boreholes intersecting a thin ore zone at an angle θ (representing a dipping zone) and at an angle of 90° to illustrate terminology used in the text. The dip angle refers to the angle between the ore zone and the normal to the borehole.

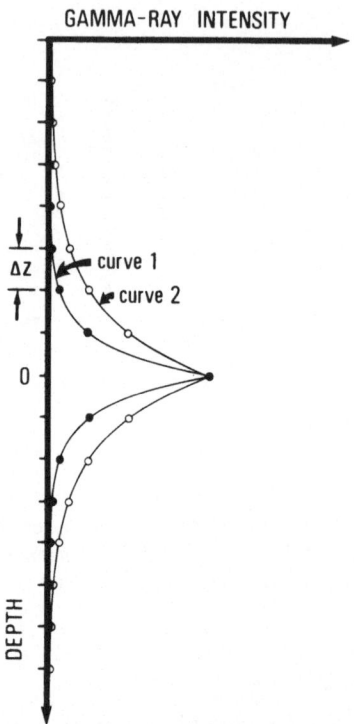

FIG. 18. The ideal gamma-ray logs past both a perpendicular zone (curve 1) and a dipping zone (curve 2). The dipping zone stretches the anomaly increasing its area and introduces errors in grade–thickness product determinations (after Conaway, 1979a).

produced curve 1 (Conaway, 1981), i.e. for the dipping bed

$$\alpha_d = \alpha \cos \theta \qquad (9)$$

The validity of eqn (9) was demonstrated by applying the appropriate dip angle correction to gamma-ray logs obtained in US Department of Energy model holes with thin ore zones dipping at angles of 0°, 30°, 45° and 60°. The shape factors α were computed from the experimental data using the semi-logarithmic plotting technique described above. These are plotted in Fig. 19 along with the predicted value of the shape factor α as computed by applying the cosine correction to the α value for the perpendicular case. The excellent agreement is apparent.

3.5.4. The Effect of Formation Parameters

The effects of variations in density, pore fluids in the rock, and the self-

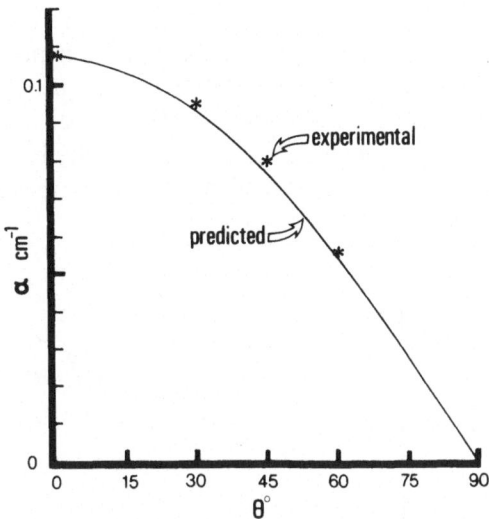

FIG. 19. Variation of the shape constant α with relative dip angle between the normal to the borehole and the ore zone, both predicted by eqn (9) and by measurements in model holes with dipping ore zones (after Conaway, 1979a).

attentuation of gamma-rays by uranium (the Z-effect) were also considered by Conaway (1980c) and Conaway et $al.$ (1980b). Experimental data to evaluate these effects are more difficult to obtain than for the previously discussed effects.

To consider the effect of density on the shape constant α, a series of blocks of rock of different density at the Geological Survey of Canada's Ottawa calibration facilities were used in conjunction with a synthetic thin radioactive zone as shown in Fig. 20(a). The gamma-ray logs

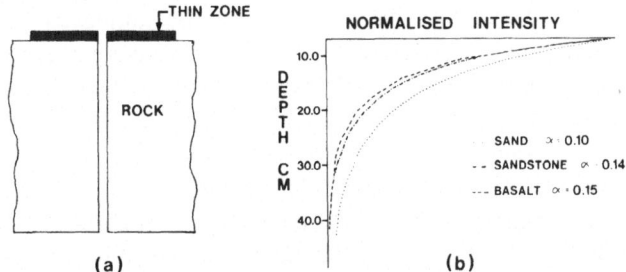

FIG. 20. (a) Position of a synthetic 'thin radioactive zone' on models (blocks of rock) of different densities. (b) Gamma-ray logs recorded for three different density models, as the detector approached the synthetic thin zone.

recorded as the detector approached the thin ore zone are shown in Fig. 20(b). It is apparent that the response is different in materials of different density. Using the semi-logarithmic plotting technique described earlier, the values of α were determined and varied from 0·10 for the low density sand (1·8 g/cm^3) to a value of 0·15 in the high density basalt (2·9 g/cm^3). The results are shown in Fig. 21. This demonstrates that α is a function

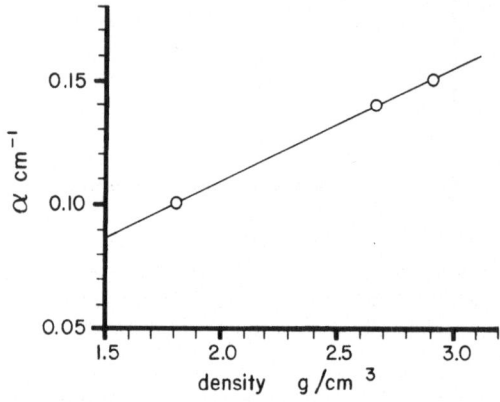

FIG. 21. Variation of the shape constant α with density from the data of Fig. 20.

of the linear attenuation coefficient which is also a function of rock density. Thus, the shape of gamma-ray log anomalies will be different in rocks of different densities. In the case of pore fluids, the variation in density of the rock caused by changes from dry rock to saturated rock will produce the same effect.

The greatest problem occurs when there is a change in formation parameter (say density) at the boundary of the radioactive zone. The shape of the anomaly will be different in rocks of different densities, thus affecting the shape of the 'tail' (or the value of the so-called 'tail factor') in computing the area under an anomaly. Figure 22 (Conaway, 1980c) illustrates this concept. The shape of the anomaly assuming α is a constant along the borehole, and assuming α is larger in the (more dense) ore zone than in the barren zone, is plotted. It is apparent that the area under the curve will be greater if the barren zone is of lower density than the ore zone. Thus, the relation $GT = KA$ will be affected. Computations of α as a function of uranium ore concentration (because it affects the

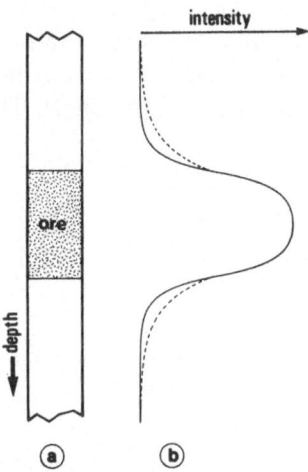

FIG. 22. Gamma-ray log recorded past an ore zone as shown in (a) for the case when the shape constant α does not change along the borehole (solid line), and the case when the shape constant is larger in the ore zone than in the barren zone (dashed line) (after Conaway, 1980c). Note the shape of the tail of the anomaly (and consequently the area) changes.

density and linear attenuation coefficient) have been done (Conaway, 1980c) using values for concrete as would be appropriate in model holes. These plots also show the effect on different gamma-ray energies. Figure 23 shows a plot of α against percentage uranium (illustrating the Z-effect)

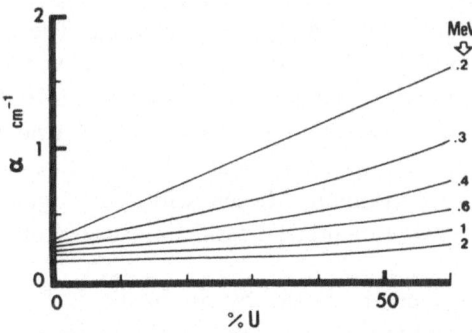

FIG. 23. Variation of the shape constant α with uranium concentration (in concrete) for different low threshold energies of a total count logging system (after Conaway, 1980c).

for different gamma-ray energies. Figure 24 shows the variation of the ratio $\alpha_{ore}/\alpha_{barren}$ for different uranium concentrations. As the uranium concentration increases, α_{ore} increases, and this invalidates the assumption of a constant α throughout the borehole. This effect is minimised

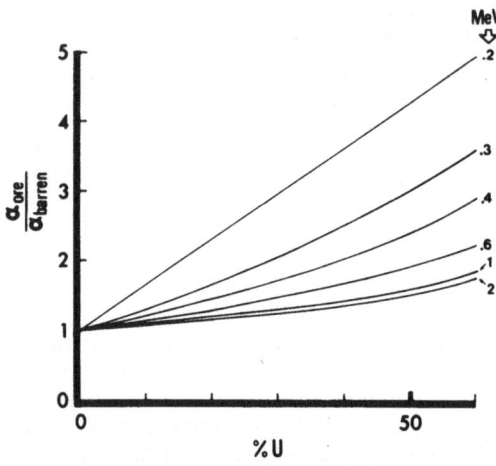

FIG. 24. Variation of ratio of $\alpha_{ore}/\alpha_{barren}$ with uranium concentration (in concrete) for a range of low threshold energies (after Conaway, 1980c).

when higher gamma-ray energies are considered. Figure 25 illustrates this effect translated into a ratio of the K-factors or sensitivities in the ore zone and in the barren zone. If only gamma-rays above $1 \cdot 0$ MeV are counted, the ratio of K_{ore}/K_{barren} is unaffected. The conventional low energy threshold of $0 \cdot 4$ MeV (for the total count window of a gamma-spectral system) introduces significant error (greater than 10%) when the ore grade reaches 10% uranium. Experimental evidence (Dodd and Eschliman, 1972) has only shown that a $0 \cdot 4$ MeV threshold is effective to ore grades of $3 \cdot 0\%$ uranium. Models with higher grades are presently unavailable to test the theoretical results illustrated in Fig. 25 on higher U ore grades.

The percentage error caused by an incorrect tail factor or shape constant α outside the ore zone will be greatest for thin zones. In effect, the area under the tail is a larger percentage of the total anomaly area for thin zones than for thick zones. The error in the computed grade–thickness product varies non-linearly with thickness of the radioactive zone, increasing rapidly as the zone thins (see Conaway, 1980c). The

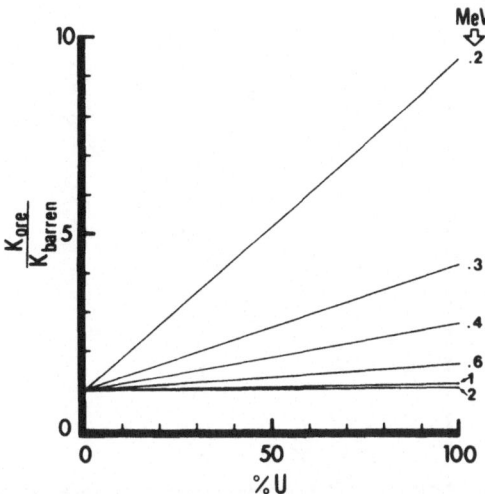

FIG. 25. Variation of the ratio of K_{ore}/K_{barren} with uranium concentration (after Conaway, 1980c). It is apparent that the error in the K-factor is minimised if a gamma-ray energy as high as 1·0 MeV is used for the low threshold. This is entirely feasible since at the high uranium concentrations involved, the gamma-ray count rate above 1·0 MeV would be quite high.

above considerations complicate any attempt to apply a simple Z-effect correction factor. The factor would be a function of both the ore grade and the zone thickness at the same time.

In an earlier paper (Conaway, 1978), the problem of measuring thickness from a gamma-ray log of thin zones was investigated. The common approach was to consider the full-width-half-maximum (FWHM) to be a good measure of the thickness. This is strictly correct only in zones which are effectively infinitely thick (which depends on α). Various rules of thumb have been applied to take care of the error in applying this FWHM rule to thin zones (e.g. full-width-four-fifths-maximum, etc.).

Using the shape factor α and consideration of various detector lengths for zones of various thickness, a set of correction curves were produced by Conaway (1978) to compute true zone thickness from measured FWHM. One of these sets of correction curves for $\alpha = 0·18$ and four different detector lengths is reproduced in Fig. 26. The corrections for detector lengths other than those in the figure can be found by interpolations. A set of curves for $\alpha = 0·14$ has also been published (Conaway, 1978). These values of α represent a typical range for bore-

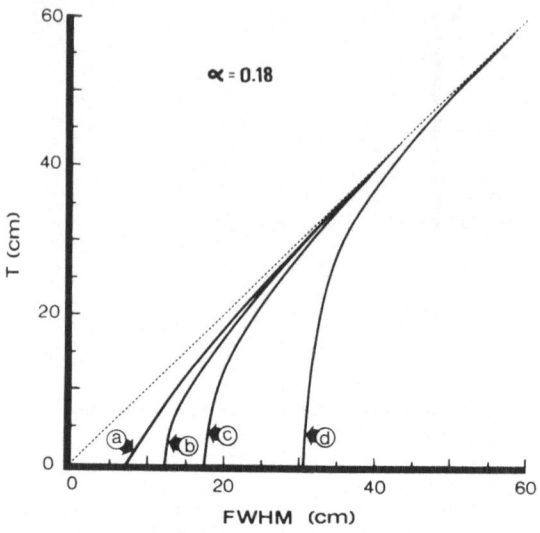

FIG. 26. Correction curves to determine true thickness of thin zones by measurement of the FWHM of anomalies on the gamma-ray log (after Conaway, 1978). True zone thickness T, for $\alpha = 0.18$, and detector lengths of (a) 1 cm, (b) 8 cm, (c) 15 cm and (d) 30 cm are given. For example, if the FWHM of an anomaly is measured as 15 cm and an 8 cm detector was used, the true thickness would be about 9.5 cm as read from the correction curve (b).

holes in the Canadian hardrock mining areas and the research logging system used by the Geological Survey of Canada (Bristow, 1979).

3.6. Application and Optimisation of the Inverse Filtering Technique

Applying many of the above considerations, the optimisation of gamma-ray logging techniques for uranium exploration was studied using multiple logging runs in a field borehole in the Bancroft uranium mining area of Canada (Conaway et al., 1980a). Variations in detector length, logging speed, sampling interval and analogue versus digital recording were all studied. Examples of inverse filtering were presented for each parameter tested, and numerous practical conclusions were drawn, including an example of deconvolution of a gamma-ray log with three values of α (0.14, 0.18 and 0.22 cm^{-1}).

For maximum definition of thin zones or complex zones of radioactivity, a short detector should be used. Figure 27(a) shows a raw gamma-ray log recorded with a 25×25 mm detector, and Fig. 27(b) shows the

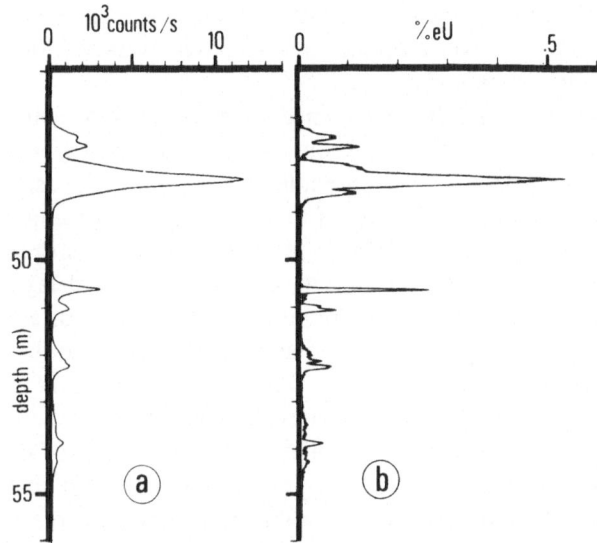

FIG. 27. (a) A portion of a raw gamma-ray log recorded with a $25 \times 25\,mm$ detector and the Geological Survey of Canada's DIGI-PROBE logging system. The log was recorded in the Bancroft uranium mining area with a logging velocity of 0.3 m/min, and $\Delta z = 1$ cm. (b) The RA-log produced from (a) by deconvolving and smoothing with the appropriate inverse filter and calibration factors. The increased resolution is readily apparent in this processed log.

deconvolved and smoothed log. The improvement in spatial resolution is apparent. Conaway *et al.* (1980*a*) suggest using a sampling interval equivalent to half the detector length or slightly shorter. They recommend continuous motion logging rather than step-wise stationary measurements and digital rather than analogue recording. If analogue recording must be used, best results are obtained with a slow logging speed, short time constant, large chart paper/depth scale, and a short digitising interval for the analogue charts.

An important consideration for users of inverse filtering is the effect of errors in the constant α (Conaway *et al.*, 1980*b*). An error in the value of α will cause the inverse filter to distort the shape of the processed log. Fortunately, since the inverse filter is linear and normalised so the sum of the filter coefficients equals 1, the area under the curve will never be changed and the accuracy of a grade–thickness product computation over an entire ore zone from background to background will be unaffected. If the value of α is too large, the high frequency component of the log will not be sufficiently amplified (e.g. thin zones will not appear as thin as they should appear). On the other hand, if α is too small, the high frequency content is over-amplified and negative overshoot or ripple will appear near bed boundaries.

A value of α should be determined as a calibration factor for each borehole probe, and ideally this should be determined in a borehole with similar properties to the field boreholes (diameter, fluid, casing, etc.).

One of the greatest advantages of the inverse filtering technique is that it can be easily implemented in a minicomputer or microprocessor to produce the radiometric assay log in almost real time (delayed only by one-half the filter length). There are no large memory requirements since it is not necessary to store the entire anomaly before it can be processed as in the iterative technique. The inverse filtering technique was implemented in the software of the research logging truck of the Geological Survey of Canada (Bristow and Killeen, 1978; Bristow, 1979; Killeen *et al.*, 1978). In October 1978 the logging system was used to conduct a series of Canadian–American intercalibration measurements in the model holes of the US Department of Energy in Grand Junction, Colorado. At that time a multiple zone model called N5 was logged and deconvolved in real time. Figure 28 shows the uranium grade distribution

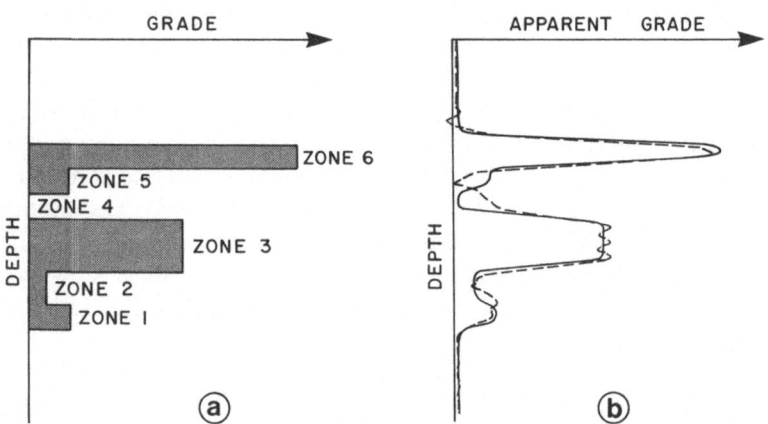

FIG. 28. (a) The uranium ore grade distribution assigned to borehole model N5 at the Grand Junction calibration facilities of the US Department of Energy. The thickness of the six zones and their assigned grades are shown to scale. (b) The deconvolved gamma-ray log of N5 produced in real time in 1978 with the DIGI-PROBE logging system plotted as an overlay of the assigned grade–thickness distribution. The assigned grade–thickness distribution has been smoothed with exactly the same smoothing used in the deconvolution of the gamma-ray log. The grade discrepancies at zone 4 and zone 5 and thickness discrepancies at zone 1 and zone 2 are apparent (see Fig. 29).

assigned to the model (Mathews *et al.*, 1978) and the deconvolved gamma-ray log. The deconvolved log shown in Fig. 28(b) indicates two discrepancies: (1) an apparent interchange of the assigned grades of zones 4 and 5 and (2) an apparent interchange of thickness of zones 1 and 2 (Bristow *et al.*, in press). Figure 29(b) shows the deconvolved log compared to the postulated necessary reassignment of grades and thickness. The US

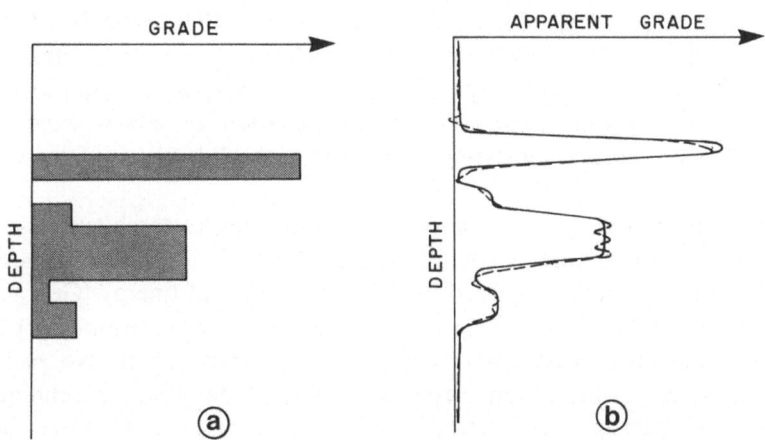

FIG. 29. (a) A postulated 'correct' grade–thickness distribution for the borehole model N5 based upon interpretation of the discrepancies shown in Fig. 28. (b) The same deconvolved log of (a) plotted as an overlay on the postulated distribution. Subsequent studies of the model have shown that the interpretation of the deconvolved log was correct.

Department of Energy reanalysed the model, and confirmed the interpretation based on the deconvolved logs. This revised grade assignment has been published (Wilson and Stromswold, 1981). Studies of log deconvolution with the inverse filter, particularly with respect to gamma-ray spectral logging, have also been published (Wilson and Stromswold, 1981).

4. GAMMA-RAY SPECTRAL LOGGING

The above discussions concerned gross count or total count logging wherein it was assumed that all gamma-rays being counted by the logging system originated from the uranium decay series alone. Where the other naturally occurring radioelements (thorium and potassium) contribute a significant portion of the recorded gamma-radiation, the above considerations regarding the application of a K-factor or sensitivity to determine uranium concentrations become invalid.

Fortunately gamma-ray spectrometry has left the realm of the laboratory and become a field technique. By about 1975 several commercial portable gamma-ray spectrometers, modified to conduct gamma-ray spectral logging, became available. The use of these early systems has been described (Killeen, 1976; Killeen and Bristow, 1976). Also an updated version was described (Killeen *et al.*, 1978) wherein digital recording of four energy windows for total count, potassium, uranium and thorium was used. They discussed the advantages of digital recording, the playback of the data, and recommendations for a new generation portable microprocessor-based system which could do inverse filtering in real time.

Descriptions of two gamma-ray spectral logging research systems are in the literature; that of the Geological Survey of Canada (Bristow, 1979), and a research logger of the US Department of Energy (George *et al.*, 1978). Stabilisation of the measured gamma-ray spectrum has been one of the main problems of gamma-ray spectral logging. No perfect solution has yet been found. The use of various stabilisation techniques has been discussed in numerous other publications (e.g. Conaway and Killeen, 1980; Stromswold and Kosanke, 1979; Conaway and Bristow, 1981).

Before discussing the application of the relation $GT = KA$ to gamma-ray spectral logging, and the inverse filtering of the spectral logs, a brief review of the principles of gamma-ray spectrometry is in order.

4.1. Physical Basis

The gamma-ray spectra for the uranium and the thorium decay series are shown in Fig. 2. The potassium spectrum (not shown) is a single gamma-ray peak at 1·46 MeV. A gamma-ray spectrometer has the ability to measure the energy of a gamma-ray and discriminate those gamma-rays associated with each of the three radioelements in three separate energy regions or windows of the spectrum and record the counts in separate memories or counters.

Due to the inherently imperfect energy measuring capabilities of scintillation detectors, and scattering of gamma-rays and degradation of gamma-ray energies in the source rock, the measured spectra for the thorium and uranium decay series will resemble those shown in Fig. 30(a) and (b) and the measured potassium spectrum will resemble that

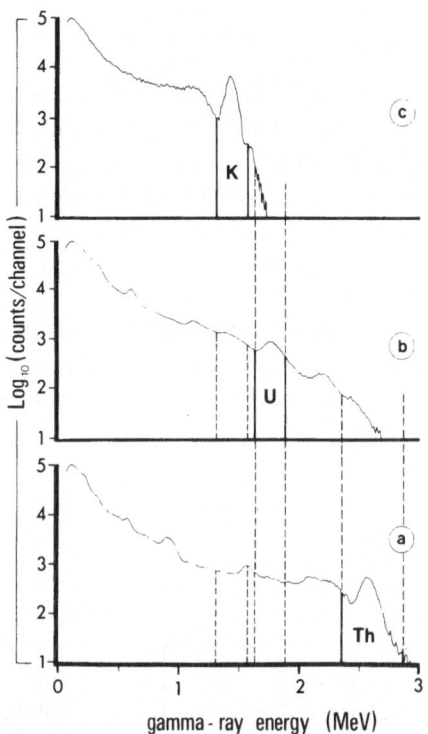

FIG. 30. (a) A gamma-ray spectrum recorded in a thorium zone of a model borehole. (b) Gamma-ray spectrum recorded in a uranium zone of a model borehole. (c) Gamma-ray spectrum recorded in a potassium zone of a model borehole. In each case, the detector was a 32 × 127 mm sodium iodide detector. Each spectrum consists of 256 energy channels spread over the range from 0 to 3·0 MeV (after Conaway and Killeen, 1980). Commonly used energy windows for K, U and Th are indicated in the respective spectrum. The dotted lines illustrate interference. For example, it is easy to see how gamma-rays from the thorium decay series can be counted in the uranium window and vice versa. (See also Fig. 31.)

shown in Fig. 30(c). Typical energy windows for a gamma-ray spectral logger are also indicated in the figure. These are 1·36–1·56 MeV (K window, 1·46 MeV peak of 40-K); 1·66–1·86 MeV (U window, 1·76 MeV peak of 214-Bi); and 2·4–2·8 MeV (Th window, 2·614 MeV peak of 208-Tl). It can be seen that there are interferences among the three spectra; some gamma-rays originating from the Th decay series will be counted in the U and K windows, some gamma-rays originating from the U decay series will be counted in the K window and to a small extent in the

Th window, and some small portion of the K gamma-rays may be counted in the U window. To determine the K-factor or sensitivity for the individual radioelements K, U and Th, the gamma-ray counts due to each individual radioelement must be determined. This means the interferences must be taken into account and the gamma-ray count rate in each window must be corrected. This is accomplished by the procedure known as spectral stripping.

4.2. The Stripping Factors

Figure 31 is a schematic representation of the interplay between the three radioelement windows (K, U and Th) identifying the stripping factors (Killeen, 1979; Conaway and Killeen, 1980). For many purposes only the

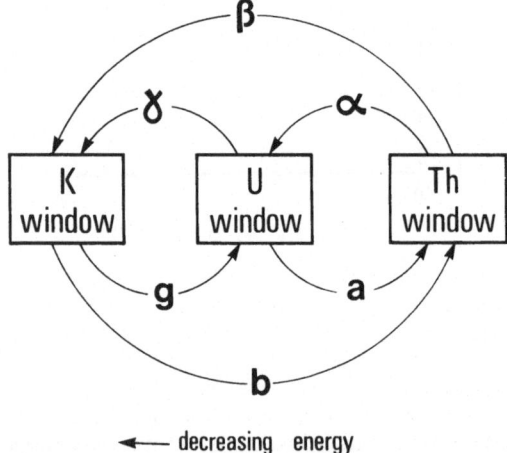

◄── decreasing   energy

FIG. 31. This figure should be studied with reference to Fig. 30. The six stripping factors are shown indicating their relationship to the gamma-ray energy windows of Fig. 30 (after Killeen, 1979). Thus, for example, the contribution of gamma-rays observed in the U window from the thorium decay series (see Fig. 30b) can be removed by subtracting from the U window the number of gamma-rays observed in the Th window multiplied by the factor α.

three stripping factors α, β and γ are used and a, b and g are assumed to be zero. For high uranium concentrations the upward stripping factor a is necessary. It has a value of approximately 0·05, but the actual value is dependent on factors such as detector size and resolution, window widths and energy settings. (Note: the stripping factor α has no relation to the shape factor discussed earlier. It is unfortunate that both factors developed in the literature using the same Greek letter for a symbol. It should be clear from the text which factor is meant.)

The stripping factors (Grasty, 1977) are applied to the digitised gamma-ray spectral log as follows:

$$Th_s = \frac{Th(1-g\gamma) + U(b\gamma - a) + K(ag - b)}{D} \tag{10}$$

$$U_s = \frac{Th(g\beta - \alpha) + U(1 - b\beta) + K(b\alpha - g)}{D} \tag{11}$$

$$K_s = \frac{Th(\alpha\gamma - \beta) + U(a\beta - \gamma) + K(1 - a\alpha)}{D} \tag{12}$$

where Th, U and K are the gamma-ray intensities recorded in the thorium, uranium and potassium windows respectively, in counts per second. Th_s, U_s and K_s are the corrected (stripped) gamma-ray intensities in those same windows. The term in the denominator, D, is given by

$$D = 1 - g\gamma - a(\alpha - g\beta) - b(\beta - \alpha\gamma) \tag{13}$$

The factors b and g are negligibly small for borehole logging detectors and may generally be eliminated from the above equations, giving

$$Th_s = \frac{Th - aU}{D} \tag{14}$$

$$U_s = \frac{U - \alpha Th}{D} \tag{15}$$

$$K_s = \frac{Th(\alpha\gamma - \beta) + U(a\beta - \gamma) + K(1 - a\alpha)}{D} \tag{16}$$

where D is now given by

$$D = 1 - a\alpha \tag{17}$$

4.3. The C-Factors or Sensitivities

To avoid confusion it has been suggested (Conway and Killeen, 1980) that the term K-factor be used to refer to total count or gross count sensitivities, while C-factors be used to refer to the sensitivities of the individual window counts. Thus the commonly used K-factor is a sensitivity for total count if all the counts are from U alone. In the spectral windows:

$$G_{Th}T_{Th}=C_{Th}A_{Th} \tag{18}$$

$$G_U T_U = C_U A_U \tag{19}$$

$$G_K T_K = C_K A_K \tag{20}$$

where A_i refers to the areas under the stripped window count log. The C-factors or window sensitivities will be in units of grade/count rate.

The stripped gamma-ray intensities from eqns (14), (15) and (16) are used in eqns (18), (19) and (20) to compute the concentrations of potassium, uranium and thorium in a similar fashion to the quantitative interpretation of gross count logs explained earlier, using either the iterative technique or the inverse filtering technique.

4.4. Derivation and Application of the Calibration Factors

The stripping factors and C-factors are determined by making measurements in model boreholes containing ore zones of known K, U and Th concentrations. The model holes are logged in the same way as field holes using the same logging parameters and the same hole size, casing, etc., as used in the field. Correction factors must be applied to adjust for any parameter which is different from field conditions. The count rates obtained by logging the model holes and the known grades and thickness of the ore zones are then used to compute all of the above calibration factors (Killeen and Conaway, 1978; Killeen, 1979; Wilson and Stromswold, 1981).

An example of a gamma-ray spectral log is shown in Fig. 32. The four windows include the total count (0·4 MeV and higher), as well as the K, U and Th windows. The log was recorded in the Bancroft uranium mining area where thorium contributes a significant component to the measured gamma-radiation. This is clear from the thorium channel log which shows four distinct anomalies. The same gamma-ray spectral log after stripping (as discussed above) and slight smoothing is shown in Fig. 33. The true contributions to the spectral windows are now evident, and it is clear, for example, that there are no potassium anomalies as were indicated by the raw spectral log, and the uranium anomalies are much smaller. Quantitative determinations may now be made on these anomalies by applying either the iterative fitting or inverse filtering technique and the appropriate C-factor as derived in model holes at calibration time.

Although gamma-ray spectral logging solves the problem of mixed radioelements in the rock, the assumption that the decay series are in

COUNTS / S

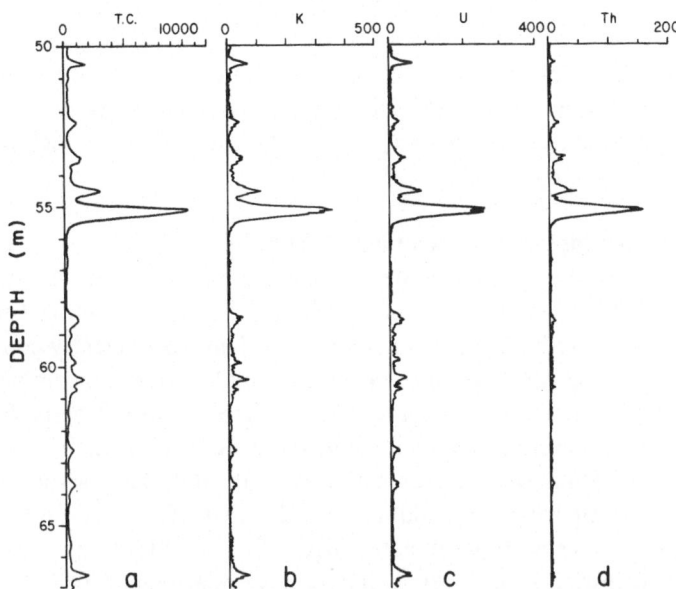

FIG. 32. A raw (uncorrected) gamma-ray spectral log showing the logs for four energy windows. These are the total count window (0·4 MeV and up), as well as the K, U and Th windows shown in Fig. 30.

COUNTS / S

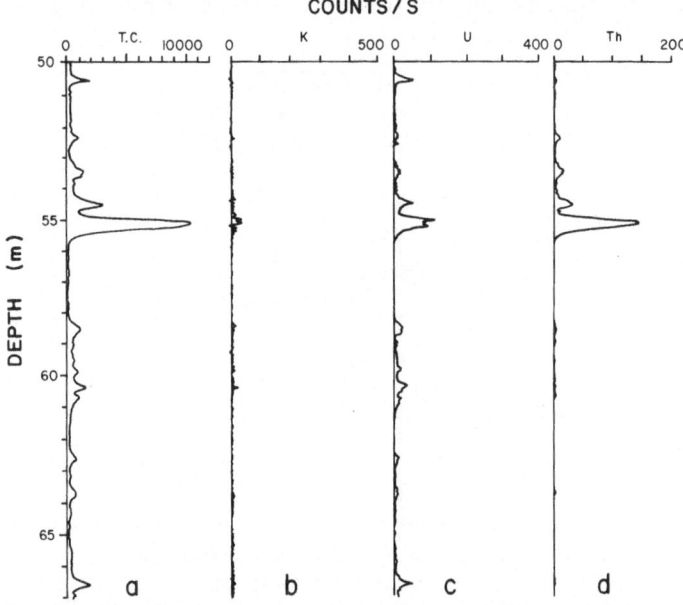

FIG. 33. The gamma-ray spectral log of Fig. 32 with stripping corrections applied, showing more accurately the true contributors to the gamma-ray spectra. The spectral window logs may now be used for quantitative measurements.

radioactive equilibrium must still be made. New developments in the use of high resolution solid-state detectors for gamma-ray spectral logging make it possible to overcome any disequilibrium problems which may be present.

4.5. New Developments in Scintillation Detectors

One of the problems of gamma-ray spectral logging with scintillation detectors is that the efficiency of the commonly used detectors (such as NaI(Tl)) is low at high gamma-ray energies. Thus to efficiently count the high energy (2·62 MeV) gamma-rays from Tl-208 in the thorium window either a larger detector or a more efficient detector would be preferable. The borehole diameter generally limits the detector size, but recently some new scintillation detector materials with high densities have been tested. A comparison of sodium iodide (NaI(Tl)), caesium iodide (CsI(Na)) and bismuth germanate ($Bi_4Ge_3O_{12}$ or 'BGO') detectors was made by Conway et al. (1980c). These have densities of 3·67, 4·51 and 7·31 g/cm^3, respectively. All three detectors were the same size (19 × 76 mm) and were tested in the same probe housing, both in the laboratory and in the model boreholes at Ottawa. To illustrate the improvement possible with the detectors of higher density, Fig. 34 shows

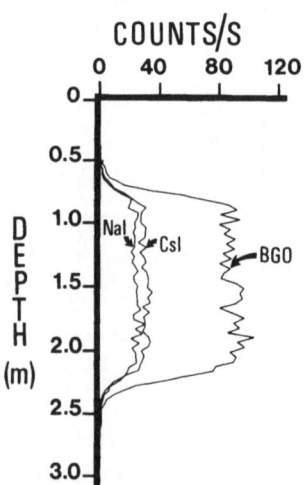

FIG. 34. Three uranium window gamma-ray spectral logs with 19 × 76 mm detectors of different materials, run through a uranium ore zone in a model hole (after Conaway et al., 1980c). The high count rate recorded by the BGO detector is a result of its high efficiency at 1·76 MeV compared to the other detectors. The CsI(Na) is also more efficient than the NaI(Tl) detector.

a uranium spectral window log through the ore zone of a model hole, for all three detectors. This development is particularly important in the case of hardrock gamma-ray spectral logging problems, where often the borehole has a diameter of 60 mm or less, and the probe and detector must be quite small. Typically a 32 mm outside diameter (OD) probe containing a 19 mm OD detector may be used. Similar favourable results of the testing of these high density detectors was reported in the USA (Wilson and Stromswold, 1981). Other promising new exotic scintillation detectors are also on the horizon (e.g. cadmium tungstate ($CdWO_4$) and mercuric iodide (HgI)).

5. HIGH RESOLUTION GAMMA-RAY SPECTRAL LOGGING

The advent of the solid-state detector, with its extremely high resolution of gamma-ray energies has opened up the possibility of detecting gamma-radiation from other daughters in the natural radioactive decay series. The bismuth-214 and thallium-208 gamma-rays used in spectrometry with scintillation detectors are usually chosen because of their abundance (intensity), their high energy, or their minimal interference from other gamma-rays of similar energies. Because of this, the scintillation gamma-ray spectral log requires radioactive equilibrium in the decay series for the indirect determination of either uranium or thorium. The use of a high resolution detector, however, makes it possible to select a number of different gamma-ray energies for counting and a great deal more information can be extracted from the gamma-ray energy spectrum.

The physical principle of operation of a solid-state detector is quite different from that of a scintillation detector. A scintillation detector produces photons detectable by a photomultiplier tube, whenever impinging gamma-rays cause excitation of electrons in the crystal lattice. In a solid-state detector, gamma-rays produce electron-hole pairs which are swept out of the sensitive region of the detector by a strong electric field. The result of both detection systems is an output pulse of amplitude proportional to the energy of the gamma-ray. Whereas a scintillating crystal may require an energy of several hundred electron-volts to generate a scintillation and an electronic pulse, a solid-state detector requires only a few electron-volts to generate a pulse. Thus gamma-rays only slightly different in energy are recognisable by the different pulse heights which they produce. An example of a portion of gamma-ray spectrum recorded with a sodium iodide (NaI(Tl)) scintillation detector and a solid-state lithium drifted germanium (Ge(Li)) detector is shown in Fig. 35 (from Śmith and Wollenberg, 1972). It is apparent from the figure how much more information is available in the Ge(Li) spectrum.

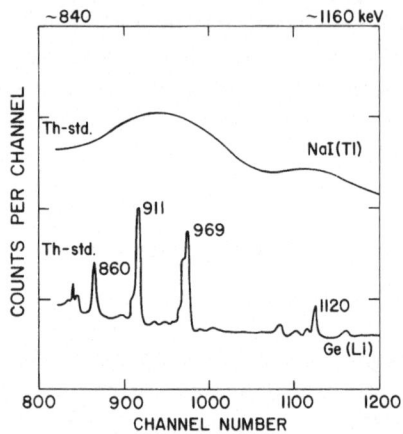

FIG. 35. A comparison of a small portion of the gamma-ray energy spectrum as measured by a scintillation detector and by a solid-state (high resolution) detector (after Smith and Wollenberg, 1972). It is easily seen that the latter contains much more information useful for interpretation.

Tables 1 and 2 list all the natural gamma-rays above 100 keV and with intensities above 0·1%. This tabulation (Smith and Wollenberg, 1972; Grasty, 1979) is a useful aid to interpretation of high resolution gamma-ray spectral logs.

TABLE 1

PRINCIPAL[a] GAMMA-RAYS OVER 100 keV EMITTED BY THE URANIUM DECAY SERIES IN EQUILIBRIUM WITH ITS DECAY PRODUCTS[b] (AFTER GRASTY, 1979)

Isotope[c]	γ-Energy (keV)	Intensity[d]	Isotope[c]	γ-Energy (keV)	Intensity[d]
Th-234	115	0·42	Th-227	236	0·51
U-235	144	0·48	Pb-214	242	7·60
Ra-223	144	0·14	Th-227	256	0·28
Ra-223	154	0·24	Pb-214	259	0·80
U-235	163	0·22	Ra-223	269	0·61
U-235	186	2·52	Rn-219	271	0·45
Ra-226	186	3·90	Pb-214	275	0·70
U-235	205	0·22	Pb-214	295	18·9

TABLE 1—*contd.*

Isotope[c]	γ-Energy (keV)	Intensity[d]	Isotope[c]	γ-Energy (keV)	Intensity[d]
Ra-223	324	0·16	Bi-214	964	0·37
Th-227	330	0·13	Pa-234M	1 001	0·83
Ra-223	338	0·12	Bi-214	1 052	0·33
Bi-211	351	0·60	Bi-214	1 070	0·26
Pb-214	352	36·3	Bi-214	1 104	0·16
Bi-214	387	0·31	Bi-214	1 120	15·0
Bi-214	389	0·37	Bi-214	1 134	0·25
Rn-219	402	0·29	Bi-214	1 155	1·70
Pb-211	405	0·18	Bi-214	1 208	0·47
Bi-214	406	0·15	Bi-214	1 238	6·10
Bi-214	427	0·10	Bi-214	1 281	1·50
Bi-214	455	0·28	Bi-214	1 304	0·11
Pb-214	462	0·17	Bi-214	1 378	4·30
Pb-214	481	0·34	Bi-214	1 385	0·80
Pb-214	487	0·33	Bi-214	1 402	1·50
Rn-222	511	0·10	Bi-214	1 408	2·60
Pb-214	534	0·17	Bi-214	1 509	2·20
Bi-214	544	0·10	Bi-214	1 539	0·53
Pb-214	580	0·36	Bi-214	1 543	0·34
Bi-214	609	42·8	Bi-214	1 583	0·73
Bi-214	666	14·0	Bi-214	1 595	0·30
Bi-214	703	0·47	Bi-214	1 600	0·34
Bi-214	720	0·38	Bi-214	1 661	1·16
Bi-214	753	0·11	Bi-214	1 684	0·24
Pa-234M	766	0·31	Bi-214	1 730	3·20
Bi-214	768	4·80	Bi-214	1 765	16·7
Pb-214	786	0·86	Bi-214	1 839	0·37
Bi-214	786	0·29	Bi-214	1 848	2·30
Bi-214	806	1·10	Bi-214	1 873	0·22
Bi-214	821	0·16	Bi-214	1 890	0·10
Bi-214	826	0·13	Bi-214	1 897	0·18
Pb-211	832	0·14	Bi-214	2 110	0·10
Bi-214	839	0·59	Bi-214	2 119	1·30
Pb-214	904	0·59	Bi-214	2 204	5·30
Bi-214	934	3·10	Bi-214	2 294	0·33
			Bi-214	2 448	1·65

[a] Photons with an intensity greater than 0·1%.

[b] *Note:* the U-235 decay series is included, but all intensities are recorded relative to 100 decays of the longest lived parent U-238. Thus, for example, for every 100 decays of U-238 there will be 16·7 gamma-rays, of energy 1765 keV, from Bi-214 and 0·83 gamma-rays from Pa-234.

[c] Decaying isotope.

[d] Decays per 100 decays of the longest lived parent.

TABLE 2

PRINCIPAL[a] GAMMA-RAYS OVER 100 keV EMITTED BY THORIUM IN EQUILIBRIUM
WITH ITS DECAY PRODUCTS (AFTER GRASTY, 1979)

Isotope[b]	γ-Energy (keV)	Intensity[c]	Isotope[b]	γ-Energy (keV)	Intensity[c]
Pb-212	115	0·61	Ac-228	675	0·11
Ac-228	129	3·03	Ac-228	702	0·20
Th-228	132	0·26	Ac-228	707	0·16
Ac-228	146	0·23	Ac-228	727	0·83
Ac-228	154	1·02	Bi-212	727	6·66
Ac-228	185	0·11	Ac-228	755	1·14
Ac-228	192	0·13	Tl-208	763	0·61
Ac-228	200	0·36	Ac-228	772	1·68
Ac-228	204	0·18	Ac-228	782	0·56
Ac-228	209	4·71	Bi-212	785	1·11
Th-228	217	0·27	Ac-228	795	5·01
Tl-208	234	0·12	Ac-228	830	0·64
Pb-212	239	44·6	Ac-228	836	1·88
Ra-224	241	3·70	Ac-228	840	10·2
Tl-208	253	0·25	Tl-208	860	4·32
Ac-228	270	3·90	Bi-212	893	0·37
Tl-208	277	2·34	Ac-228	904	0·90
Ac-228	279	0·24	Ac-228	911	30.0
Bi-212	288	0·34	Ac-228	944	0·11
Pb-212	300	3·42	Ac-228	948	0·13
Ac-228	322	0·26	Bi-212	952	0·18
Bi-212	328	0·14	Ac-228	959	0·33
Ac-228	328	3·48	Ac-228	965	5·64
Ac-228	332	0·49	Ac-228	969	18·1
Ac-228	338	12·4	Ac-228	988	0·20
Ac-228	341	0·44	Ac-228	1 033	0·23
Ac-228	409	2·31	Ac-228	1 065	0·15
Ac-228	440	0·15	Bi-212	'1 079	0·54
Bi-212	453	0·37	Tl-208	1 094	0·14
Ac-228	463	4·80	Ac-228	1 096	0·14
Ac-228	478	0·25	Ac-228	1 111	0·36
Ac-228	504	0·22	Ac-228	1 154	0·17
Ac-228	510	0·51	Ac-228	1 247	0·59
Tl-208	511	8·10	Ac-228	1 288	0·12
Ac-228	523	0·13	Ac-228	1 459	1·08
Ac-228	546	0·23	Ac-228	1 496	1·09
Ac-228	562	1·02	Ac-228	1 502	0·60
Ac-228	571	0·19	Bi-212	1 513	0·31
Ac-288	572	0·17	Ac-228	1 557	0·21
Tl-208	583	31·0	Ac-228	1 580	0·74
Ac-228	651	0·11	Ac-228	1 588	3·84

TABLE 2—contd.

Isotope[b]	γ-Energy (keV)	Intensity[c]	Isotope[b]	γ-Energy (keV)	Intensity[c]
Bi-212	1 621	1·51	Ac-228	1 666	0·22
Ac-228	1 625	0·33	Ac-228	1 686	0·11
Ac-228	1 630	2·02	Bi-212	1 806	0·11
Ac-228	1 638	0·56	Ac-228	1 887	0·11
			Tl-208	2 614	36·0

[a] With more than 0·1 decays per 100 decays of the longest lived parent.
[b] Decaying isotope.
[c] Decays per 100 decays of the longest lived parent, Th-232.

At present the main problem associated with the use of high resolution detectors is their cooling requirements. In order to operate, they must be kept at temperatures below $-150°C$. This requires liquid nitrogen or an equivalent coolant. The Ge(Li) detector requires these low temperatures at all times, even during storage and transportation. Also, large detectors are difficult to manufacture and are therefore very costly. A Ge(Li) borehole probe was used for gamma-ray spectral logging in a uranium mine in Sweden (Lauber and Landstrom, 1972). Their cryostat kept the probe cool for 10 h under working conditions, after which time the liquid nitrogen had to be replenished. A natural gamma-ray spectrum recorded in the Ranstad uranium mine was published to show the possibilities of the method. The counting times, however, were fairly long due to the small size of the detector ($22 \, cm^3$) and the need for a large number of channels to utilise the high resolution of the detector. The authors suggested that a 4–6 channel analyser, with its channels centred on peaks of interest, should be a viable arrangement. The use of a borehole Ge(Li) detector was suggested (Gorbatyuk et al., 1973) to determine the uranium content of ore using the 186 keV gamma-ray peak which is a combination of the 185·7 and 186·2 keV gamma-rays from 235-U and 226-Ra respectively. They also tried the 1·001 MeV peak of 234-Pa. This is a low intensity peak, but has the advantage of being high energy and relatively free of interference and also is high in the decay series avoiding disequilibrium problems. A simplification of the detector cooling problem was developed by Boynton (1975). It consists of using canisters or cartridges of solid propane 3·7 cm diameter by 57 cm long instead of liquid nitrogen. The solid propane melts to a liquid during cooling, without much change in volume, unlike the liquid nitrogen which converts to a gas, increasing in volume drastically and thus requiring venting of the probe. The most recent designs in the USA involve the use of a solid copper bar, pre-cooled to liquid nitrogen temperatures. No literature is yet available on these developments. In a review of nuclear geophysics in Sweden, borehole gamma-ray spectral logging measurements with both NaI(Tl) and Ge(Li) detectors were described (Christell et al., 1976). The use of intrinsic germanium (also called hyperpure Ge) in borehole probes used for uranium exploration was also described in 1976 (Senftle et al., 1976). The intrinsic Ge has the advantage of only requiring cooling to operate, but not during storage or transportation as is the case with Ge(Li) detectors.

Senftle et al. (1976) discussed several gamma-ray peaks which may be utilised for analysis of uranium such as the 63·3 keV peak of 234-Th, first daughter of 238-U. In 1977 the measurement of disequilibrium using a solid-state detector in a borehole probe was developed (Tanner et al., 1977). They utilised two probes, one with a Ge(Li) detector, the other with an intrinsic Ge detector. The latter is suitable for low energy gamma-ray measurement whereas the former, being of large volume (45 cm³), is used for high energy gamma-ray measurements. Their procedure is to first delineate zones for detailed investigation by logging continuously at about 1·0 m/min. The interesting zones are then analysed with 10 min counting times to determine their uranium content and also their state of radioactive equilibrium or disequilibrium. In situ assaying is a direct measurement based on the 63·3 keV gamma-ray of 234-Th and also the 1001·4 keV gamma-ray of 234-Pa. These are always in equilibrium with the parent 238-U because of their short half-life and proximity to the parent in the decay series. Six isotopes or groups of isotopes are evaluated in a single measurement of disequilibrium. Comparisons of scintillation detector logs and solid-state detector logs were also shown by Tanner et al. (1977). A series of holes drilled through a roll front uranium deposit were logged, and the state of equilibrium was displayed as a ratio of equivalent uranium (measured indirectly) to uranium measured directly. Through the use of the high resolution detector they were able to present logs of Pa–234, Th–234, Th–230, U–235 + Ra–226, Pb–214, Pb–210, Th–227, Ra–223 and K–40. Note some of these logs include data from the U–235 decay series (see Fig. 1 and Table 1). Sensitivity of the method is about 70 ppm uranium for the 10 min counting time.

The use of the solid-state detector has a number of advantages over scintillation detectors along with the disadvantages caused by the detector cooling problems. However, a commercially available technique is presently offered in the USA by at least one logging service company.

6. FUTURE DEVELOPMENTS

It is apparent that digital logging systems will eventually completely replace analogue logging systems. *In situ* data processing capability will be greatly expanded as speed, memory capacity, and ruggedness of the mini- or microcomputer-based field logging systems evolve. Computer libraries of characteristic radioelement ratios and other information will be made available (stored in memory) to be scanned for correlation with the measured logging data, and preliminary on-site interpretations will be made by the computer. These will eventually include correlation with data obtained simultaneously by the same logging tool with respect to other measured parameters.

High resolution solid-state detectors which can operate at near room temperature will likely be developed, eliminating the present complexity of the cooling problems. With further developments in detector efficiency of these solid-state detectors, even the scintillation detector could become as little-used as the Geiger-Müller detector is today, although this is not likely for some considerable time. In uranium exploration particularly, the reliability of both the data acquisition and the data processing techniques will be improved so as to become acceptable as a quantitative measurement, without question in any geologic situation including multi-radioelements, disequilibrium and variable formation parameters.

ACKNOWLEDGEMENTS

I would like to thank my co-workers at the Geological Survey of Canada for their contributions to these developments, especially Quentin Bristow who designed and constructed the GSC's DIGI-PROBE logging system, and John G. Conaway who developed the application of inverse filtering techniques to gamma-ray logs for uranium exploration.

I am also grateful to K. A. Richardson and C. J. Mwenifumbo for their helpful comments and suggestions regarding the manuscript.

REFERENCES

BOYNTON, G. R. (1975). Canister cryogenic system for cooling germanium semiconductor detectors in borehole and marine probes, *Nuclear Instruments and Methods*, **123**, 599–603.

BRISTOW, Q. (1979). Airborne and vehicle mounted geophysical data acquisition system controlled by Nova minicomputers, in *Proceedings of the 6th Annual Data General User's Group Meeting, New Orleans*, Dec. 4–7.

BRISTOW, Q. and KILLEEN, P. G. (1978). A new computer-based gamma ray spectral logging system, in *Geophysics Golden Gateway to Energy*, 49th Annual Meeting of the SEG (Abstract), pp. 117–18.

BRISTOW, Q., CONAWAY, J. G. and KILLEEN, P. G. (in press). Application of inverse filtering to gamma-ray logs: A case study, *Geophysics*.

CHRISTELL, R., LJUNGGREN, K. and LANDSTROM, O. (1976). Brief review of developments in nuclear geophysics in Sweden, *Nuclear Techniques in Geochemistry and Geophysics*, Proc. Series IAEA, Vienna, p. 21–46.

CONAWAY, J. G. (1978). Problems in gamma-ray logging: thin zone correction factors, in *Current Research, Part C, Geol. Surv. Can.*, Paper 78–1C, pp. 19–21.

CONAWAY, J. G. (1979a). Problems in gamma-ray logging: the effect of dipping beds on the accuracy of ore grade determinations, in *Current Research, Part A, Geol. Surv. Can.*, Paper 79–1A, pp. 41–4.

CONAWAY, J. G. (1979b). Computer processing of gamma-ray logs: a program for the determination of radioelement concentrations, in *Current Research, Part B, Geol. Surv. Can.*, Paper 79–1B, pp. 27–32.

CONAWAY, J. G. (1980a). Exact inverse filters for the deconvolution of gamma-ray logs, *Geoexploration*, **18**, 1–14.

CONAWAY, J. G. (1980b). Direct determination of the gamma-ray logging system response function in field boreholes, *Geoexploration*, **18**, 187–99.

CONAWAY, J. G. (1980c). Uranium concentrations and the system response function in gamma-ray logging, in *Current Research, Part A, Geol. Surv. Can.*, Paper 80–1A, pp. 77–87.

CONAWAY, J. G. (1981). Deconvolution of gamma-ray logs in the case of dipping radioactive zones, *Geophysics*, **46**(2), 198–202.

CONAWAY, J. G. and BRISTOW, Q. (1981). Pitfalls in quantitative gamma-ray logging: Calibration sleeves and Am–241 temperature stabilization, *Society of Professional Well Log Analysts 22nd Annual Logging Symposium Transactions, Mexico City*.

CONAWAY, J. G. and KILLEEN, P. G. (1978a). Quantitative uranium determinations from gamma-ray logs by application of digital time series analysis, *Geophysics*, **43**(6), 1204–21.

CONAWAY, J. G. and KILLEEN, P. G. (1978b). Computer processing of gamma-ray logs: iteration and inverse filtering, in *Current Research, Part B, Geol. Surv. Can.*, Paper 78–1B, pp. 83–8.

CONAWAY, J. G. and KILLEEN, P. G. (1980). Gamma-ray spectral logging for uranium, *Can. Inst. Mining Metall. Bull.*, **73**(813), 115–23.

CONAWAY, J. G., ALLEN, K. V., BLANCHARD, Y. B., BRISTOW, Q., HYATT, W. G. and KILLEEN, P. G. (1979). The effects of borehole diameter, borehole fluid, and steel casing thickness on gamma-ray logs in large diameter boreholes, in *Current Research, Part C, Geol. Surv. Can.*, Paper 79–1C, pp. 37–40.

CONAWAY, J. G., BRISTOW, Q. and KILLEEN, P. G. (1980a). Optimization of gamma-ray logging techniques for uranium, *Geophysics*, **45**(2), 292–311.

CONAWAY, J. G., KILLEEN, P. G. and BRISTOW, Q. (1980b). Variable formation parameters and nonlinear errors in quantitative borehole gamma-ray log interpretation, in *Technical Program, Abstracts and Biographies*, 50th Annual International Meeting of the SEG (Abstract), pp. 121–2.

CONAWAY, J. G., KILLEEN, P. G. and HYATT, W. G. (1980c). A comparison of bismuth germanate, cesium iodide, and sodium iodide scintillation detectors for gamma-ray spectral logging in small diameter boreholes, in *Current Research, Part B, Geol. Surv. Can.*, Paper 80–1B, pp. 173–7.

CZUBEK, J. A. (1971). Differential interpretation of gamma-ray logs: I. Case of the static gamma-ray curve, Report No. 760/1, Nuclear Energy Information Center, Polish Government Commissioner for Use of Nuclear Energy, Warsaw, Poland.

CZUBEK, J. A. (1972). Differential interpretation of gamma-ray logs: II. Case of the dynamic gamma-ray curve, Report No. 793/1, Nuclear Energy Information Center, Polish Government Commissioner for Use of Nuclear Energy, Warsaw, Poland.

DAVYDOV, Y. B. (1970). Odnomernaya obratnaya zadacha gamma-karotazha skvazhin (One dimensional inversion problem of borehole gamma logging), *Izvestiya Vysshoya Uchebnoye Zavedeniya Geologiya i Razvedka*, No. 2, 105–9 (in Russian).

DODD, P. H. and ESCHLIMAN, D. H. (1972). Borehole logging techniques for uranium exploration and evaluation, in *Uranium Prospecting Handbook*, Ed. S. H. U. Bowie, M. Davis and D. Ostle. Inst. Min. Metal., London, pp. 244–76.

GEORGE, D. C., EVANS, H. B., ALLEN, J. W., KEY, B. N., WARD, D. L. and MATHEWS, M. A. (1978). A borehole gamma-ray spectrometer for uranium exploration, US Dept. of Energy, Grand Junction Office, Report GJBX–82(78).

GORBATYUK, O. V., KADISOV, E. M., MILLER, V. V. and TROITSKII, S. G. (1973). Possibilities of determining uranium and radium content of ores by measuring gamma radiation in a borehole using a spectrometer with a Ge(Li) detector, Translated from *Atomnaya Energiya*, **35**(5), 355–7, Nov. 1973: Consultants Bureau, Plenum Publishing Co., New York.

GRASTY, R. L. (1977). A general calibration procedure for airborne gamma-ray spectrometers, in *Report of Activities, Part C, Geol. Surv. Can.*, Paper 77–1C, pp. 61–2.

GRASTY, R. L. (1979). Gamma-ray spectrometric methods in uranium exploration: Theory and operational procedures, in *Geophysics and Geochemistry in the Search for Metallic Ores*, Geol. Surv. Can., Econ. Geol. Rep. 31, pp. 147–61.

HSU, H. P. (1970). *Fourier analysis*. Simon and Schuster, New York.

KANASEWICH, E. R. (1973). *Time Sequence Analysis in Geophysics*. University of Alberta Press, Calgary.

KILLEEN, P. G. (1976). Portable borehole gamma-ray spectrometer tests, in *Report of Activities, Part A, Geol. Surv. Can.*, Paper 76–1A, pp. 487–9.

KILLEEN, P. G. (1978). Gamma-ray spectrometric calibration facilities—a preliminary report, in *Current Research, Part A, Geol. Surv. Can.*, Paper 78–1A, pp. 243–7.

KILLEEN, P. G. (1979). Gamma-ray spectrometric methods in uranium exploration—application and interpretation, in *Geophysics and Geochemistry in the Search for Metallic Ores*, Geol. Surv. Can., Econ. Geol. Rep. 31, pp. 163–229.

KILLEEN, P. G. and BRISTOW, Q. (1976). Uranium exploration by borehole gamma-ray spectrometry using off-the-shelf instrumentation, in *Exploration for Uranium Ore Deposits*, Proc. Series IAEA, Vienna, pp. 393–414.

KILLEEN, P. G. and CONAWAY, J. G. (1978). New facilities for calibrating gamma-ray spectrometric logging and surface exploration equipment, *Can. Inst. Mining Metall. Bull.*, **71**(793), 84–7.

KILLEEN, P. G., CONAWAY, J. G. and BRISTOW, Q. (1978). A gamma-ray spectral logging system including digital playback, with recommendations for a new generation system, in *Current Research, Part A, Geol. Surv. Can.*, Paper 78–1A, pp. 235–41.

LAUBER, A. and LANDSTROM, O. (1972). A Ge(Li) borehole probe for *in situ* gamma-ray spectrometry, *Geophys. Prospect.*, **20**, 800–13.

MATHEWS, M. A., KOIZUMI, C. J. and EVANS, H. B. (1978). DOE—Grand Junction logging model data synopsis, US Dept. of Energy, Grand Junction Office, Report GJBX–76(78).

ROBINSON, E. A. (1967). *Multichannel time series analysis with digital computer programs*, Holden-Day, San Francisco.

ROSHOLT, J. N., JR. (1959). Natural radioactive disequilibrium of the uranium series, *US Geol. Surv. Bull.*, 1084–A, pp. 1–30.

SCOTT, J. H. (1963). Computer analysis of gamma-ray logs, *Geophysics*, **28**(3), 457–65.

SCOTT, J. H. (1980). Pitfalls in determining the dead time of nuclear well-logging probes, *Society of Professional Well Log Analysts 21st Annual Logging Symposium Transactions*, Paper H, pp. 1–11.

SCOTT, J. H., DODD, P. H., DROULLARD, R. F. and MUDRA, P. J. (1961). Quantitative interpretation of gamma-ray logs, *Geophysics*, **26**(2), 182–91.

SENFTLE, F. E., MOXHAM, R. M., TANNER, A. B., BOYNTON, G. R., PHILBIN, P. W. and BAICKER, J. A. (1976). Intrinsic germanium detector used in borehole sonde for uranium exploration, *Nuclear Instruments and Methods*, **138**, 371–80.

SMITH, A. R. and WOLLENBERG, H. A. (1972). High-resolution gamma-ray spectrometry for laboratory analysis of the uranium and thorium decay series, *Proceedings of the Second International Symposium on the Natural Radiation Environment*, Houston, Texas, Ed., J. A. S. Adams, W. M. Lowder and T. F. Gesell. US Dept. of Commerce, Springfield, Virginia, pp. 181–231.

STROMSWOLD, D. C. and KOSANKE, K. L. (1979). Spectral gamma-ray logging I: Energy stabilization methods, *Society of Professional Well Log Analysts 20th Annual Logging Symposium Transactions*, Paper DD, pp. 1–10.

SUPPE, S. A. and KHAIKOVICH, I. M. (1960). Resheniye pryamoi zadachi gamma-karotazha v sluchaye slozhnogo raspredeleniya radioaktivnogo elementa v aktivykh plastakh (Solution of the linear problem of gamma logging in the case of a complex distribution of the radioactive element in the active strata), *Voprosy Rudnoi Geofiziki*, No. 1 (in Russian).

TANNER, A. B., MOXHAM, R. M. and SENFTLE, F. E. (1977). Assay for uranium and determination of disequilibrium by means of *in situ* high resolution gamma-ray spectrometry, US Geol. Surv. Open File Report 77–571.

WILSON, R. D. and STROMSWOLD, D. C. (1981). Spectral gamma-ray logging studies, Report prepared for US Dept. of Energy by Bendix Field Engineering Corporation, Grand Junction, Colorado, under contract No. DE–AC13–76GJ01664.

Chapter 5

ACOUSTIC LOGGING: THE COMPLETE
WAVEFORM AND ITS INTERPRETATION

DENNIS RADER

Teleco Oilfield Services Inc., Meriden, Connecticut, USA

SUMMARY

The waveform generated in acoustic logging contains significantly more information than the traditional compressional interval travel time measurement. Four components of the wavetrain, viz. shear and compressional lateral waves, the Stoneley and pseudo-Rayleigh waves, are discussed. A description of experimental studies of the acoustic wavetrain in homogeneous and inhomogeneous scale models is given followed by analytical results pertaining to each of the four component waves. The analytical results are presented in three parts: First, individual results for the pseudo-Rayleigh and Stoneley waves are discussed. Secondly, full waveform studies are presented which are based on numerical integration schemes. Finally, the results of branch cut integration techniques applied to the shear and compressional lateral waves are given. The lateral wave impulse responses and the responses to a realistic source function are presented for a variety of borehole parameters. It is also shown that the shear arrival should be viewed as a composite of both the shear and pseudo-Rayleigh signals.

1. INTRODUCTION

Continuous well logging measurements fall into three technique groups: electrical, nuclear and acoustic (or sonic). When non-continuous, i.e. point by point, devices such as the formation tester or core sampler are considered, a fourth, purely mechanical category is added.

Most wireline logging measurements are aimed at the goal of formation evaluation, a term which refers to the determination of lithology, geometry of structure (porosity and microstructure), pore fluid identification, density, permeability and mechanical properties. Among the important exceptions is the cement bond log, an acoustic measurement for evaluating the quality of the cement bond achieved in cased holes. In almost all cases, formation evaluation objectives cannot be achieved unambiguously without using two or more logging techniques. Even when all relevant logs are available, several formation properties cannot be directly determined. In particular, some key parameters such as formation permeability and mechanical strength can be inferred but not deduced with existing wireline logging technology.

Acoustic logging is unique in that it is the only continuous logging measurement which is directly sensitive to some of the mechanical properties of formation rock. Thus, when the formation density is known, from the gamma–gamma log, acoustic wave interval transit time measurements provide a means for determining the elastic moduli of the rock. While the traditional application of the acoustic log is for porosity determination, it has received increasing attention in recent years for its potential in several other applications including mechanical property determination.

In conventional open hole acoustic logging a single measurement is made as a function of depth in a borehole, viz. the compressional wave transit time in the formation over a fixed spatial interval, typically 1 ft. In conjunction with the appropriate additional logging data, this measurement can be applied, with varying degrees of validity, to porosity determination, lithology identification and seismic correlation. As we shall see, however, this one measurement is not sufficient to generate substantive information about the elastic moduli or other mechanical properties of the formation, an important but elusive set of parameters. To make further progress, more detailed evaluation of the full acoustic waveform is required, which includes a determination of the shear wave velocity in the formation.

It has been known for some time that the conventional acoustic waveform does indeed contain usable information in addition to the compressional wave velocity. The problem is one of data processing and appropriate tool design to optimise information retrieval. Acquisition of the shear wave speed, for example, provides a deductive means for describing the elastic properties of the formation and also offers the potential for shear–seismic correlation. Conventional seismic sections are based only on compressional wave speeds. Future seismic studies will very likely include additional data based on shear wave initiation and detection.

The acoustic waveform contains two other prominent components known as the pseudo-Rayleigh and Stoneley waves. These are guided interface waves, both of which are strongly influenced by the shear modulus of the formation. They will be discussed in Sections 3 and 4.

Until recently, adequate analytical and experimental results, which must be the starting point for developing data processing schemes to extract the desired information, were not available to describe the full acoustic waveform. In the balance of this paper the status of recent analytical studies together with supporting experimental work will be reviewed. First, a background review is provided.

2. ELASTIC WAVE PROPAGATION AND ACOUSTIC LOGGING

The fluid saturated porous rocks found in nature cannot actually be described as elastic solids, i.e. they exhibit dissipation, due to several mechanisms, when subjected to external loading cycles. Accordingly, acoustic waves which propagate through rock experience attenuation and dispersion (i.e. phase velocity is a function of frequency). Moreover, rock is in general anisotropic. In order to make the borehole acoustic logging problem tractable, however, rock is assumed to be elastic and isotropic. This simplification is generally not as serious as may appear because most *in situ* rocks typically exhibit modest attenuation, particularly at the high stresses which obtain at depth in the logging environment (see, for example, Wyllie *et al.*, 1962; Knopoff, 1964; Walsh, 1966; Gordon and Davis, 1968; Gordon and Rader, 1970). The degree to which anisotropy may be important is less clear cut since the acoustic logging measurement is generally made normal to the bedding plane while some transverse properties are often of interest, i.e. in the bedding plane. (For a further discussion of seismic waves in transversely isotropic formations, see Levin, 1980.) Nonetheless, we expect to gain access to the major features of the acoustic waveform with an idealised model.

A fundamental result in elastic wave propagation in isotropic solids is the existence of two and only two modes of propagation when the medium is unbounded. The first is a longitudinal mode which is geometrically irrotational (i.e. the curl of the displacement vector is zero) and is alternately known as the dilatational, extensional, P, or compressional wave. The second is a transverse mode which is geometrically equivoluminal (i.e. the divergence of the displacement vector is zero) and is commonly known as the S or shear wave. A detailed discussion of the subject will be found in Kolsky (1953). All other propagation modes, and their characteristic dispersion curves, occur as a consequence of the presence of boundaries either at a free surface or at an interface with another medium.

The compressional speed (v_c) and the shear speed (v_s) can be expressed in terms of two elastic constants (λ and μ) and the density (ρ), which completely define the solid, as follows:

$$v_c = \{(\lambda + 2\mu)/\rho\}^{\frac{1}{2}} \tag{1}$$

$$v_s = (\mu/\rho)^{\frac{1}{2}} \tag{2}$$

The constants λ and μ are known as Lamé constants. Alternatively, the wave speeds can be expressed in terms of the 'engineering' constants E, μ and v where E is Young's modulus, μ is the shear modulus (identical to the Lamé constant) and v is Poisson's ratio:

$$v_c = \{E(1 - v)/\rho(1 + v)(1 - 2v)\}^{\frac{1}{2}} \tag{3}$$

$$v_s = \{E/2\rho(1 + v)\}^{\frac{1}{2}} \tag{4}$$

We also have the convenient relationship between v, v_c and v_s:

$$v = \{\tfrac{1}{2}(v_c/v_s)^2 - 1\}/\{(v_c/v_s)^2 - 1\} \tag{5}$$

The two sets of elastic constants are related by the expressions

$$\mu = E/\{2(1+v)\}, \quad \lambda = Ev/\{(1+v)(1-2v)\} \tag{6}$$

Thus, if both v_c and v_s are measured, together with ρ, the elastic properties of the solid are completely defined. Since in conventional acoustic logging only v_c is measured, the elastic moduli cannot be determined.

A schematic description of the basic method of acoustic logging is illustrated in Fig. 1. Further discussion of the conventional logging measurement can be found in several sources such as Pirson (1963). In Fig. 1, 'ray' paths are indicated for shear and compressional waves. The transmitter, T (regarded as an omni-directional point source), emits an acoustic pulse (the 'source function') with a centre frequency in the

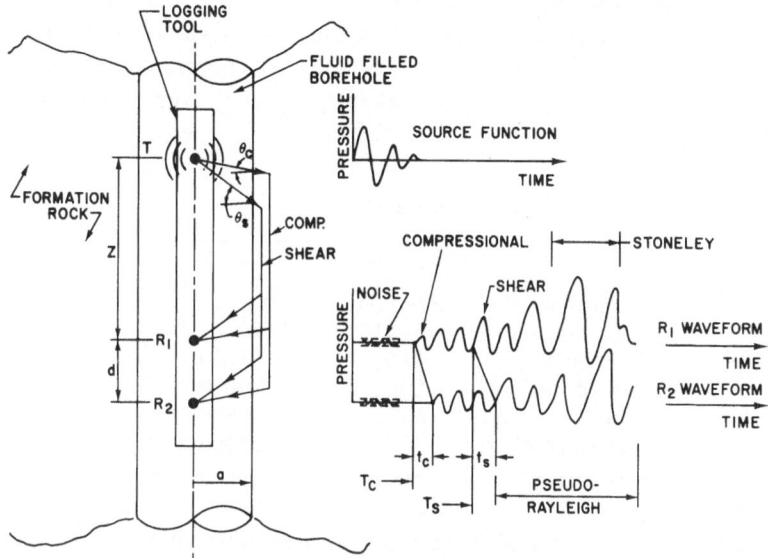

FIG. 1. Schematic illustration of the acoustic logging technique with simulated source function and waveforms.

10–30 kHz range. A pressure wave propagates through the borehole fluid at a speed v and is refracted at the borehole wall. The reflected wave is ignored here. At one critical angle of incidence, θ_c, a 'lateral' or 'head' wave is generated parallel to the borehole wall and propagates at the compressional speed in the formation. At the critical angle θ_s, another lateral wave is generated which propagates at the shear speed in the formation.

The critical incidence angles are determined from Snell's law when the refraction angle is 90°:

$$\theta_c = \sin^{-1} v/v_c; \quad \theta_s = \sin^{-1} v/v_s \tag{7}$$

The existence of these waves is not obvious and cannot be described starting from Snell's law. Rather, a solution to the full wave propagation problem must be obtained which leads to a description of the lateral wave propagation and the results stated above for θ_c and θ_s. See, for example Ewing *et al.* (1957).

The lateral wave amplitude diminishes along the propagation path as energy continuously radiates back into the borehole. Re-radiated signals are detected at the receivers R_1 and R_2 with a time delay corresponding to the propagation time over distance d in the formation. The *interval transit times* t_c and t_s, for the compressional and shear respectively, are generally normalised to a unit distance; they are commonly expressed in microseconds per foot. Typical values of t_c and v_c in common formation rocks and steel casing are given in Table 1 (Schlumberger, 1972).

TABLE 1

	$t_c(\mu s/ft)$	$v_c(ft/s)$
Sandstone	55·5–51·0	18 000–19 600
Limestones	47·6–43·5	21 000–23 000
Dolomites	43·5	23 000
Anhydrite	50·0	20 000
Salt	66·7	15 000
Casing (steel)	57·0	17 500

The shear speeds depend on the value of v but are typically about 60% of v_c. More precisely,

$$t_s = \{2(1 - v)/(1 - 2v)\}^{\frac{1}{2}} t_c \tag{8}$$

where $0 < v < \frac{1}{2}$.† The interval transit time in borehole fluids such as drilling mud and water varies but is typically about $190\,\mu s/ft$.

Interval transit times and propagation velocities given for the rocks in Table 1 pertain to the *matrix* material. The presence of the fluid-filled porosity increases the mechanical compliance of the rock structure and the corresponding values of t_c. The values of t_c and v_c reduce to that of the matrix when the porosity is zero.

The classic Wyllie equation between the observed interval travel time and the porosity ϕ is the linear relationship

$$t_c = \phi t_f + (1 - \phi)t_{ma} \tag{9a}$$

† In theory, the allowable range for v is $-1 < v < \frac{1}{2}$. For an incompressible solid $v = \frac{1}{2}$. Negative values of v have not been observed in isotropic solids but some examples exist for non-isotropic solids.

or

$$\phi = \frac{t_c - t_{ma}}{t_f - t_{ma}} \tag{9b}$$

where t_c = log reading, t_f = transit time in the pore fluid and t_{ma} = transit time in the matrix. This relationship is a consequence of the assumption that the wave responds to the two-phase material as if it passed through the separate phases in sequence. In practice, this relationship works fairly well in clean, consolidated formations with uniform intergranular porosity. In such cases an acoustic log can be used to determine porosity when the lithology has been identified (which gives t_{ma}) and t_f is taken as $190\,\mu s/ft$. A more comprehensive discussion can be found in Schlumberger (1972). More elaborate treatments of the wave propagation problem in porous rock have been given by Toksoz et al. (1976) and O'Connell and Budiansky (1974), among others.

Given the nature of the acoustic signal as a lateral wave, the log measurement has an apparently shallow depth of investigation. Indeed, it is qualitatively different from most nuclear and electrical logging measurements which respond to the properties of a significant volume of formation rock. In acoustic logging we are therefore concerned about the problems of formation invasion and alteration in a zone near the borehole wall. Fortunately, these effects invariably result in a reduction in acoustic impedance to a value between that of the borehole fluid and the undisturbed formation. Thus, there will be a positive radial gradient of acoustic impedance with the consequences illustrated in Fig. 2. Two lateral wave arrivals are shown. One of these (labelled 1) propagates near the borehole wall, as before, but in a region where the formation acoustic properties have been modified. A second lateral wave (labelled 2) is shown at a hypothetical boundary between the extremity of the invaded/altered zone and the undisturbed formation. A curved refraction path is shown through the invaded zone to the hypothetical boundary. At short transmitter to receiver spacings the near-wall arrival may be detected first since its path length is the lesser of the two waves. However, the propagation speed in the near-wall zone will be anomalously low (i.e. t_c will be high) giving erroneous results. At sufficiently long spacings, the signal from the undisturbed formation will be detected first because its higher velocity will more than compensate for the added path length. Moreover, the intrinsic attenuation in the undisturbed formation will typically be less than that in the invaded/altered zone, thereby enhancing the relative amplitude of the

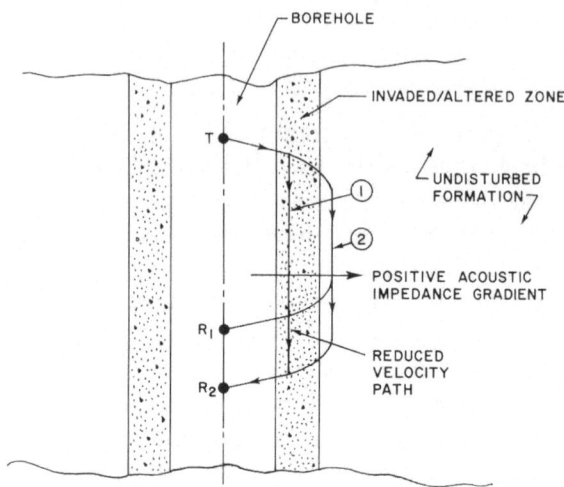

F IG. 2. Effect of the invaded/altered zone on the acoustic logging problem.

desired signal. Examples of the benefits derived from a long-spacing acoustic logging tool are given by Aron *et al.* (1978).

As with all logging tools, acoustic devices of the type discussed here, known as 'mandrel tools', have operating limitations. In cases where the formation propagation speed is less than the borehole fluid speed, there can be no lateral wave arrival. This is a rare occurrence for the compressional wave. The shear speed does drop below the wave speed in the borehole fluid with greater frequency and in such cases no direct determination of the formation shear properties can be obtained. An example of this occurrence will be discussed in Section 4.

In practice, care must be taken to ensure that the logging tool structure does not provide a direct acoustic path to the receivers since such a signal would precede the formation compressional arrival in the waveform, making the detection of the latter more difficult. This is generally accomplished by elongating the acoustic path in the tool using slots or other geometric patterns. When the spacing between the transmitter and receiver is Z (Fig. 1) the *total* transit times of the compressional and shear arrivals, T_c and T_s respectively, are

$$T_c = \frac{Z}{v_c} + \frac{2a \cos \theta_c}{v} \tag{10a}$$

and

$$T_s = \frac{Z}{v_s} + \frac{2a \cos \theta_s}{v} \tag{10b}$$

where a is the borehole radius.

As suggested by the simulated waveforms in Fig. 1, the compressional arrival will be readily detected when the noise level is small compared to the compressional signal amplitude. The shear arrival will, however, be 'buried' in the wavetrain leading to a difficult signal detection problem. In fact, given the complex structure of the waveform, it is not immediately clear whether a distinct shear arrival actually exists. As we shall see later, the shear signal cannot in general be isolated; rather a composite of the shear and pseudo-Rayleigh waves can be identified and will be termed the shear–pseudo-Rayleigh arrival following Tsang and Rader (1979). The Stoneley arrival follows and partially overlaps the shear–pseudo-Rayleigh. These waves will be discussed in the next two sections.

The wavetrain illustrated in Fig. 1 can be thought of as the output of a system (consisting of the borehole geometry, borehole fluid and formation rock) which is excited by an input function in the form of the source waveform. It is the transfer function of this system which we wish to study. In the following two sections, experimental and theoretical efforts toward this end are reviewed.

3. EXPERIMENTAL STUDIES

Before embarking on a discussion of the analytical problem (Section 4), some experimental studies of acoustic logging will be reviewed in order to develop a physical appreciation of this complex problem. The results described in this section were obtained by the author at Schlumberger-Doll Research Center in the early 1970s and were previously reported at a meeting of the Acoustical Society of America (Rader, 1975).

Although significant progress on the analysis of the borehole acoustic problem began some three decades ago (Biot, 1952), more complete analytical treatments did not appear until the 1960s and 1970s (White, 1962; White and Zechman, 1968; Peterson, 1974; Roever et al., 1974; Rosenbaum, 1974; Tsang and Kong, 1979; Tsang and Rader, 1979; Cheng and Toksoz, 1981). Several key aspects of the wave propagation problem have thus been at least partially understood for some time but controlled experimental testing had only been performed to a limited extent in the field. Although the depth of penetration of the acoustic measurement is generally small, full size laboratory test facilities are impractical because the effects of boundary reflections would require prohibitively large test formations. Control of physical (acoustic) properties would also be difficult in such a facility. An experimental approach based on modelling techniques was therefore undertaken.

A schematic description of a cylindrical test block for studying acoustic logging is shown in Fig. 3. This facility was built to simulate the

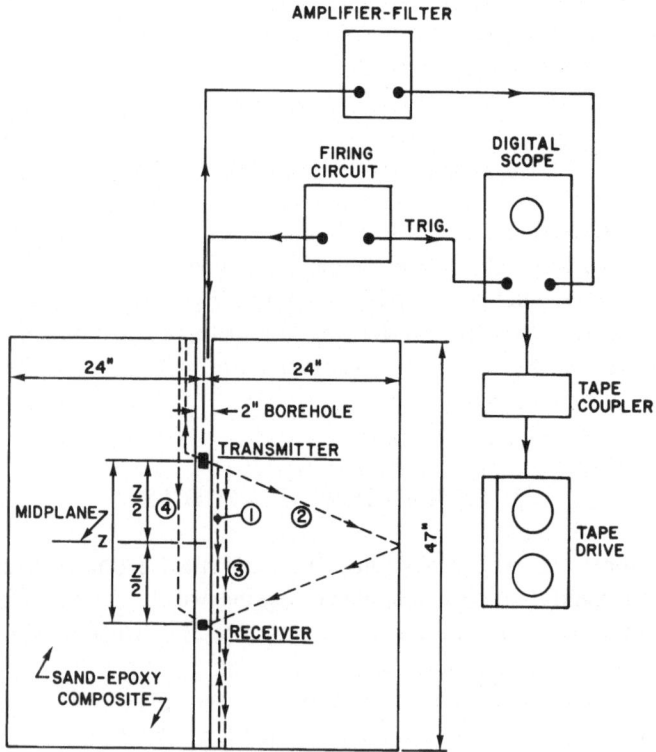

FIG. 3. Experimental set-up for scale model studies of borehole acoustic waveforms.

logging problem in a 'typical' mid-range porous sandstone for which $t_c = 88\ \mu s/ft$ ($v_c = 11\ 400\ ft/s$), penetrated by an 8 in diameter borehole. A scaling ratio of 4:1 was chosen and applied to the borehole diameter and transmitter frequency. Thus a 10 kHz source waveform suitable for a field tool was modelled with a 40 kHz transmitter and the borehole diameter was set at 2 in. In this way the ratio of acoustic wavelength to borehole diameter, the key non-dimensional parameter, is held constant.

The formation material was an epoxy–sand mixture cast in place around a borehole moulding tube which was subsequently withdrawn. Great care was taken to assure material homogeneity. The borehole fluid was a medium weight oil chosen for its electrical insulation properties. This fluid is somewhat lower in density and wave speed than borehole fluids encountered in the field and is the one respect in which modelling fidelity was not maintained. This, however, has a small effect on the results in relation to field waveforms. The external dimensions of the formation were chosen to minimise the effect of boundary reflections. The model parameters and equivalent field conditions are summarised in Table 2.

TABLE 2

	Model	Equivalent field parameter
Formation		
Borehole diameter (in)	2	8
t_c (μs/ft)	88	88
t_s (μs/ft)	139	139
Source frequency (kHz)	40	10
Z	variable	4 × model value
Ratio of wavelength in borehole fluid to borehole radius	0·71 (at 40 kHz)	0·71 (at 10 kHz)
Density (g/cm^3)	2·2	2·2
Borehole fluid		
Interval transit time (μs/ft)	212	212
Density (g/cm^3)	0·88	0·88

The transmitter was a hollow piezo-electric cylinder made of lead zirconate titanate $\frac{3}{4}$ in diameter, $\frac{3}{4}$ in long and $\frac{1}{16}$ in thick. It was driven in the simple radial mode at up to 500 V by the pulsing circuit. The receiver element was a solid cube of lead metaniobate approximately $\frac{1}{10}$ in on a side. In order to have the experiments closely simulate analytical treatments, the borehole was kept as free as possible from materials which would distort the acoustic fields. Thus, the transducer assembly was mounted in a nylon and teflon structure which provided minimal acoustic impedance mismatch relative to the borehole fluid. Teflon control rods were brought out at the top surface for manual or motorised operation. The necessary insulated conductors were carried along the control rods from the pulsing circuit to the transducer elements.

Acoustic waveforms detected by the receiver were displayed on an analogue or digital oscilloscope and, in the latter case, recorded digitally on magnetic tape. The open field output waveform, or 'source function', of the transmitter was observed in a separate oil-filled tank and is shown in Fig. 4. The source waveform is a relatively wideband function which is desirable for this kind of experiment.

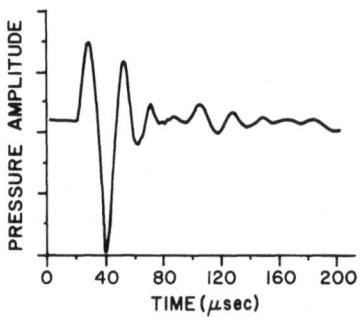

FIG. 4. Open field transmitter waveform used in experimental studies.

Some of the acoustic paths in the model which are indicated in Fig. 3 will be discussed for later reference. To simplify the discussion, consider only compressional waves. Lateral waves, i.e. the 'logging' signals, propagate along path 1. The direct side-boundary reflection occurs along path 2. The lateral wave signals which are reflected from the bottom and top surfaces of the model propagate along paths 3 and 4 respectively. For the parameters of this model, the first interfering reflection occurs along path 2 when $Z < 9.6$ in and along path 3 when $Z > 9.6$ in. The unobstructed waveform duration, ΔT, in microseconds, at any value of Z, expressed in inches, is given approximately by

$$\Delta T = 329 - 7Z \qquad (11)$$

when the transducers are centred with respect to the midplane of the model as shown in Fig. 3.

Typical waveforms observed in the test facility are shown in Fig. 5 at spacings of $Z = 9, 10, 11$ and 12 in. Except for the time scale which is compressed by a factor of 4, these waveforms are the same as would be

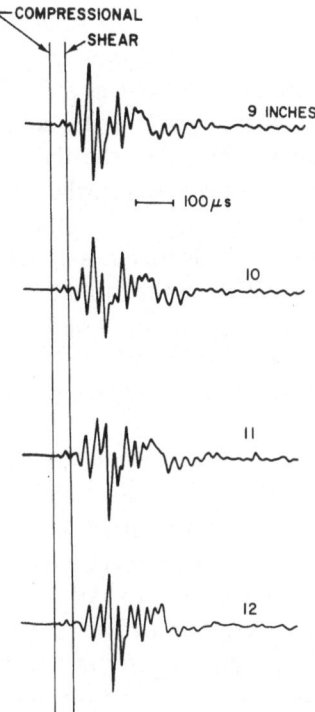

FIG. 5. Waveforms observed at $Z=9$, 10, 11 and 12 in in the scale model test facility of Fig. 3.

observed under equivalent field conditions. In practice, field waveforms will be affected by the presence of the logging tool, borehole irregularities and formation inhomogeneities which are not modelled in the experiments.

The first $1\frac{1}{2}$ cycles of the waveform clearly represent the compressional arrival. A line representing the theoretical position of the shear arrival is shown to the right of the compressional but cannot be clearly identified with a feature of the wavetrain. Further to the right, a feature having a consistently dominant amplitude is evident. Two high amplitude components of the acoustic wavetrain which arrive in the time range beyond the shear arrival are the Stoneley and pseudo-Rayleigh waves (see Cheng and Toksoz, 1981, for example). The Stoneley wave is a coupled mode between the borehole fluid and the formation which propagates at a speed slightly below the borehole fluid speed, v. It exhibits only modest dispersion. The pseudo-Rayleigh wave is a modified classical Rayleigh wave in which the effect of fluid on the solid boundary is to produce attenuation in the solid due to radiation into the fluid medium. It is a strongly dispersive mode in borehole geometry wherein the phase velocity is equal to the shear speed in the solid at the low frequency cutoff and asymptotically approaches the fluid acoustic speed at high frequencies. These waves are discussed further in Section 4.

It is only in the last waveform ($Z = 12$ in) that a reasonable amount of 'separation' of the various arrivals becomes apparent. This, together with enhanced depth of investigation in the formation, is the main motivation for implementing so-called long-spaced acoustic logging tools. However, attenuation of the compressional arrival due to both geometric spreading and intrinsic material dissipation is one factor which limits the maximum practical spacing.

Evidently, few general conclusions can be drawn about the structure of the acoustic waveform or the information which can be obtained from it without further analysis. Experimentally, it is possible, however, to plot large numbers of waveforms in a compact seismic-type display in which amplitude information is converted into varying tones of darkness on the white to black scale. Such variable density presentations are frequently helpful in elucidating the general features of a set of complex waveforms.

In order to achieve this presentation with the test model, the control rods on the transmitter and receiver were motorised in such a way that the two transducers moved in opposite directions centred about the midplane of the model. While in motion at a separation rate of about 1 in/min, the transmitter was pulsed 20 times per second. The resulting waveforms were digitised and stored on magnetic tape. In subsequent processing they were displayed in the form shown in Fig. 6. In this figure the entire structure of the acoustic problem in borehole geometry can be seen. Along the vertical (time) scale each waveform is represented by a variation in grey tone related to the wave amplitude at each point in time. Three fundamental arrivals are clearly observable, the compressional, shear–pseudo-Rayleigh (labelled pseudo-Rayleigh in the figure) and the Stoneley. The group velocity of each signal can be determined from the figure as the reciprocal of the slope of the respective bands.

It is also possible to deduce some semi-quantitative information. The compressional and shear–pseudo-Rayleigh arrivals are clearly attenuated with propagation distance whereas the Stoneley arrival shows very little attenuation. These observations are consistent with analytical results.

The strong horizontal bands in Fig. 6 may lead to some confusion

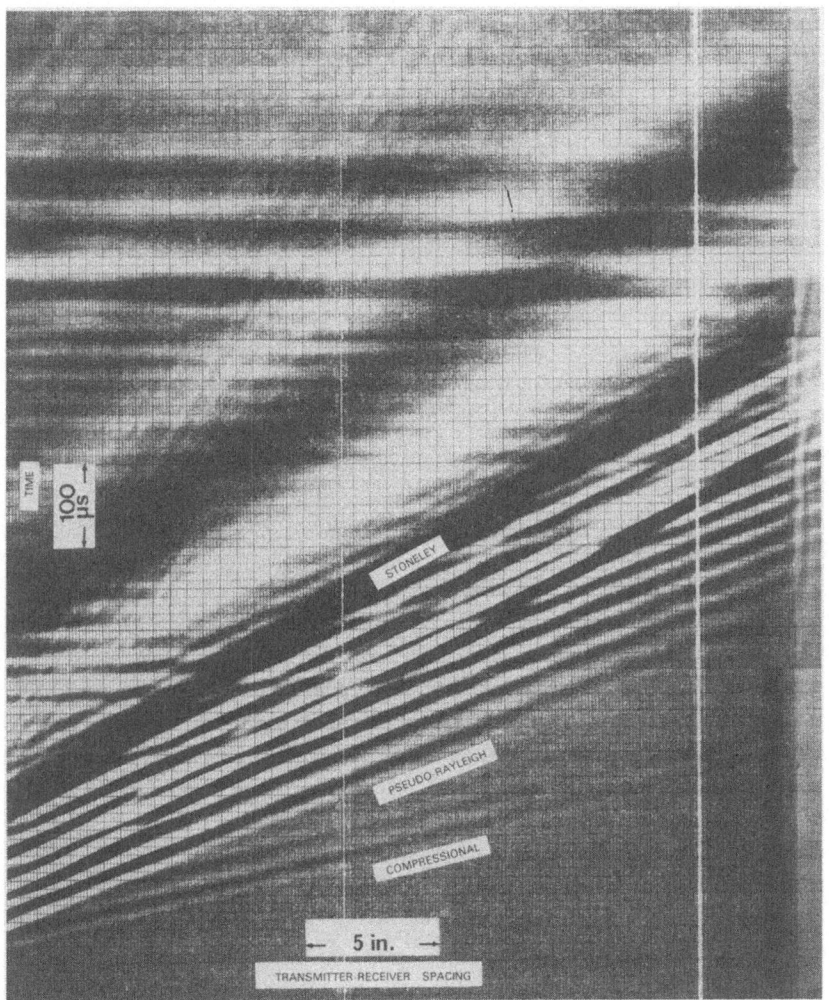

FIG. 6. Variable density display of model formation waveforms.

since they suggest the presence of a signal propagating at infinite speed! This is actually an artefact of the experimental set-up. All signals which propagate along paths 3 and 4 (Fig. 3) propagate over the same distance for any value of the spacing Z, and thus arrive at the receiver at the same time, when the transmitter and receiver move in opposite directions centred about the midplane. These signals therefore appear as horizontal bands in Fig. 6. The other sets of sloping bands can also be related to artefacts of the technique.

It is important to note that in all the experiments described in this section, the transmitter and receiver were very accurately centred in the borehole. In other experiments carried out with the test formation it was shown that severe reduction in signals occurs, particularly for the compressional, when either or both of the transducers are moved off centre. With the receiver one-quarter borehole radius off centre, for example, signal reduction was of the order of 6 dB. Similar results were reported by Roever et al. (1974).

Modelling techniques are also fruitful for studying additional aspects of the acoustic logging problem. In particular we would like to observe the response of the logging tool to a change in formation properties and to determine the influence of dip in the bedding plane as the tool passes from one formation to the next. As we shall see in the next section, an analytical description of the acoustic problem can be achieved in the idealised case where the formation is isotropic, elastic and homogeneous. When the beds dip with respect to the borehole axis (i.e. the bedding plane intersects the borehole axis at an angle between 0° and 90°), the acoustic logging problem is much more complex and has not been treated analytically.

A three-layer model test facility designed to investigate the effects of changes in formation properties and the orientation of bed boundaries on tool response is shown schematically in Figs. 7(a) and (b). Also shown are series of waveforms obtained in the neighbourhood of the bed boundaries.

The model incorporates a limestone bed sandwiched between upper and lower beds made of the same epoxy–sand composite used in the first model (Fig. 3). One of the bed boundaries is normal to the borehole axis (0° dip) and the other is at 45°. The model was 'logged' using a transmitter–receiver pair spaced at 11 in. Waveforms were recorded at 0·5 in intervals and displayed in a 'wiggle' trace format to facilitate observation of the compressional arrival and the early portion of the shear–pseudo-Rayleigh arrival. The transducers and instrumentation were identical to their counterparts for the model of Fig. 3.

In Fig. 7(a), the model was logged in the upward direction across the non-dipping boundary. Successive positions of the receiver are indicated alongside the traces and on the sketch of the model. When the transmitter is in its initial position in the limestone at x_0, the receiver is also in the limestone. When the transmitter is at x_1, the receiver is just at

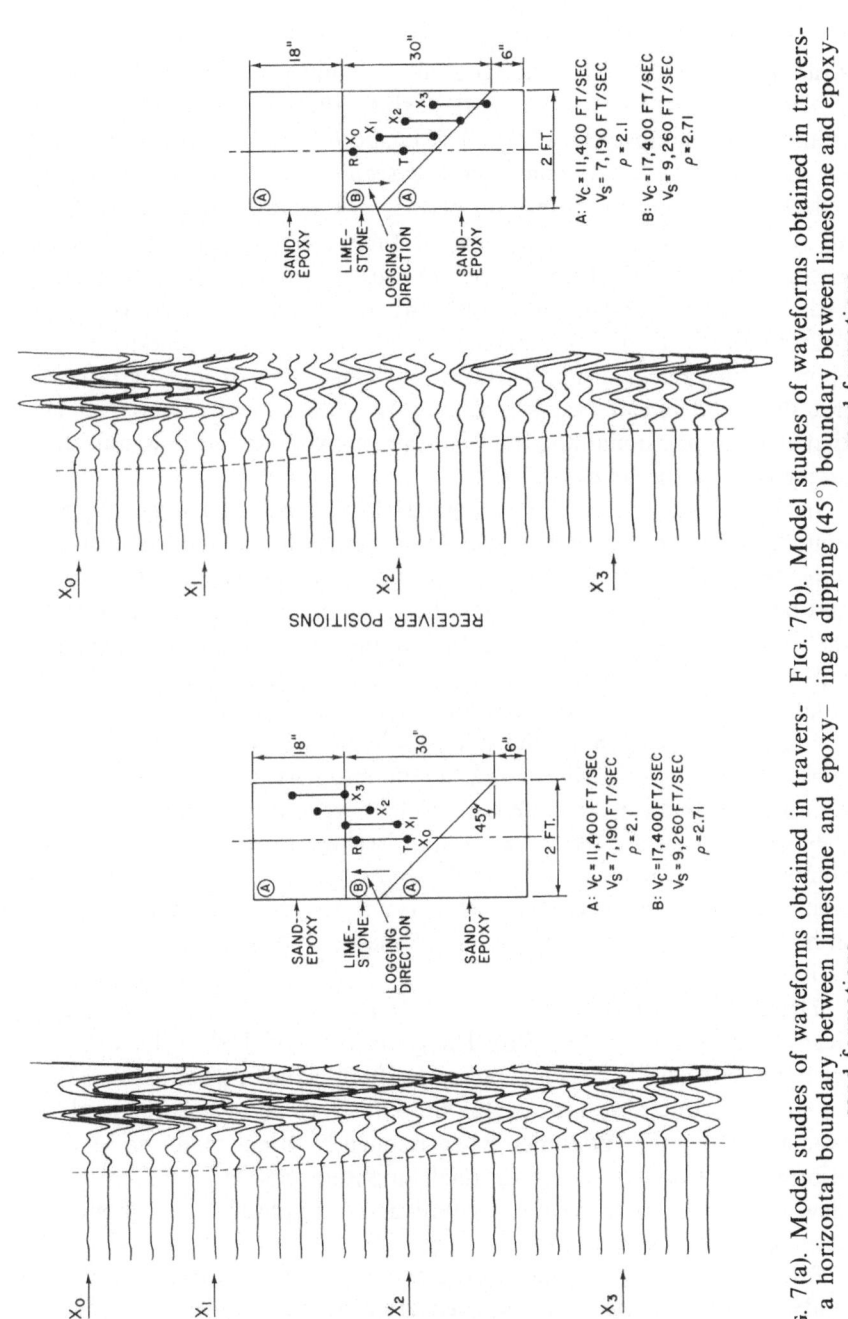

FIG. 7(a). Model studies of waveforms obtained in traversing a horizontal boundary between limestone and epoxy—sand formations.

FIG. 7(b). Model studies of waveforms obtained in traversing a dipping (45°) boundary between limestone and epoxy—sand formations.

the boundary, and in the successive positions the transducers evenly straddle the boundary (x_2) and are fully within the epoxy–sand bed (x_3).

The waveforms shown in Fig. 7(a) exhibit the expected characteristics. Initially, the compressional signal is unaffected by the epoxy–sand bed except for some interference on the second cycle due to boundary reflections. During the transition period, as the transmitter moves from x_1 to x_3, the compressional arrival time increases linearly. Beyond position x_3, the arrival time is the appropriate value for the epoxy–sand bed. Similarly a logging tool with two (or more) receivers would, of course, show a gradual change in t_c between the limestone and epoxy–sand values during the transition. Thus, acoustic logging tools do not sharply delineate bed boundaries. We also note for later reference that the compressional amplitude in the epoxy–sand bed is significantly larger than in the limestone bed.

In Fig. 7(b), the model is logged downward across the dipping boundary. In this case the successive positions of the receiver are indicated. As in Fig. 7(a), the waveforms shown in the ranges x_0 to x_1 and beyond x_3 are as expected for the respective homogeneous cases. In the transition region, however, the results are dramatically different. Substantial distortion of the compressional arrival is clearly evident. It would be difficult to extract an accurate value of t_c under such conditions. Distortion of the compressional arrival in Fig. 7(b) is a consequence of destructive phase interference. As signals from various azimuths around the circumference of the borehole arrive at the receiver, they will be shifted in time as a result of having propagated over different combinations of path lengths in the two formation materials. Somewhat similar effects occur when the receiver is not centred in the borehole even in homogeneous formations leading to sharply reduced signal amplitude as indicated earlier.

There is no effective means for dealing with the type of signal degradation shown in Fig. 7(b). The geometry modelled in this example is an extreme case in that it represents a high acoustic contrast and very high dip. Fortunately, this extreme situation does not occur frequently in practice. However, attempts have frequently been made to extract information from the compressional amplitude and its attenuation. In view of the foregoing results, compressional amplitude data should be used with considerable care.

Both the three-layer and homogeneous models are useful tools for studying other waveform details such as the Stoneley wave. The Stoneley arrival has attracted considerable attention in recent years because of its prominence in the waveform and its sensitivity to formation shear modulus (White, 1965; Ingram, 1978). There are three obvious features which characterise the Stoneley wave arrival in typical logging waveforms:

1. The signal is relatively low in frequency, typically 2–5 kHz when the transmitter centre frequency is 10–25 kHz.
2. Its apparent amplitude is generally significantly greater than the compressional arrival.
3. Its propagation speed is slightly less than the acoustic speed in the borehole fluid.

Stoneley waves are generalised Rayleigh waves which propagate along the interface between two media and decay exponentially away from the boundary on both sides. They characteristically propagate at a speed lower than the slowest body wave in the two media and are non-dispersive when the interface is planar. In the non-planar interface geometry of the borehole, Stoneley waves are dispersive, i.e. their phase and group velocities are functions of the ratio of wavelength to borehole diameter. As discussed in Section 4, this dispersion is very slight, ranging from about 75% of the borehole fluid speed at low frequencies in soft formations to a value approaching the fluid speed at high frequency. However, the kinematic field of the wave motion changes significantly with frequency.

At high frequencies (small wavelength to diameter ratio) the disturbance is confined to a narrow region near the borehole well. In this regime, the wave motion is analogous to the plane boundary case. At low frequencies, the pressure distribution is nearly uniform across the borehole.

In the low frequency limit, the phase velocity v_{st} is given by (e.g. White, 1965)

$$v_{st} = \frac{v}{(1 + B/\mu)^{1/2}} \tag{12}$$

where μ is the formation shear modulus and B and v are the bulk modulus and acoustic speed respectively of the borehole fluid. This low frequency wave is often termed the 'tube' wave and is the same mode commonly observed in 'water hammer' phenomena.

The low frequency regime for borehole Stoneley waves begins when the wavelength is about three times the borehole diameter or greater. At a typical propagation speed of 4500 ft/s, a 2 kHz Stoneley wave would be considered low frequency in an 8 in borehole. This corresponds to 8 kHz in the model test facilities.

In general, logging tools primarily detect the low frequency Stoneley wave because the Stoneley excitation function increases sharply as frequency decreases (Ingram, 1978) and the source function (transmitter output) spectrum is generally rich in low frequency components. Moreover, at low frequencies the pressure amplitude is nearly uniform across the borehole diameter. Thus, since the disturbance is confined to an increasingly narrow range near the borehole wall with increasing frequency, the pressure amplitude on the borehole axis is only significant at low frequencies. Stoneley waves have the further property that they do not attenuate with distance unless one or both of the two media are dissipative. The dispersion curves do not exhibit a low frequency cutoff.

A series of experiments were carried out in the homogeneous model (Fig. 3) to study Stoneley waves in borehole geometry. Some typical results are shown in Fig. 8. In Fig. 8(a), the full sonic waveform is shown for a spacing of $Z = 12$ in. An enlargement of the compressional arrival is shown in Fig. 8(b), where the first three peaks are designated E_1, E_2 and E_3 respectively. A bandpass filter (4–10 kHz) was used to extract the signal shown in Fig. 8(c). In further experiments this arrival was shown to have all the characteristics of the low frequency Stoneley wave, namely

1. The propagation speed was given by eqn (12)
2. There was minimal attenuation of the signal (less than 2 dB/ft).
3. The signal was unchanged as the receiver was moved across the borehole diameter.

The signal of Fig. 8(c) models the result which would be obtained in the frequency range of 1–2·5 kHz in an 8 in borehole. Its amplitude is about one-fifth of the peak amplitude of the full waveform, but about four times the peak compressional amplitude (E_3) shown in Fig. 8(b). These data confirm that the low frequency Stoneley wave is a significant component of the high amplitude portion of the acoustic waveform. Additional Stoneley wave energy outside of the 4–10 kHz band and some higher frequency components of the pseudo-Rayleigh wave also contribute to the high amplitude region of the waveform.

In Fig. 9, similarly obtained Stoneley arrivals are shown at spacings of 6 in in both the epoxy–sand and limestone beds of the three layer model formation. The two waveforms are evidently quite similar even though the shear modulus of the limestone is about double that of the epoxy–sand (for the epoxy–sand $\mu/B = 5\cdot8$, for the limestone $\mu/B = 11\cdot8$). As discussed in the next section the Stoneley amplitude is a

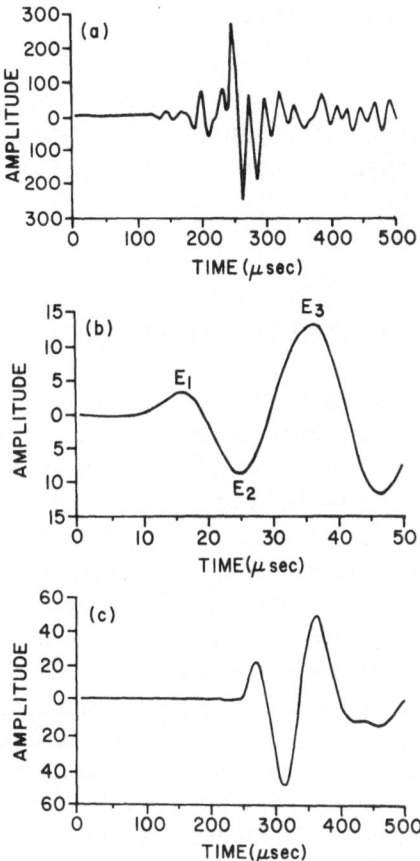

FIG. 8. (a) The full acoustic waveform at $Z = 12$ in in the model formation of Fig. 3. (b) Expanded view of the compressional arrival (time delay $= 106\,\mu$s). (c) Stoneley wave extracted by means of a 4–10 kHz bandpass filter.

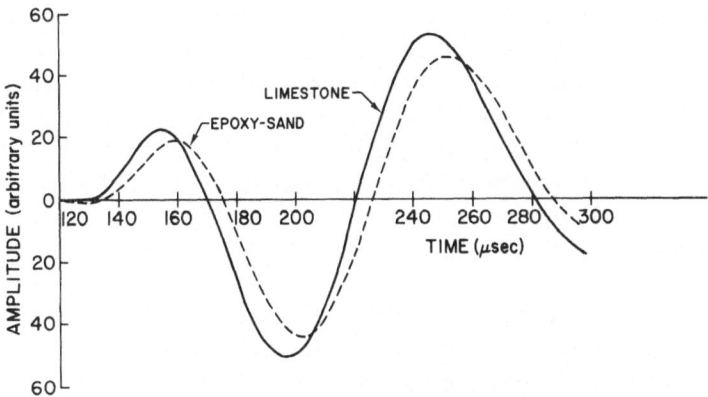

FIG. 9. Comparison of Stoneley waves in limestone ($\mu/B = 11\cdot8$) and epoxy–sand ($\mu/B = 5\cdot8$) at $Z = 6$ in.

strongly increasing function of μ/B for $\mu/B < 4$, but increases slowly above that value. Thus, the amplitude of the Stoneley arrival is not a good indicator of formation shear modulus for medium-hard to hard lithologies.

4. ANALYTICAL STUDIES

In the preceding sections we have described the borehole acoustic waveform in terms of two body waves, the critically refracted compressional and shear arrivals, and two guided interface modes, the pseudo-Rayleigh and Stoneley arrivals. Most of the published analytical work to date approaches the mathematical problem either in terms of a numerical evaluation of the full waveform or provides a description of the individual waveform components through closed form or numerical techniques. A brief summary of the key analytical studies will be useful.

Biot (1952) obtained dispersion curves for the two guided waves. White and Zechman (1968) used numerical techniques to compute the full acoustic waveform for a case in which the formation exhibited dissipation. The introduction of dissipation eliminates the need to deal with mathematical singularities in the numerical integration. Roever *et al.* (1974) and Peterson (1974) studied the normal mode solutions by residue analysis of poles in the complex wavenumber plane and computed the body wave arrivals by means of branch cut integration employing ordinary asymptotic methods.

Tsang (1978) and Tsang and Kong (1979) observed that certain poles, previously regarded as 'extraneous' but which actually influence the branch cut evaluations of lateral waves, were not taken into account in earlier work. Tsang and Rader (1979) used an improved numerical technique, 'real axis integration', to compute the full acoustic waveform and a 'modified' asymptotic technique which accounts for the 'extraneous' poles to obtain the shear and compressional head wave arrivals through branch cut integration. With this method it was possible to isolate and study the impulse response of the lateral waves for the first time.

The interested reader is referred to the articles by Tsang (1978) and Tsang and Kong (1979) for a detailed description of the modified asymptotic technique required for carrying out the branch cut integrations which generate the lateral wave arrivals. Briefly, however, the borehole problem has an 'extraneous' pole associated with the Rayleigh wave (more properly the pseudo-Rayleigh wave when the borehole is fluid-filled). This is analogous to the extraneous pole which exists in the classical Lamb problem as described by Tsang (1978). The position of this extraneous pole is strongly dependent on Poisson's ratio at a given frequency. For large values of v the pole lies on the upper Riemann sheet in the complex wavenumber plane, as shown in Tsang and Rader (1979). For small values of v the pole lies on the lower Riemann sheet. The pole's proximity to the compressional branch point is such that its influence cannot be neglected even on the lower Riemann sheet. As shown in the foregoing references, the influence of the extraneous poles and the modified asymptotic technique can be quite large, not merely a perturbation.

In recent work, Cheng and Toksoz (1981) have carried out more exhaustive studies of the guided wave dispersion curves. They have also generated complete acoustic waveforms starting with the formulation of Tsang and Rader (1979).

In reviewing recent analytical results, it is convenient to first consider Stoneley and pseudo-Rayleigh waves followed by a description of full waveform numerical calculations. Finally, a description of the branch cut integration results for lateral waves will be given.

4.1. Stoneley and Pseudo-Rayleigh Waves

Typical group and phase velocity dispersion curves for Stoneley and pseudo-Rayleigh waves (fundamental mode), computed by Cheng and Toksoz (1981) are shown in Fig. 10. The dashed curves are for a 5·28 in diameter borehole and the solid curves are for an 8 in diameter borehole containing a 3·6 in diameter logging tool. The Stoneley wave exhibits very little dispersion with both phase and group velocities remaining in a narrow band just below the borehole fluid acoustic speed. Pseudo-Rayleigh waves, however, are strongly dispersive beginning with the formation shear velocity at the low frequency cutoff and asymptotically approaching the fluid acoustic speed at high frequency. The group velocity curve exhibits an Airy phase at an intermediate frequency. As noted by Cheng and Toksoz (1981), the effect of a 'hard' tool is to reduce the apparent hole diameter. In other words, referring to Fig. 10, the dispersion curves 'with tool' in a large diameter hole are similar to those 'without tool' in a smaller hole in the same frequency range.

We note that the low frequency cutoff occurs in the frequency range commonly used for acoustic logging. Thus, energy which arrives at the

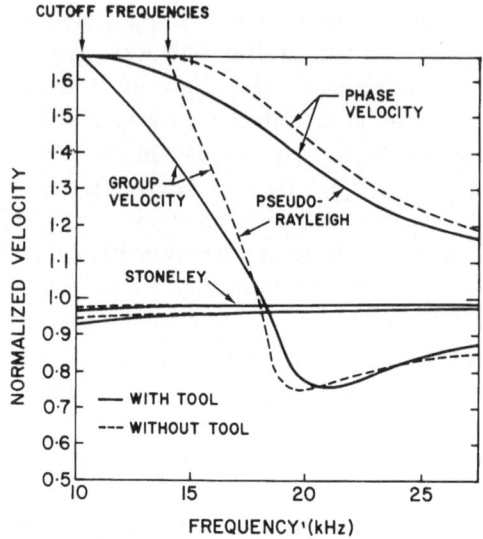

FIG. 10. Pseudo-Rayleigh and Stoneley wave dispersion curves for $v_c = 5.94 \times 10^5$ cm/s and $v_s = 3.05 \times 10^5$ cm/s. Dashed curves: 5.28 in diameter borehole with no tool. Solid curves: 8 in diameter borehole with a 3.6 in diameter tool. Velocities on the ordinate are normalised to the borehole fluid acoustic speed, $v = 1.83 \times 10^5$ cm/s. (After Cheng and Toksoz, 1981. Reprinted with permission of the Society of Exploration Geophysicists.)

shear speed in the acoustic waveform contains contributions from both the shear and pseudo-Rayleigh waves. A mathematical description of this phenomenon develops naturally in performing the shear branch cut integration as shown by Tsang and Rader (1979).

It is also clear from Fig. 10 that the Stoneley and pseudo-Rayleigh waves will overlap (at around 18 kHz in this case). This is consistent with our previous discussion in connection with Fig. 8.

Below the cutoff frequency, the pseudo-Rayleigh curve has a branch in the complex wavenumber plane. This branch is a decaying surface wave with energy radiating away from the surface. Mathematically this corresponds to one of the extraneous poles of the pseudo-Rayleigh wave mentioned earlier.

Physically, the pseudo-Rayleigh wave is analogous to the ordinary Rayleigh wave at an interface bounded by a vacuum. The presence of a fluid on the boundary leads to some decay due to radiation into the fluid medium.

Figure 11 illustrates the relationships between formation stiffness, i.e.

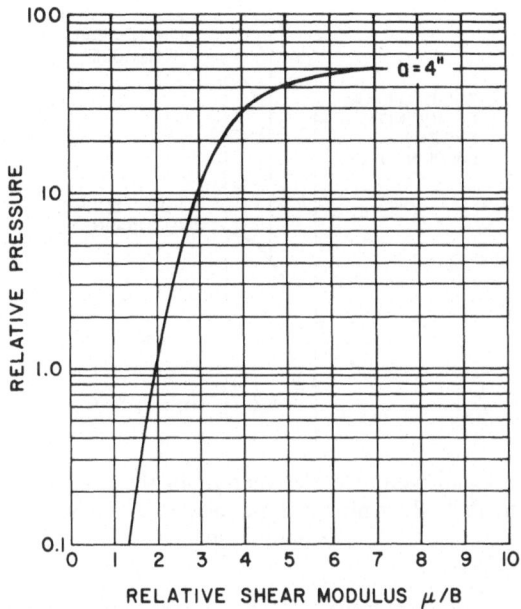

FIG. 11. Representative Stoneley wave amplitude on the axis of an 8 in diameter borehole as a function of relative formation shear modulus. (After Ingram, 1978.)

relative shear modulus, and induced Stoneley wave amplitude in a fluid-filled 8 in borehole (Ingram, 1978). The first portion of the curve, up to $\mu/B \simeq 4$, shows a very strong dependence of amplitude on formation shear modulus. For $\mu/B > 4$, the curve is relatively insensitive to the shear modulus. Thus, the Stoneley amplitude offers a means, in principle, for deducing the shear modulus in 'soft' formations where a shear lateral wave is small or does not exist. The relative amplitudes of the Stoneley waves shown in Fig. 9 are consistent with the curve of Fig. 11. From eqn (12) it is also clear that the Stoneley velocity offers a means in principle for obtaining formation shear modulus.

4.2. Full Waveform Computations

An integral solution for the pressure at (r, z, t), P_r, due to a point source at the origin in a borehole of radius a (see Fig. 12) is given by Tsang and Rader (1979). Using a Laplace Contour L, which is above and parallel to the real ω axis in the complex frequency plane, P_r is given by

$$P_r = \frac{iP_0 R_0}{4\pi} \int_L \mathrm{d}\omega X(\omega)\exp(-i\omega t) \int_{-\infty}^{\infty} \mathrm{d}k_z A J_0(k_r r) \exp(ik_z z) \quad (13)$$

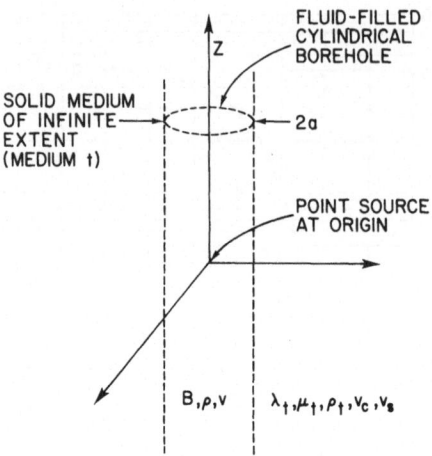

FIG. 12. Geometry and parameter designation for the analytical problem. (After Tsang and Rader, 1979. Reprinted with permission of the Society of Exploration Geophysicists.)

where P_0 is the peak pressure of the source waveform observed at a distance R_0 in an infinite fluid medium, J_0 is the zero-order Bessel function, k_z is the z direction wavenumber, k_r is the radial component of the fluid wavenumber, and $X(\omega)$ is the frequency spectrum of the source function. When the source is the decaying sinusoid illustrated in Fig. 13,

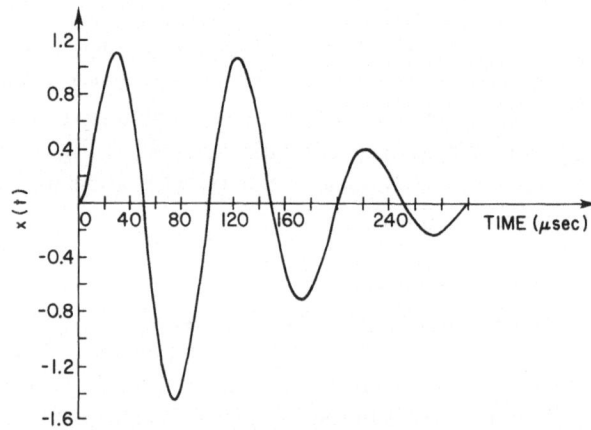

FIG. 13. Source function $x(t)$ for $\omega_0 = 2\pi \times 10^4$ rad/s and $\alpha = 0.79\,\omega_0/\pi$. (After Tsang and Rader, 1979. Reprinted with permission of the Society of Exploration Geophysicists.)

i.e.

$$x(t) = 4\alpha t \, \exp(-\alpha t) \, \sin \omega_0 t \, u(t) \qquad (14)$$

where $u(t)$ is the unit step function, we have

$$X(\omega) = \frac{8 \alpha \omega_0 (\alpha - i\omega)}{\{(\alpha - i\omega)^2 + \omega_0^2\}^2} \qquad (15)$$

which is illustrated in Fig. 14. Tsang and Rader chose $\omega_0 = 2\pi \times 10^4$ rad/s and $\alpha = 0.79 \, \omega_0/\pi$.

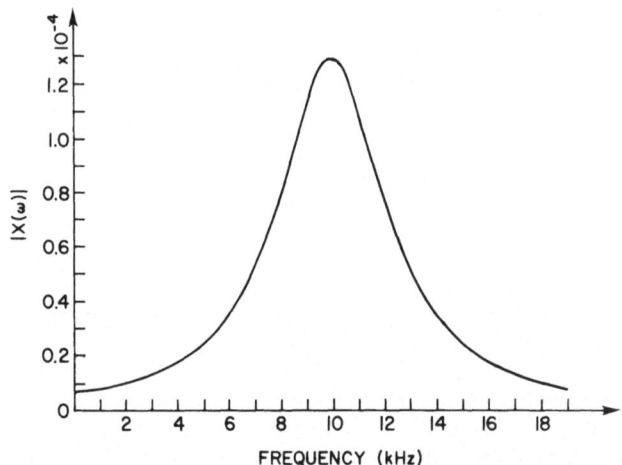

FIG. 14. Frequency spectrum $X(\omega)$ of the source function. (After Tsang and Rader, 1979. Reprinted with permission of the Society of Exploration Geophysicists.)

We have the further definitions:

$$k_r^2 = k^2 - k_z^2 \quad (k = \omega/v) \qquad (16)$$

and

$$A = -\frac{f_1 k_r a H_1^{(1)}(k_r a) + f_2 H_0^{(1)}(k_r a)}{f_1 k_r a J_1(k_r a) + f_2 J_0(k_r a)} \qquad (17)$$

where

$$f_1 = \mu_t \left[\frac{2k_s^2}{a^2} - (k_s^2 - 2k_z^2)^2 \, \frac{H_0^{(1)}(k_r^{(c)} a)}{k_r^{(c)} a H_1^{(1)}(k_r^{(c)} a)} \right.$$

$$\left. - 4k_z^2 k_r^{(s)} \, \frac{H_0^{(1)}(k_r^{(s)} a)}{a H_1^{(1)}(k_r^{(s)} a)} \right] \qquad (18)$$

$f_2 = Bk^2 k_s^2$, $H_1^{(1)}$ $(k_r a)$ and $J_1 (k_r a)$ are the first-order Hankel and Bessel functions, $k_r^{(c)} = (k_c^2 - k_z^2)^{\frac{1}{2}}$, $k_r^{(s)} = (k_s^2 - k_z^2)^{\frac{1}{2}}$, $k_c = \omega/v_c$, and $k_s = \omega/v_s$.

With this integral formulation, and the real axis numerical integration technique described in Tsang and Rader (1979), any desired time duration of the received waveform can be obtained. Figure 15 shows the

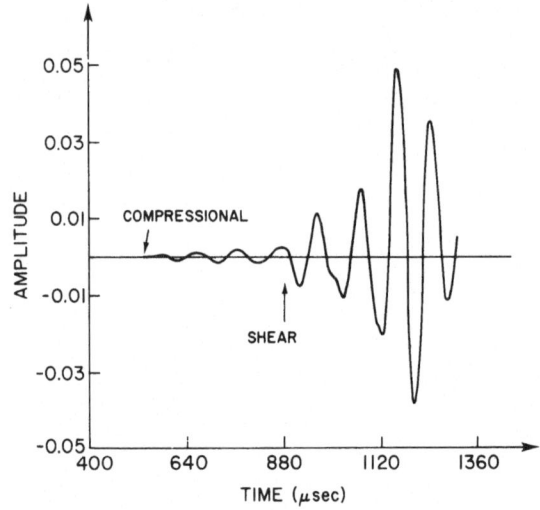

FIG. 15. Computed waveform for $v = 1\cdot44 \times 10^5$ cm/s, $v_c = 3\cdot5 \times 10^5$ cm/s, $v_s = 1\cdot95 \times 10^5$ cm/s, $\rho = 0\cdot88$, $\rho_t = 2\cdot1$, $a = 10$ cm, $r = 0$ and $z = 150$ cm. (After Tsang and Rader, 1979. Reprinted with permission of the Society of Exploration Geophysicists.)

results of such a computation for a 10 cm radius borehole penetrating a formation having properties similar to the epoxy–sand model discussed earlier. The compressional and shear arrival times are in agreement with values computed from eqns (10a) and (10b). In Fig. 15 and subsequent figures, eqn (13) is normalised by setting $P_0 R_0 = 1/\pi$.

In Fig. 16, results are shown for a formation which is softer (v_c and v_s are lower than in Fig. 15) but with higher fluid and formation densities. The shear velocity in this case is only 6% higher than the borehole fluid speed. This waveform may be compared with that shown in Fig. 17 for which only the formation shear speed has been changed to a value 6% *below* the borehole fluid speed. The first portion of the waveforms in Fig. 16 and 17 are similar, but the shear and pseudo-Rayleigh arrivals are absent in the latter as expected. The Stoneley arrival is evident, however, in Fig. 17.

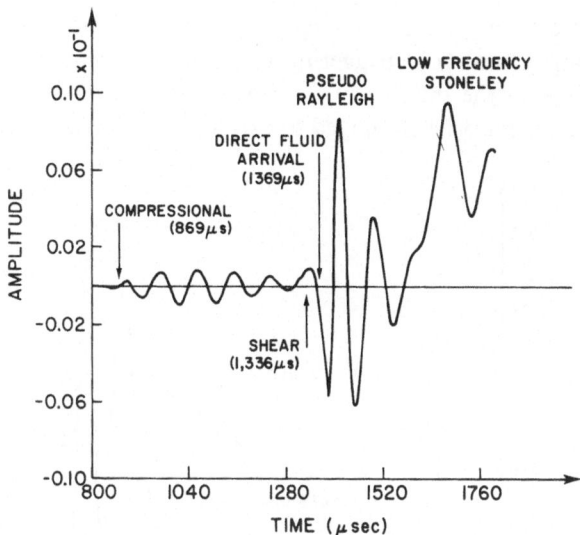

FIG. 16. Computed waveform for $v = 1 \cdot 54 \times 10^5$ cm/s, $v_c = 2 \cdot 77 \times 10^5$ cm/s, $v_s = 1 \cdot 63 \times 10^5$ cm/s, $\rho = 1 \cdot 2$, $\rho_t = 2 \cdot 4$, $a = 10$ cm, $r = 0$ and $z = 210 \cdot 8$ cm. (After Tsang and Rader, 1979. Reprinted with permission of the Society of Exploration Geophysicists.)

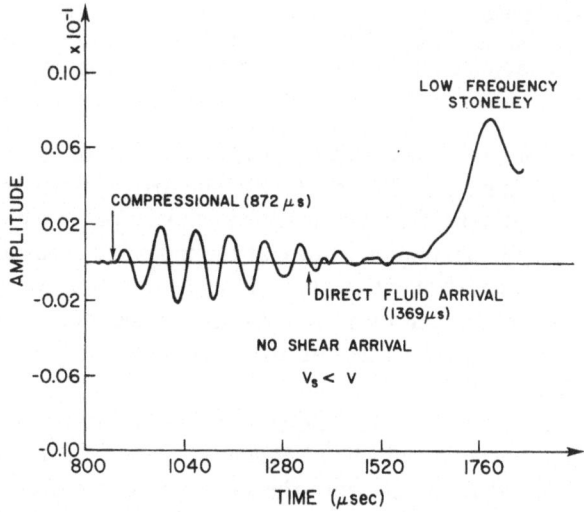

FIG. 17. Computed waveform for $v = 1 \cdot 54 \times 10^5$ cm/s, $v_c = 2 \cdot 77 \times 10^5$ cm/s, $v_s = 1 \cdot 45 \times 10^5$ cm/s, $\rho = 1 \cdot 2$, $\rho_t = 2 \cdot 4$, $a = 10$ cm, $r = 0$ and $z = 210 \cdot 8$ cm. (After Tsang and Rader, 1979. Reprinted with permission of the Society of Exploration Geophysicists.)

When noise levels are low enough, the compressional arrival can be detected using a threshold triggering technique, for example. Since the shear arrival is in the interior of the waveform it is obvious that some additional 'logic' must be employed to extract t_s. Figures 18–21 exhibit a

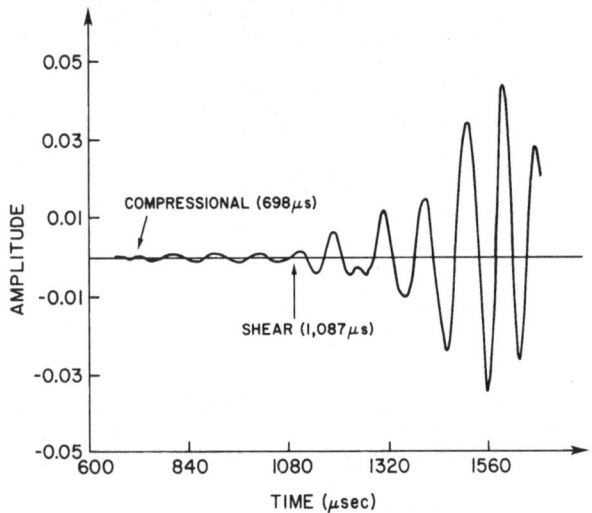

FIG. 18. Computed waveform for $z = 200\,\text{cm}$, $v = 1.44 \times 10^5\,\text{cm/s}$, $v_c = 3.5 \times 10^5\,\text{cm/s}$, $v_s = 2.02 \times 10^5\,\text{cm/s}$, $\rho = 0.88$, $\rho_t = 2.1$, $a = 10\,\text{cm}$ and $r = 0$. (After Tsang and Rader, 1979. Reprinted with permission of the Society of Exploration Geophysicists.)

series of four computed waveforms spaced at intervals of 30·4 cm (approximately 1 ft) beginning at 200 cm (6·6 ft). The arrival times indicated on the figures are obtained from the specified acoustic parameters. Although there is no clearly identifiable feature at the indicated shear arrival time, there is a waveform segment of about $1\frac{1}{2}$ cycle duration which is clearly similar in shape for all form waveforms immediately following the indicated 'onset' time. This feature overlaps the trailing portion of the compressional.

In the article by Aron *et al.* (1978) a 'four-fold correlation' scheme is described by means of which a sequence of four waveforms can be digitally processed to extract both the shear and compressional arrivals. That technique was applied to the waveforms of Fig. 18–21 and produced interval transit times corresponding to the specified velocities ($v_c = 3.5 \times 10^5\,\text{cm/s}$ and $v_s = 2.02 \times 10^5\,\text{cm/s}$) to three significant figures. Thus, at least one automatic procedure has been successfully applied to

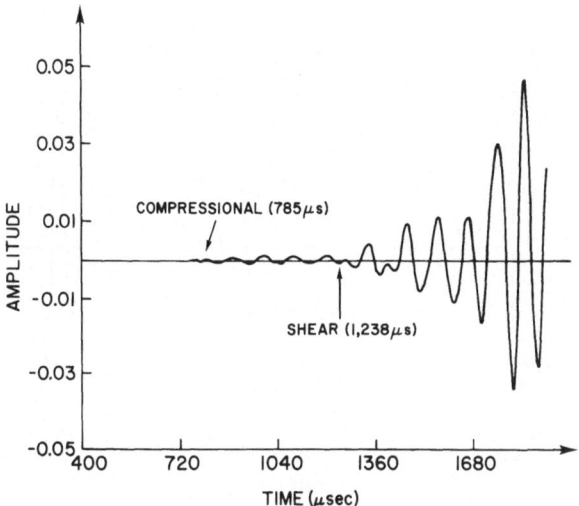

FIG. 19. Computed waveform for $z = 230 \cdot 4$ cm; other parameters the same as in Fig. 18. (After Tsang and Rader, 1979. Reprinted with permission of the Society of Exploration Geophysicists.)

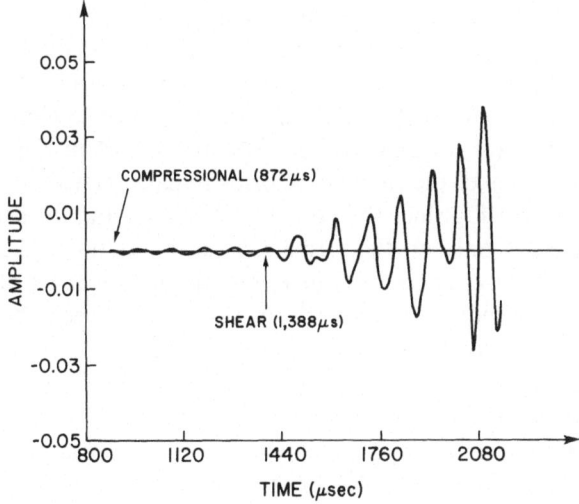

FIG. 20. Computed waveform for $z = 260 \cdot 8$ cm; other parameters the same as in Fig. 18. (After Tsang and Rader, 1979. Reprinted with permission of the Society of Exploration Geophysicists.)

182   DENNIS RADER

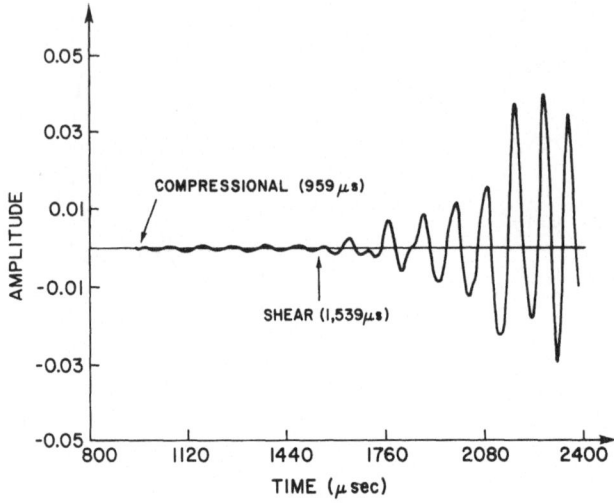

FIG. 21. Computed waveform for $z = 291 \cdot 2$ cm; other parameters the same as in Fig. 18. (After Tsang and Rader, 1979. Reprinted with permission of the Society of Exploration Geophysicists.)

the multiple arrival time detection problem. Although the conditions in this example are ideal (the waveforms are noise-free), good results were demonstrated by Aron *et al.* (1978) on actual field data.

Figure 22 is an example of a set of full waveforms calculated by Cheng and Toksoz (1981) to observe the influences of Poisson's ratio. In each of Fig. 22(a), (b) and (c) only the compressional speed is changed giving $v = 0 \cdot 32$ in Fig. 22(a), $v = 0 \cdot 26$ in Fig. 22(b) and $v = 0 \cdot 1$ in Fig. 22(c). The prominent portions of the three waveforms remain relatively unchanged, which is not surprising since the Stoneley amplitude and the shear arrival time both depend only on the formation shear modulus. The compressional arrival time changes of course, but the interesting feature is the decrease in amplitude of the waveform segment between the compressional and shear arrivals as Poisson's ratio decreases.

The dependence of compressional amplitude on Poisson's ratio is discussed below in connection with branch cut integration of the first compressional lateral wave. Mathematically, the effect is related in part to the position of the previously discussed 'extraneous' pole which moves from the lower to the upper Riemann sheet as v increases and has a more pronounced effect on the compressional branch cut integration. As suggested by Tsang and Rader (1979) and Cheng and Toksoz (1981), the

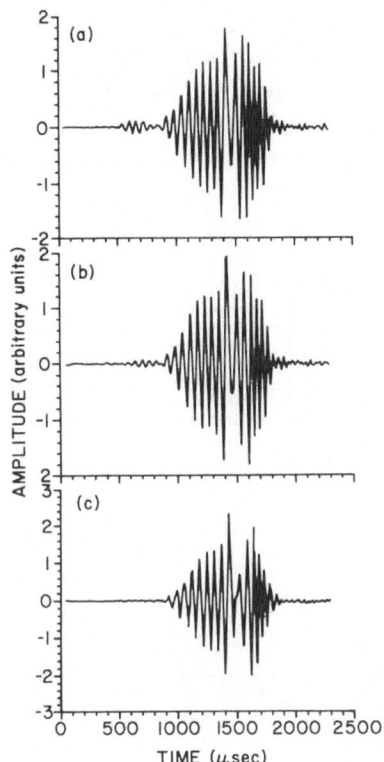

FIG. 22. Computed waveforms calculated for three different values of Poisson's ratio: (a) $v=0.32$; (b) $v=0.26$; (c) $v=0.1$. Compressional velocities: (a) $v_c=5.94 \times 10^5$ cm/s, (b) $v_c=5.33 \times 10^5$ cm/s, (c) $v_c=4.57 \times 10^5$ cm/s. Other parameters: $v_s=3.05 \times 10^5$ cm/s, $\rho=1.2$, $\rho_t=2.3$, $a=6.7$ cm, $r=0$ and $z=244$ cm. (After Cheng and Toksoz, 1981. Reprinted with permission of the Society of Exploration Geophysicists.)

compressional amplitude could therefore provide a means for determining v or v_c/v_s in principle.

4.3. Branch Cut Integration

The foregoing method of numerical integration is a useful tool for studying the full waveform under any prescribed set of conditions. Branch cut integration, however, affords an opportunity to isolate and study the compressional and shear lateral waves in detail. Tsang and Rader (1979) employed a ray expansion of the parameter A (eqn 17), the first term of which characterises the first radial ray as depicted in the inset of Fig. 23. Associated with the integration are three branch points in the complex Z-wavenumber plane. The compressional arrival is obtained by a branch cut integration associated with the compressional branch point in which the effect of the 'extraneous' Rayleigh pole is properly included using the modified asymptotic method described by Tsang (1978).

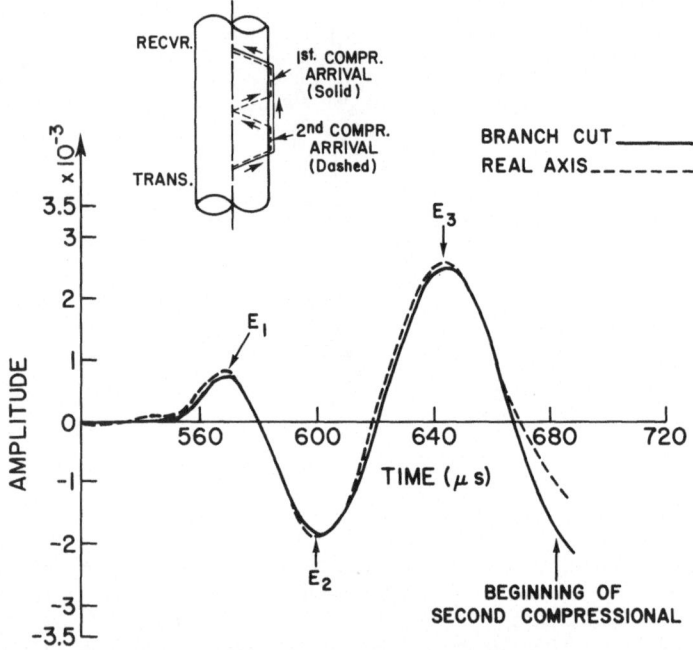

FIG. 23. Comparison of compressional arrival computed with real axis and branch cut integrations for $v = 1.44 \times 10^5$ cm/s, $v_c = 3.5 \times 10^5$ cm/s, $v_s = 2.2 \times 10^5$ cm/s, $\rho = 0.88$, $\rho_t = 2.1$, $a = 10$ cm, $r = 0$ and $z = 150$ cm. (After Tsang and Rader, 1979. Reprinted with permission of the Society of Exploration Geophysicists.)

The first three peaks of the compressional arrival (E_1, E_2 and E_3) are included in the waveform computed by this technique as illustrated in Fig. 23. (The source function is that of Fig. 13.) The method cannot be carried further in time without including additional terms in the ray expansion since the 'second' (and subsequent) compressional rays arrive at later times. However, the first $1\frac{1}{2}$ cycles of the waveform is in excellent agreement with the corresponding result from the real axis integration method described in Section 4.2. Tsang (1978) has shown that there are serious discrepancies between the two waveform calculations when ordinary asymptotic techniques are employed in the branch cut integration.

This branch cut integration method is a powerful tool for studying the amplitude characteristics as well as the impulse response function of the lateral wave. Figures 24–26 illustrate the amplitude dependence (of E_2)

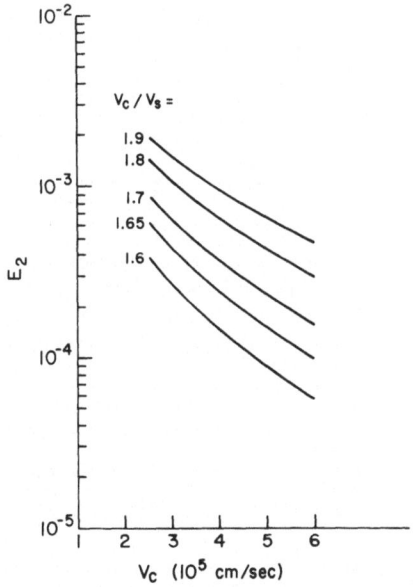

FIG. 24. Compressional amplitude E_2 as a function of compressional velocity v_c for various v_c/v_s, with $v = 1·44 \times 10^5$ cm/s, $\rho = 0·88$, $\rho_t = 2·5$, $a = 10$ cm, $r = 0$ and $z = 150$ cm. (After Tsang and Rader, 1979. Reprinted with permission of the Society of Exploration Geophysicists.)

on the compressional velocity, v_c/v_s ratio and formation density, respectively, for the source waveform of Fig. 13. Figure 24 exhibits a clear declining trend in amplitude response as a function of v_c for any given v_c/v_s (or equivalently, any v). The opposite trend obtains as a function of v_c/v_s for given v_c as exhibited in Fig. 25. This is consistent with the results shown in Fig. 22 discussed previously. Compressional amplitude is a comparatively weak function of formation density (Fig. 26). Referring back to Fig. 7(a) and (b) we note that the compressional amplitude is significantly higher in the epoxy–sand formation than in the limestone formation. This is consistent with the results shown in Fig. 24–26.

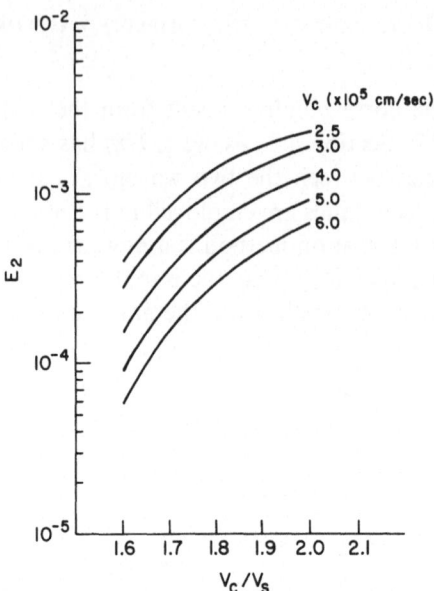

FIG. 25. Compressional amplitude E_2 as a function of v_c/v_s for $v_c = 2\cdot5$, $3\cdot0$, $4\cdot0$, $5\cdot0$ and $6\cdot0$ ($\times 10^5$ cm/s), with $v = 1\cdot44 \times 10^5$ cm/s, $\rho = 0\cdot88$, $\rho_t = 2\cdot5$, $a = 10$ cm, $r = 0$ and $z = 150$ cm. (After Tsang and Rader, 1979. Reprinted with permission of the Society of Exploration Geophysicists.)

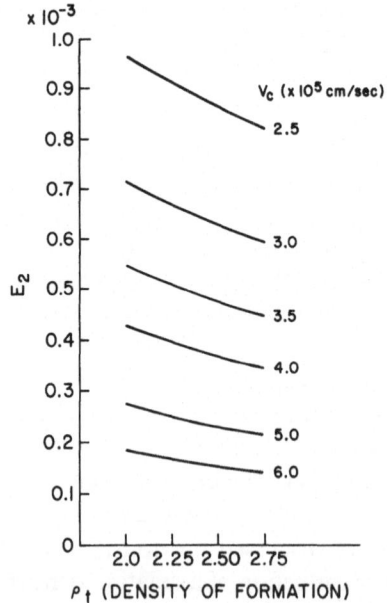

FIG. 26. Compressional amplitude E_2 as a function of the density of the formation ρ_t for $v_c = 2\cdot5$, $3\cdot0$, $3\cdot5$, $4\cdot0$, $5\cdot0$ and $6\cdot0$ ($\times 10^5$ cm/s), with $v = 1\cdot44 \times 10^5$ cm/s, $v_s = v_c/1\cdot7$, $\rho = 0\cdot88$, $a = 10$ cm, $r = 0$ and $z = 150$ cm. (After Tsang and Rader, 1979. Reprinted with permission of the Society of Exploration Geophysicists.)

Two impulse response functions are shown in Fig. 27. For the typical parameters specified, the response is monotonically increasing to frequencies well beyond the bandwidth of transmitters used in conventional acoustic logging. When the source spectrum is of the general shape shown in Fig. 14, and the impulse response is monotonically increasing, the compressional arrival will be richer in high frequency components than the source function. A quick estimate based on zero-crossing intervals in Fig. 23, for example, illustrates this point.

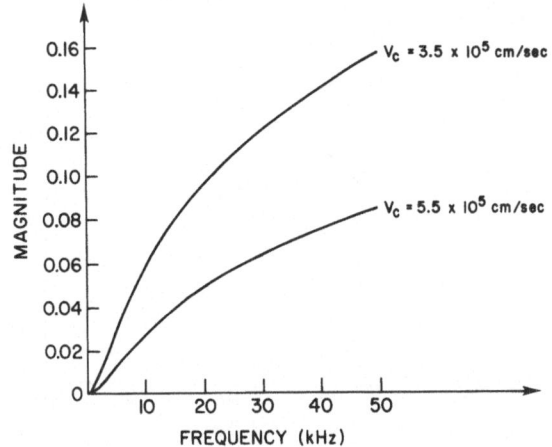

FIG. 27. Frequency spectrum of the first compressional arrival for $v_c/v_s = 1.6$, $v_c = 3.5$ and 5.5 ($\times 10^5$ cm/s), with $v = 1.44 \times 10^5$ cm/s, $\rho = 0.88$, $\rho_t = 2.25$, $a = 10$ cm, $r = 0$ and $z = 150$ cm. (After Tsang and Rader, 1979. Reprinted with permission of the Society of Exploration Geophysicists.)

In evaluating the shear branch cut integration, the pseudo-Rayleigh pole, which is in the neighbourhood of the shear branch point, must be taken into account. The wave arrival has therefore been termed 'composite shear–pseudo-Rayleigh wave' by Tsang and Rader (1979). The location of the pseudo-Rayleigh pole is a function of frequency and its coordinates in general include both real and imaginary components in the complex Z–wavenumber plane (Tsang and Rader, 1979).

The position of the pseudo-Rayleigh pole off the real axis corresponds physically to attenuation via radiation into the fluid medium. Its movement with frequency reflects the dispersive character of the wave which, as we have seen, is significant in borehole geometry.

Figure 28 provides further insight into the behaviour of the pseudo-Rayleigh wave in relation to the shear body wave. The loss tangent† and ratio of shear to pseudo-Rayleigh phase velocity, v_s/v_R, are plotted versus frequency for a typical set of borehole conditions. In the 10–20 kHz range the phase velocity of the pseudo-Rayleigh wave is within 2% of the shear velocity and its loss tangent is no more than 0·02. Thus, the pseudo-Rayleigh wave cannot be distinguished from the shear in practical logging systems. Moreover, since the shear amplitude drops off rapidly, approximately at the rate $1/Z^2$ (Tsang and Rader, 1979), the pseudo-Rayleigh will generally be the dominant component of the composite arrival.

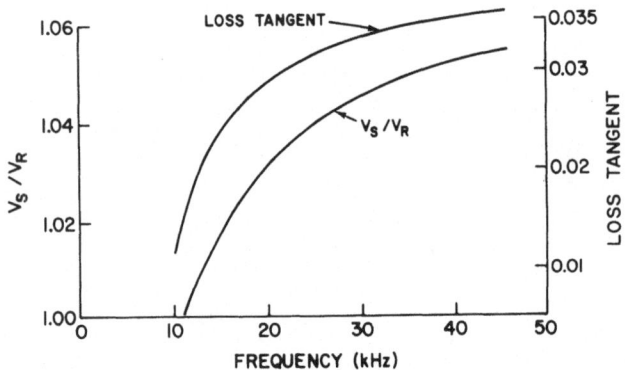

FIG. 28. The ratio v_s/v_R and loss tangent of the pseudo-Rayleigh wave as a function of frequency. Parameters: $v = 1·44 \times 10^5$ cm/s, $v_c = 5·0 \times 10^5$ cm/s, $v_s = v_c/1·9$, $\rho = 0·88$, $\rho_t = 2·25$ and $a = 10$ cm. The vertical scale on the left applies to v_s/v_R while the vertical scale on the right applies to the loss tangent. (After Tsang and Rader, 1979. Reprinted with permission of the Society of Exploration Geophysicists.)

Figures 29 and 30 illustrate the composite arrival impulse response function for various values of v_c/v_s and various values of v_c respectively. The curves in both figures exhibit peaks roughly in the 10–20 kHz range, suggesting that a source function covering this bandwidth is appropriate for exciting strong shear–pseudo-Rayleigh arrivals, at least for the parameter ranges of Figs. 24 and 25.

Using notation analogous to the compressional arrival for identifi-

---

† The loss tangent is the ratio of the imaginary to real part of the pseudo-Rayleigh wavenumber. Energy dissipation increases with the imaginary part of the wavenumber.

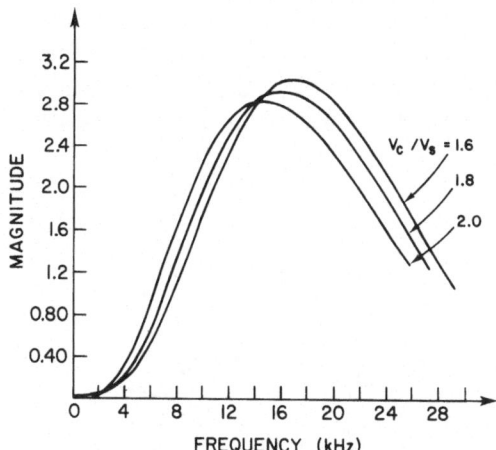

FIG. 29. Frequency spectrum of first shear arrival for three different v_c/v_s ratios (1·6, 1·8 and 2·0), with $v = 1·44 \times 10^5$ cm/s, $v_c = 4·5 \times 10^5$ cm/s, $\rho = 0·88$, $\rho_t = 2·5$, $a = 10$ cm, $r = 0$ and $z = 150$ cm. (After Tsang and Rader, 1979. Reprinted with permission of the Society of Exploration Geophysicists.)

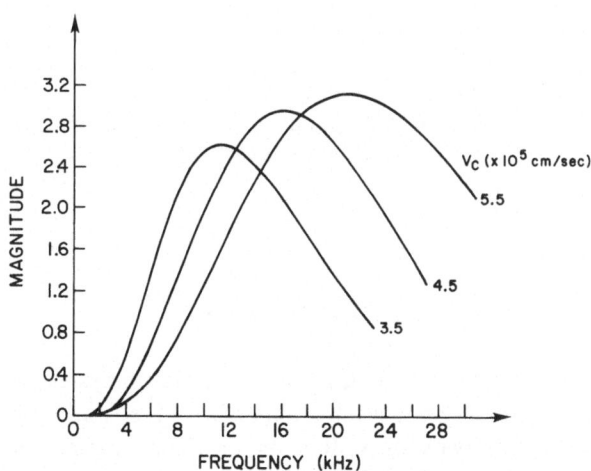

FIG. 30. Frequency spectrum of first shear arrival for $v_c = 3·5$, 4·5 and 5·5 ($\times 10^5$ cm/s), with $v_c/v_s = 1·8$, $v = 1·44 \times 10^5$ cm/s, $\rho = 0·88$, $\rho_t = 2·5$, $a = 10$ cm, $z = 150$ cm and $r = 0$. (After Tsang and Rader, 1979. Reprinted with permission of the Society of Exploration Geophysicists.)

cation, the E_2 shear–pseudo-Rayleigh peak amplitude is plotted in Fig. 31 for various values of v_c versus v_c/v_s (the source function is again that of Fig. 13). As in the case of the compressional (Fig. 25), the amplitude increases as v_c increases for fixed v_c/v_s and increases with v_c/v_s for fixed v_c.

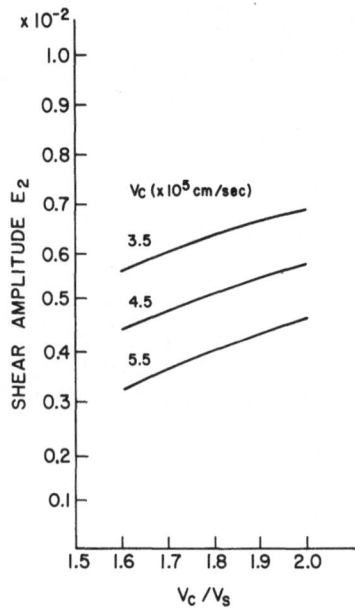

FIG. 31. Shear amplitude E_2 as a function of v_c/v_s for $v_c = 3{\cdot}5$, $4{\cdot}5$ and $5{\cdot}5$ ($\times 10^5$ cm/s), $v = 1{\cdot}44 \times 10^5$ cm/s, $\rho = 0{\cdot}88$, $\rho_t = 2{\cdot}5$, $a = 10$ cm, $r = 0$ and $z = 150$ cm. (After Tsang and Rader, 1979. Reprinted with permission of the Society of Exploration Geophysicists.)

5. CONCLUSIONS

We have shown that scale model experimental studies of the acoustic logging problem can be fruitfully applied to the study of the main features of the full waveform. Results of these studies are in good agreement with the idealised analytical results for homogeneous, isotropic and elastic formations. Model studies were also used to elucidate the behaviour of the compressional arrival in the presence of dip and high contrast beds.

The results of analytical studies have been summarised for the four major components of the acoustic waveform, i.e. compressional, shear, pseudo-Rayleigh and Stoneley waves. Stoneley waves are dominant arrivals propagating just below the borehole fluid velocity and virtually without dispersion. In non-dissipative media they also propagate without attenuation. Pseudo-Rayleigh waves are highly dispersive in borehole geometry, overlapping both the shear and Stoneley waves. They attenuate modestly due to radiation into the borehole fluid. Under normal logging conditions, this arrival appears as a composite with the shear. The full waveform has been studied using numerical techniques to carry out the solution which is available in integral form. The results of shear and compressional lateral wave analyses based on a branch cut integration technique have been discussed. This technique has been used to isolate the individual arrivals for a specific source function and to study their impulse response characteristics.

Means for extracting the shear and compressional interval transit times have been reported in the literature and discussed here. Bandpass filtering has been used to extract the Stoneley wave which typically predominates as a low frequency signal in the acoustic wavetrain. A possible application of the Stoneley amplitude for formation shear modulus has been described for 'soft' formations.

Finally, we may conclude that the waveform of acoustic logging is now well understood and sophisticated means for analysing any aspect of interest are available. Analytical results are, however, less complete when complications such as inelastic, anisotropic and inhomogeneous formation structures are of interest.

ACKNOWLEDGEMENTS

The author wishes to express his appreciation to the Society of Exploration Geophysicists for granting permission to use figures from the papers by Tsang and Rader (1979) and Cheng and Toksoz (1981). Professor Toksoz's kindness in providing a manuscript of the latter paper prior to its publication is also appreciated.

Support by Schlumberger-Doll Research for the author's experimental and analytical work, and associated previous publications, in the period 1973–1978, is gratefully acknowledged.

A special note of thanks is due to Dr L. Tsang with whom the author previously collaborated. His work on the analysis of the borehole

DENNIS RADER

acoustic problem constitutes a significant contribution to an important area of geophysics.

Thanks are due to Ms B. Brown for her care and effort in typing the manuscript and to Mr G. Daigle for preparing the illustrations.

REFERENCES

ARON, J., MURRAY, J. and SEEMAN, B. (1978). Formation compressional and shear interval transit-time logging by means of long spacing and digital techniques, SPE Paper 7446, 53rd Ann. Tech. Conf. and Exhib., SPE of AIME.

BIOT, M. A. (1952). Propagation of elastic waves in a cylindrical bore containing a fluid, *J. Appl. Phys.*, **23**, 997–1005.

CHENG, C. H. and TOKSOZ, M. N. (1981) Seismic wave propagation in a fluid filled borehole and synthetic acoustic logs, *Geophysics*, **46**, 1042–53.

EWING, M., JARDETSKY, W. and PRESS, F. (1957). *Elastic Waves in Layered Media.* McGraw-Hill Book Co., Inc., New York.

GORDON, R. B. and DAVIS, L. A. (1968). Velocity and attenuation of seismic waves in imperfectly elastic rock, *J. Geophys. Res.*, **73**, 3917–35.

GORDON, R. B. and RADER, D. (1970). Imperfect elasticity of rock: Its influence on the velocity of stress waves, in *Structure and Physical Properties of the Earth's Crust*, Ed. J. G. Heacock. American Geophysical Union Monograph 14, pp. 235–42.

INGRAM, J. D. (1978). Method and Apparatus for Acoustic Logging of a Borehole, United States Patent No. 4 131 875.

KNOPOFF, L., (1964). Q, *Review of Geophysics*, **2**, 625–60.

KOLSKY, H. (1953). *Stress Waves in Solids.* Clarendon Press, London (Reprinted 1964, Dover, New York).

LEVIN, F. K. (1980). Seismic velocities in transversely isotropic media, II, *Geophysics*, **45**, 3–17.

O'CONNELL, R. J. and BUDIANSKY, B. (1974). Seismic velocities in dry and saturated cracked solids, *J. Geophys. Res.* **79**, 5412–26.

PETERSON, E. W. (1974). Acoustic waves propagating along a fluid filled cylinder, *J. Appl. Phys.*, **45**, 3340–50.

PIRSON, S. J. (1963). The elastic wave propagation logs, in *Handbook of Well Log Analysis*, Prentice Hall, Inc., Englewood Cliffs, New Jersey.

RADER, D. (1975). Acoustic logging of oil wells, presented at the 89th Meeting of the Acoustical Society of America, Austin, Texas.

ROEVER, W. L., ROSENBAUM, J. H. and VINING, T. F. (1974). Acoustic waves from an impulsive source in a fluid-filled borehole, *J. Acoust. Soc. Am.*, **55**, 1144–57.

ROSENBAUM, J. H. (1974). Synthetic microseismograms: logging in porous formations, *Geophysics*, **39**, 14–32.

SCHLUMBERGER (1972). *Log Interpretation, Volume I—Principles.* Schlumberger Limited, New York.

TOKSOZ, M. N., CHENG, C. H. and TIMUR, A. (1976). Velocities of seismic waves in porous rocks, *Geophysics*, **41**, 621–45.

TSANG, L. (1978). Time harmonic solution of the elastic head wave problem incorporating the influence of Rayleigh poles, *J. Acoust. Soc. Am.*, **63**, 1302–9.

TSANG, L. and KONG, J. A. (1979). Asymptotic methods for the first compressional head wave arrival in a fluid-filled borehole, *J. Acoust. Soc. Am.*, **65**, 647–54.

TSANG, L. and RADER, D. (1979). Numerical evaluation of the transient acoustic waveform due to a point source in a fluid filled borehole, *Geophysics*, **44**, 1706–20.

WALSH, J. B. (1966). Seismic wave attenuation in rock due to friction, *J. Geophys. Res.*, **71**, 2591–9.

WHITE, J. E. (1962). Elastic waves along a cylindrical bore, *Geophysics*, **27**, 327–33.

WHITE, J. E. (1965). *Seismic Waves: Radiation, Transmission and Attenuation.* McGraw-Hill, New York.

WHITE, J. E. and ZECHMAN, R. E. (1968). Computed response of an acoustic logging tool, *Geophysics*, **33**, 302–10.

WYLLIE, M. R. J., GARDNER, G. H. F. and GREGORY, A. R. (1962). Studies of elastic wave attenuation in porous media, *Geophysics*, **27**, 569–89.

Chapter 6

ELECTRICAL ANISOTROPY:
ITS EFFECT ON WELL LOGS

J. H. MORAN

Briarcliff 840, Spicewood, Texas, USA

and

S. GIANZERO

Gearhart Industries, Inc., Austin, Texas, USA

SUMMARY

The influence of electrical anisotropy on the response of a variety of electrode and induction devices is studied. Standard logging tools are considered where anisotropy can lead to some ambiguity in the interpretation of the measurements. Other tools are considered that are designed to be sensitive to anisotropy and yield a measurement of it. Our conclusions appear as the effects of anisotropy on standard logging tools and a proposal for a new tool that appears most promising for the measurement of anisotropy. This new tool is a micro-resistivity sidewall device similar to a micro-SFL, proximity or micro-laterolog. The standard tools are somewhat insensitive to the vertical resistivity unless the dip is large except near bed boundaries. Among the standard tools, both the electrode and induction devices tend to yield the horizontal resistivity unless the spacings of the electrode device are small, of the order of the borehole diameter.

NOMENCLATURE

$a(1)$	borehole radius
$a(2)$	width of rectangular pad
A	electrode position
A	vector potential

A_z	component of vector potential
A_R	area of receiver coil
$\mathbf{E}$	electric field (in Maxwell's equations)
G_i	generalised conductivity in a dipping anisotropic medium
h_{mc}	mudcake thickness
$\mathbf{H}$	magnetic field (in Maxwell's equations)
$I(1)$	current into and out of electrode source
$I(2)$	current flowing in a wire
$I_0(\ ),\ I_1(\ )$	Bessel functions
$J_0(\ ),\ J_1(\ )$	Bessel functions
$\mathbf{J}$	current density
$\mathbf{J}_s$	current source density
k_h	horizontal propagation constant
k_m	propagation constant of borehole fluid
k_v	vertical propagation constant
K	tool coefficient
$K_0(\ ),\ K_1(\ )$	Bessel functions
$L(1)$	normal AM spacing
$L(2)$	distance of A electrode from boundary
$L(3)$	length of current-carrying wire
$L(4)$	I-L two coil T–R spacing
$\mathbf{M}_s,\ M$	strength of source magnetic dipole (corresponding to induction log transmitter coil)
$M_x,\ M_y,\ M_z$	Cartesian components of $\mathbf{M}_s$
N_R	number of receiver coil turns
$\mathbf{R}$	rotational operator
R_a	apparent resistivity as measured by logging tool
R_h	resistivity parallel to the bedding planes
R_m	mud resistivity
R_{mc}	mudcake resistivity
R_v	resistivity normal to the bedding planes
$R_x,\ R_y,\ R_z$	receiver coil
$s,\ s',\ s_L$	contracted distance
$T_x,\ T_y,\ T_z$	transmitter coils
$V(1)$	scalar potential
$V(2)$	voltage induced in a receiver coil
$V_1,\ V_2$	potential induced in anisotropic medium
$V_h,\ V_v$	measured voltages for horizontal and vertical spacings of pad device

V_s	source voltage in an anisotropic medium
x, y, z	Cartesian coordinates
ρ, z	cylindrical coordinates
z	normal spacing
α, α_i	bed inclination
β, β_i	dip azimuth
β_h, β_v	horizontal and vertical longitudinal propagation constants
β_m	longitudinal propagation constant of mud
δ_h, δ_v	horizontal and vertical skin depths
Γ	reflection coefficient
λ	coefficient of anisotropy
λ_a	apparent anisotropy coefficient
μ	variable of integration
μ_0	magnetic permeability of free space
π, π_x, π_y, π_z	Hertz vector
σ	conducting tensor
σ_a	apparent conductivity from logging tool
σ_h	conductivity parallel to bedding planes
σ_m	borehole conductivity
σ_r	resistive part of induction log conductivity signal
σ_x	reactive part of induction logging signal
σ_{xx}, etc.	elements of conducting tensor
σ_v	conductivity perpendicular to bedding planes
T	transmission coefficient
ϕ	scalar potential
ω	angular frequency

1. INTRODUCTION

In many sedimentary strata, electric current flows more easily parallel to the bedding planes than transversely to them. One reason for this anisotropy is that the solid particles possess a flat elongated shape and take an orientation parallel to the plane of disposition. This results in a pore structure allowing current to flow more easily parallel to the bedding plane in the saturating mineralised water than transverse to the bedding plane. This electrical anisotropy is usually more pronounced in shales. Under some conditions of sedimentation the resistivity in the bedding plane varies with direction. This effect, however, is usually relatively small and will be ignored here. Our only concern in this chapter is with electrical anisotropy but we should recognise that the same features of sedimentation that cause electrical anisotropy also result in anisotropy in other physical parameters. Permeability, for example, is usually much more affected than resistivity and the elastic and sonic properties may be expected to exhibit anisotropy. The elastic properties in fact will be anisotropic as a result of the stress distribution in the sediments, even without consideration of the pore structure.

If a sample is cut from the sediment parallel to the bedding plane, the resistivity measured with current flowing along this plane is referred to as the longitudinal resistivity R_h. On the other hand, the resistivity with current flowing normal to the bedding plane is referred to as the transverse resistivity R_v. The coefficient of anisotropy λ is defined as

$$\lambda = \sqrt{R_v/R_h}$$

and is generally equal to or greater than unity. Laboratory measurements (Hill, 1972; Keller and Frischknecht, 1966) have found λ in the range 1–2·5. The geometric mean resistivity R, defined as

$$R = \sqrt{R_v R_h}$$

will also be used. Hence in terms of λ and R

$$R_h = R/\lambda$$

and

$$R_v = \lambda R$$

In addition to anisotropy occurring as a result of the orientation of the rock grains, anisotropy may also occur when a formation is made up of a sequence of thin isotropic beds of differing lithologic characteristics (e.g. sequences of shales and sands.) The logging system may have dimensions such that these thin beds appear as a homogeneous anisotropic formation. This is described as macroscopic anisotropy and it can act in addition to the microscopic anisotropy due to the grain shape. Macroscopic anisotropy has been considered by Runge and Hill (1971), Habberjam (1975) and Cambell (1977).

If a current source I is placed at the origin of a cylindrical coordinate system (ρ, z) in a homogeneous formation with resistivity R_v along z and

R_h in planes normal to z, then the potential $V(\rho, z)$ at any point ρ, z is (Smythe, 1950)

$$V(\rho,\, z) = \frac{IR}{4\pi} \frac{1}{\sqrt{\rho^2 + \lambda^2 z^2}}$$

Equipotentials are therefore ellipsoids. Measuring the potential at a fixed distance L from the source, such that $\rho = L\sin\alpha$ and $z = L\cos\alpha$ where α is the angle the 'tool axis' makes with the z-axis, then

$$V(L\sin\alpha,\, L\cos\alpha) = \frac{IR}{4\pi L} \frac{1}{\sqrt{\sin^2\alpha + \lambda^2 \cos^2\alpha}}$$

The so-called 'apparent resistivity' $R_a = \dfrac{4\pi V L}{I}$ is

$$R_a = R / \sqrt{\sin^2\alpha + \lambda^2 \cos^2\alpha}$$

In the most common case $\alpha = 0$ and $R_a = R_h$. This perhaps surprising result is referred to as the 'paradox' of anisotropy. The other extreme is $\alpha = \pi/2$ and $R_a = R$. No orientation of the tool yields $R_a = R_v$.

Electrical anisotropy was recognised by Schlumberger (1920) early in the history of electrical prospecting. A method for computing the potential in anisotropic formations was presented by Maillet and Doll (1932). The anisotropy of formations has been used to advantage for the determination of the directions of strike and dip both in surface prospecting and well logging (Schlumberger and Schlumberger, 1930; Schlumberger et al., 1933; Schlumberger et al., 1934). In a recent paper by Moran and Gianzero (1979) it was shown that the paradox of anisotropy remains valid even for high frequency alternating currents.

Features observed on resistivity logs stimulated the work of Kunz and Moran (1958). One such feature was the generally higher reading of the 16 in normal than the induction in shales. Another feature was the low readings of the lateral device just before the measuring electrodes entered a shale from a sand. The specific features associated with anisotropy tend to be highly localised 'boundary' effects associated with the transition from one formation to another or from the mud column to the formation. This behaviour will be explained in detail later in this chapter.

The importance of anisotropy for logging arises in several ways. For thick sands the anisotropy is sufficiently small that it may safely be ignored. On the other hand, if the reservoir consists of thin layers of sand and shale, interpretation of the logs for oil saturation can be in considerable error unless the anisotropy or shale content is accounted for. Currently this is done using auxiliary logs such as the neutron, density and SP. A direct resistivity method for the determination of λ would be highly desirable. Another important occurrence of anisotropy is in the interpretation of ULSEL (Runge and Hill, 1971). Here the goal is to determine the distance from the well bore to a nearby salt dome. Since the spacings of the normal devices used range up to several thousand feet, both microscopic and macroscopic anisotropy are important. Uncertainty in λ is directly reflected into uncertainty in the distance to the salt dome. A similar technique (Mitchell *et al.*, 1972) is often used in locating a cased hole relative to a relief hole in the case of a blowout. Again the spacings, 75 ft or so, are so large that macroscopic anisotropy must be considered and again λ is of first-order importance in the interpretation. Also, the previously mentioned use of anisotropy to determine dip and strike is of some interest as a possible basis for a more effective dipmeter. Finally, it is possible that a routine reliable method for the determination of λ could be a useful lithological parameter in logging.

Unfortunately even today there is no reliable direct method for measuring λ with a logging tool. A few methods that have been suggested will be considered, leading to the conclusion that a pad-type micro-resistivity device shows the greatest promise at least for the determination of micro-anisotropy.

Recently a great deal of relevant theoretical material has appeared, most of it motivated by studies of magnetised plasmas (Brown, 1967; Felsen and Marcuwitz, 1973). A number of papers have appeared that are related to geophysical problems (Chetayev, 1962, 1966 *a,b*; Sinha and Battacharya, 1967; O'Brien and Morrison, 1967; Sinha, 1968; Al'tgauzen, 1969; Davydov, 1970; Chetayev and Belen'kaya, 1971; Asten, 1974; Tsang *et al.*, 1974; Moran and Gianzero, 1979). Except for the last paper, these papers are directed to surface geophysical methods. The last paper is principally concerned with borehole methods.

The problems studied in this chapter use a number of different mathematical representations. This leads to complications in the presentation which is further complicated by the detailed algebraic manipulations required. In a reasonable space it is only possible to outline the analyses. This will not trouble one familiar with the techniques used but may unfortunately leave questions in the mind of one not familiar with these techniques.

Our treatment will proceed by considering three broad categories of problems:

(a) D.C. resistivity devices;
(b) A.C. resistivity devices;
(c) A.C. induction devices.

For the most part it is assumed that the borehole penetrates the formation normal to the bedding planes. All of our studies are based on analytical solutions—no numerical methods are considered.

2. D.C. RESISTIVITY DEVICES

2.1. General Equations for No Formation Dip

On the assumption that all the beds are horizontal and the borehole is vertical, one may speak of horizontal (longitudinal) conductivity σ_h parallel to the bedding planes and vertical (transverse) conductivity σ_v perpendicular to these planes. If the current sources lie on the axis of the borehole, the potential surfaces and the lines of current flow possess cylindrical symmetry. Consequently any anisotropy can be described mathematically via a conductivity tensor σ, viz.

$$\sigma = \begin{bmatrix} \sigma_h & 0 & 0 \\ 0 & \sigma_h & 0 \\ 0 & 0 & \sigma_v \end{bmatrix} \tag{1}$$

Ohm's law is

$$\mathbf{J} = \sigma \mathbf{E} \tag{2}$$

where $\mathbf{J}$ is the current density and $\mathbf{E}$ is the electric field strength.

Electrode tools have electric fields which can be derived from the gradient of a scalar potential. This fact coupled with the equation of continuity leads to the following generalisation of Laplace's equation:

$$\frac{1}{\rho} \frac{\partial}{\partial \rho} \left(\rho \sigma_h \frac{\partial V}{\partial \rho} \right) + \frac{\partial}{\partial z} \left(\sigma_v \frac{\partial V}{\partial z} \right) = 0 \tag{3}$$

where $V(\rho, z)$ is the scalar potential in a cylindrical coordinates system (ρ, ϕ, z) whose z axis coincides with the borehole axis.

Assuming that the variation of the coefficient of anisotropy with ρ and

z is expressible in the form

$$\lambda = f(\rho)g(z) \tag{4}$$

then from eqn (3)

$$\frac{f}{\rho}\frac{\partial}{\partial \rho}\left(\frac{f}{R}\frac{\partial V}{\partial \rho}\right) + \frac{1}{g}\frac{\partial}{\partial z}\left(\frac{1}{Rg}\frac{\partial V}{\partial z}\right) = 0 \tag{5}$$

where $R = \sqrt{R_h R_v}$ is the geometric mean of the resistivity. The equation is written in terms of resistivity since D.C. measurements are proportional to this parameter. By introducing the change of variables

$$\rho' = \int_0^\rho \frac{d\rho}{f(\rho)} \text{ and } z' = \int_0^z g(z)\,dz \tag{6}$$

and letting

$$V(\rho, z) = V'(\rho', z') \tag{7}$$

eqn (5) becomes

$$\frac{\partial}{\partial \rho'}\left(\frac{\rho}{R}\frac{\partial V'}{\partial \rho'}\right) + \frac{\partial}{\partial z'}\left(\frac{\rho}{R}\frac{\partial V'}{\partial z'}\right) = 0 \tag{8}$$

Finally, by introducing a new value of resistivity at each point, given by

$$R' = \frac{\rho'}{\rho} R \tag{9}$$

eqn (8) becomes

$$\frac{1}{\rho'}\frac{\partial}{\partial \rho'}\left(\frac{\rho'}{R'}\frac{\partial V'}{\partial \rho'}\right) + \frac{\partial}{\partial z'}\left(\frac{1}{R'}\frac{\partial V'}{\partial z'}\right) = 0 \tag{10}$$

A comparison of eqn (10) with eqn (3) shows that in terms of the primed coordinates (ρ', z') the differential equation for V' is that required for an isotropic medium of resistivity R'. Therefore, with the conditions imposed by eqn (4), any medium involving anisotropic formations may be replaced by an equivalent isotropic one. This is done by mapping each point (ρ, z) in the original space with a corresponding point (ρ', z') of a new space. In this mapping the potential remains the same at corresponding points, but the mean resistivity changes from R to R' $(=\rho'R/\rho)$. Kunz and Moran (1958) have shown that the boundary conditions are invariant under this same transformation.

2.2. Effect of the Borehole

Subsurface resistivity measurements with an electrode device are generally made by lowering the device into a drilled hole filled with drilling mud. The presence of the electrically conductive (and isotropic) column of drilling mud containing the device can be taken into account in eqn (4) by noting that

$$\lambda = \begin{cases} 1 & \text{for } \rho < a \\ \lambda_0 & \text{for } \rho > a \end{cases}$$

where a is the radius of the borehole. Setting $f(e) = \lambda$ and $g(z) = 1$, since λ is a function of ρ, from eqn (6)

$$z' = z$$

and

$$\rho' = \begin{cases} \rho & \text{for } \rho < a \qquad (11a) \\ a + (\rho - a)/\lambda_0 & \text{for } \rho > a \qquad (11b) \end{cases}$$

The equivalent isotropic resistivity after the transformation is, from eqn (9),

$$R' = \begin{cases} R_m & \text{for } \rho < a \\ \dfrac{R\rho'}{\lambda_0\rho' - (\lambda_0 - 1)a} & \text{for } \rho > a \end{cases} \qquad (12)$$

where R_m is the resistivity of the mud filling the hole and R is the geometric mean resistivity of the anisotropic formation.

The radial resistivity profile as given by eqn (12) is shown in Fig. 1. Note that as $\rho' \to \infty$ the resistivity R' of the equivalent isotropic formation approaches asymptotically $R_h = R/\lambda_0$, the horizontal resistivity. At the edge of the borehole ($\rho' = a$), as seen from eqn (12), R' equals R; but at $\rho' = (1 + 1/\lambda_0)a$, which is less than twice the borehole radius, $R' = \frac{1}{2}(R + R_h)$. At this distance R' has returned halfway to its asymptotic value R_h.

From this analysis the departure curves for normal and lateral devices (Fig. 2) should lie above the isotropic curves for a formation of resistivity R_h and below for a formation of conductivity $R = \lambda_0 R_h$, and they should approach the former for large spacings.

An exact treatment of the problem as exemplified by Fig. 1 shows that the characteristic features are indeed borne out. As mentioned earlier,

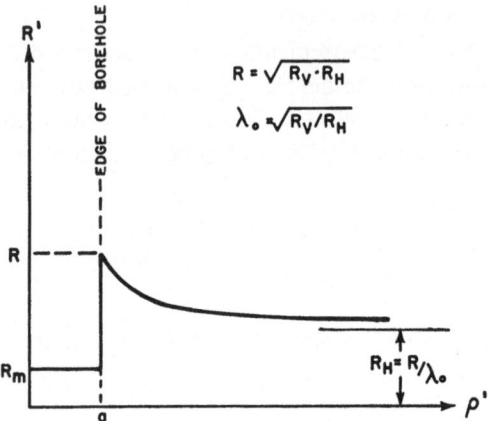

FIG. 1. Radial plot of the resistivity R' of an isotropic medium equivalent to a homogeneous anisotropic medium pierced by an isotropic homogeneous mud column of resistivity R_m. (Plotted for $\lambda_0 = 2$.)

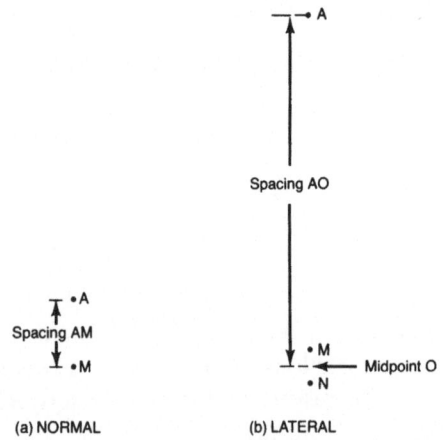

FIG. 2. Configuration of normal and lateral devices

one is led to believe that the transformation to an equivalent isotropic formation will often serve to gain a qualitative prediction of the apparent resistivity to be expected.

The results of the computations for the normal device are shown in Fig. 3 for various values of R_h/R_m (ratio of horizontal formation resistivity to mud resisitivity) and for four degrees of anisotropy ($\lambda = 1$, 1·5, 2 and 2·5). Each curve shows the variation of R_a/R_m (ratio of apparent resistivity to mud resistivity) versus AM/d (ratio of normal

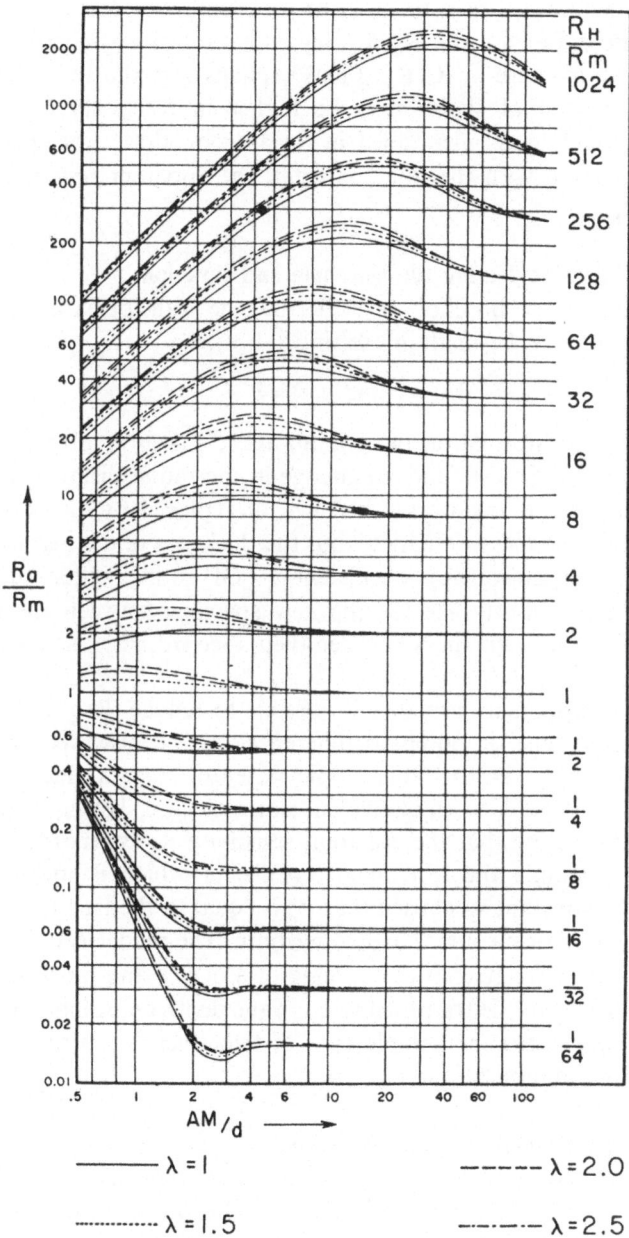

FIG. 3. Normal device (no invasion).

spacing to hole diameter). The comments on an infinite, homogeneous, anisotropic medium are seen to describe the qualitative features of the curves accurately. For a given R_h/R_m, increased anisotropy corresponds in general to an increase in R_a/R_m. (This explains why the 16 in normal reading at the level of shales may be higher in resistivity than the induction log reading, as mentioned in the Introduction.) For spacings large in comparison with the hole diameter, the apparent resistivity R_a approaches R_h.

2.3. General Case Neglecting the Borehole and Invasion

If the influence of the borehole is neglected, the transformation to an isotropic medium can be made by setting

$$f(\rho) = 1 \text{ and } g(z) = \lambda \qquad (13)$$

in eqn (4). The transformation of eqns (6) and (7) then involves only a stretching in the z-direction and no change in the radial distance or the mean resistivity. Each anisotropic bed will be replaced by an isotropic bed having the same mean resistivity, but the thickness of the bed will be multiplied by the anisotropy coefficient λ. Of course, after such a transformation one must take for the new spacings between the electrodes of a sonde the differences between their z'-coordinates, as given by eqn (6).

The above transformation makes it possible to reduce the problem of finding the potentials, or potential differences, to be observed on the measuring electrodes in the anisotropic case to that of finding the corresponding potentials in an equivalent isotropic case. The equivalence is completed by noting that the apparent resistivity R_a as recorded on a log must also include the instrument coefficient which is merely the proportionality constant so chosen that R_a is equal to the true resistivity for an infinite, homogeneous, isotropic medium. The reader is cautioned that this constant is related to the spacing of a device and must also be stretched according to the transformation in the depth coordinate. Let us see how this works for the sample case of two thick anisotropic beds separated by a common boundary.

2.4. Two-Media Problem Without Borehole

In the first bed the horizontal and vertical resistivities are R_{h_1} and R_{v_1} respectively, with similar definition in the second bed (i.e. R_{h_2} and R_{v_2}). If the original problem is converted to the equivalent isotropic problem, and if L is the distance from electrode A to the boundary and z is the vertical coordinate of M with reference to A (positive downward), the potential in the two beds can be found by the method of images:

$$V = \frac{IR_1}{4\pi\lambda_1}\left[\frac{1}{|z|} + \frac{R_2 - R_1}{R_2 + R_1}\left(\frac{1}{|2L - z|}\right)\right] \quad \text{for A and M in medium no. 1}$$

$$V = \frac{I}{4\pi}\frac{2R_1 R_2}{(R_1 + R_2)}\frac{1}{|\lambda_1 L + \lambda_2(z - L)|} \quad \begin{array}{l}\text{for A in}\\ \text{medium no. 1} \quad (14)\\ \text{M in medium no. 2}\end{array}$$

$$V = \frac{IR_2}{4\pi\lambda_2}\left[\frac{1}{|z|} + \frac{R_1 - R_2}{R_1 + R_2}\left(\frac{1}{|2L + z|}\right)\right] \quad \text{for A and M in medium no. 2}$$

where R_1 and R_2 are the mean resistivities. Equations (14) are demonstrated graphically in Fig. 4.

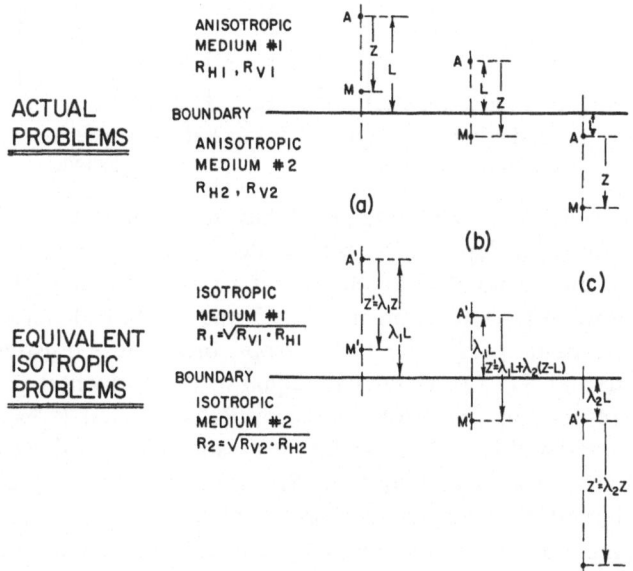

FIG. 4. Two semi-infinite anisotropic media, borehole effect neglected. (Drawn for $\lambda_1 = 1\cdot25$, $\lambda_2 = 2$.)

2.4.1. Special Case $R_2 = R_1$

This is a case of special interest because the equivalent isotropic problem corresponds to a homogeneous medium. The apparent resistivity profiles versus depth for such a case are illustrated by the solid curves in Fig. 5 for the normal device and in Fig. 6 for the lateral. In each figure the upper anisotropic bed (region 1) has a horizontal resistivity R_{h1} of 2, and

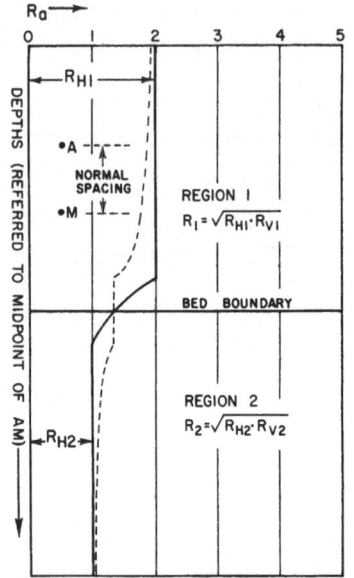

FIG. 5. Normal sonde, borehole effect neglected.

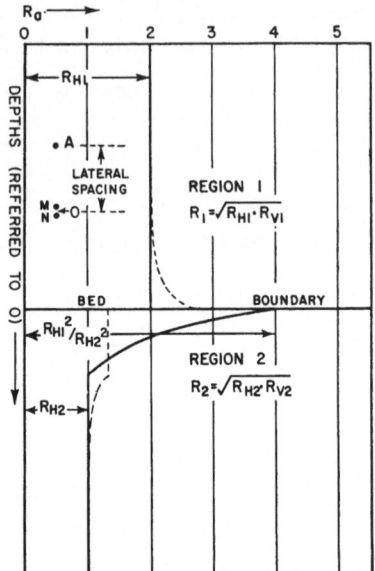

FIG. 6. Lateral sonde, borehole effect neglected.

the lower anisotropic bed (region 2) has R_{h_2} equal to 1. (Note the reference points for depths have been shifted to the midpoint of AM for the normal, and to point O for the lateral, i.e. the average tool spacing.)

Also shown for comparison in Figs. 5 and 6 (dashed lines) are the apparent resistivity profiles for two isotropic beds whose resistivities $R_1{}^*$ (upper bed) and $R_2{}^*$ (lower bed) are equal to the horizontal resistivities of the corresponding anisotropic beds, i.e. $R_1{}^* = R_{h_1}$ and $R_2{}^* = R_{h_2}$.

Both normal and lateral curves for the anisotropic case show values of R_a which are constant and equal to R_{h_1}, until the measuring electrodes cross the boundary. Then, for the normal, while the A electrode is in one medium and the measuring electrode is in the other, the value of R_a makes a smooth transition from R_{h_1} to R_{h_2}. This transition from an apparent resisitivity R_{h_1} to R_{h_2} is completed in the distance AM which is the sonde spacing. The influence of anisotropy is a 'boundary' effect as mentioned in the Introduction. For the lateral, when the measuring electrodes cross the boundary, the apparent resisitivity makes a jump such that the readings on the two sides of the boundary are in the ratio λ_1/λ_2. This sudden jump is followed by a smooth transition to the value R_{h_2} which takes place in the distance of one sonde spacing AO.

This example illustrates some of the novelty due to anisotropy. If we attempted to interpret the apparent resistivity curve in this case as though the media were isotropic, we would have assigned resistivities R_{h_1} and R_{h_2} to match the curve at a distance from the boundary. Then the behaviour at the boundary would be misleading (compare dashed and solid curves in Figs. 5 and 6).

2.4.2. Special Case $R_{h_1} = R_{h_2}$, but $R_1 \neq R_2$

The apparent resistivity R_a will be the same in both media except in the immediate neighbourhood of the boundary. Suppose $\lambda_2 > \lambda_1$. Then, as indicated in Fig. 7, the reading of the normal device will increase as the boundary is approached from above reaching a value $2\lambda_2 R_h/(\lambda_1 + \lambda_2)$ when the measuring electrode arrives at the boundary. After the measuring electrode crosses the boundary, the apparent resisitivity decreases smoothly, becoming equal to $2\lambda_1 R_h/(\lambda_1 + \lambda_2)$ when the current electrode reaches the boundary. The reading then increases to the value R_h as the normal device continues to move downward.

The lateral device under the same circumstances will show a decreasing reading, as shown in Fig. 8, arriving at a value $2\lambda_1 R_h/(\lambda_1 + \lambda_2)$ when

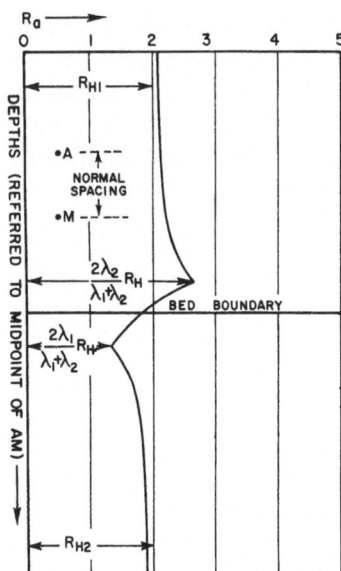

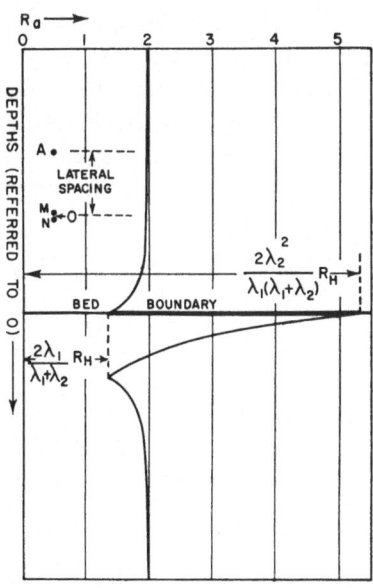

FIG. 7. Normal sonde, borehole effect neglected. Special case of $R_{h_1} = R_{h_2}$ $= R_h$. (Plotted for $\lambda_1 = 1$, $\lambda_2 = 2$, $R_h = 2$.)

FIG. 8. Lateral sonde, borehole effect neglected. Special case of $R_{h_1} = R_{h_2}$ $= R_h$. (Plotted for $\lambda_1 = 1$, $\lambda_2 = 2$).

the measuring electrodes are just above the boundary. When the measuring electrodes cross the boundary, the apparent resisitivity jumps up by a factor of $(\lambda_2/\lambda_1)^2$. It then decreases, reaching a second minimum when the current electrode passes the boundary. Thereafter the apparent resisitivity increases towards its final value of R_h.

The resistivity profiles of Figs. 5–8 indicate that the response of the lateral device is more strikingly affected by the presence of anisotropy than the response of the normal.

The general case has been considered in detail by Kunz and Moran (1958) who show how λ can be determined under ideal conditions.

2.5. Micro-Resistivity Devices for Measuring Anisotropy
It should be apparent from the previous discussion that because of the paradox of anisotropy, resistivity tools oriented perpendicular to the bedding planes are influenced by R_v only in the vicinity of a boundary separating media of different resistivity properties. The best way to circumvent the paradox is to destroy the cylindrical symmetry by means of a micro-resistivity (pad) device which, of course, is completely eccentric.

Consider the following idealisation of a pad device, exemplified in Fig. 9. The essential idea is to drive a single voltage electrode on a grounded pad and monitor the return currents at two electrodes separated electrically by means of insulation from the rest of the pad. One monitor electrode is displaced vertically and the second horizontally from the voltage electrode. The principal measurement is the ratio of the

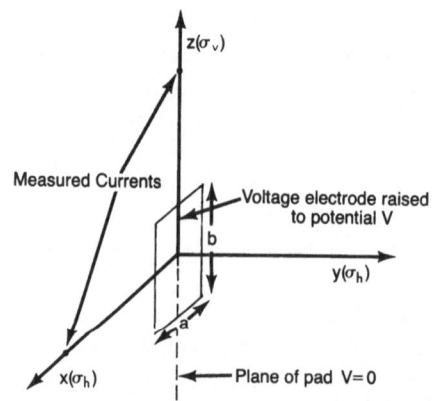

FIG. 9. Idealisation of a conducting pad device.

return currents. Assuming that the borehole fluid is sufficiently conducting that the pad is an infinite conducting plane, then the collected current
densities of the horizontal $J(x, 0)$ and vertical $J(0, z)$ spacings can be
closely approximated by

$$J(x, 0) \equiv J_h = -\frac{\sigma_h V ab\lambda}{2\pi}\left(\frac{1}{x}\right)^3 \tag{15}$$

$$J(0, z) \equiv J_v = -\frac{\sigma_h V ab\lambda}{2\pi}\left(\frac{\lambda}{z}\right)^3 \tag{16}$$

where a and b are the length and width of the electrode. If the spacings x
and z are chosen to be identical, then the anisotropy coefficient is found
from

$$\lambda = \sqrt[3]{\frac{J_h}{J_v}} \tag{17}$$

It should be understood that the fall-off of current density as the cube
of the distance is brought about by the severe 'shunt effect' of the nearby
ground. One might think that this shunt effect might result in a device
with a severe mudcake effect because of shallow penetration in the
formation. The results of a theoretical investigation which included a
mudcake are shown in Fig. 10. Clearly, up to 0·5 in of the mudcake error in
the anisotropy is negligible.

2.6. Comparison With a Modified Micro-Normal

A comparison of the proposed device with a conventional tool like the
micro-normal can only be achieved by modifying the micro-normal for
an anisotropy measurement.

A current electrode on an insulating pad with a return at infinity
serves as an idealisation of the micro-normal. In addition to measuring
the voltage in the vertical direction, the voltage is measured in a
horizontal direction by means of a second spacing, as shown in Fig. 11.

The chief difference in the measuring capability of this tool compared
with the conducting pad device is the sensitivity to λ. It can be shown
that, for equal horizontal and vertical spacings, the anisotropy coefficient
is given by

$$\lambda = \frac{V_h}{V_v} \tag{18}$$

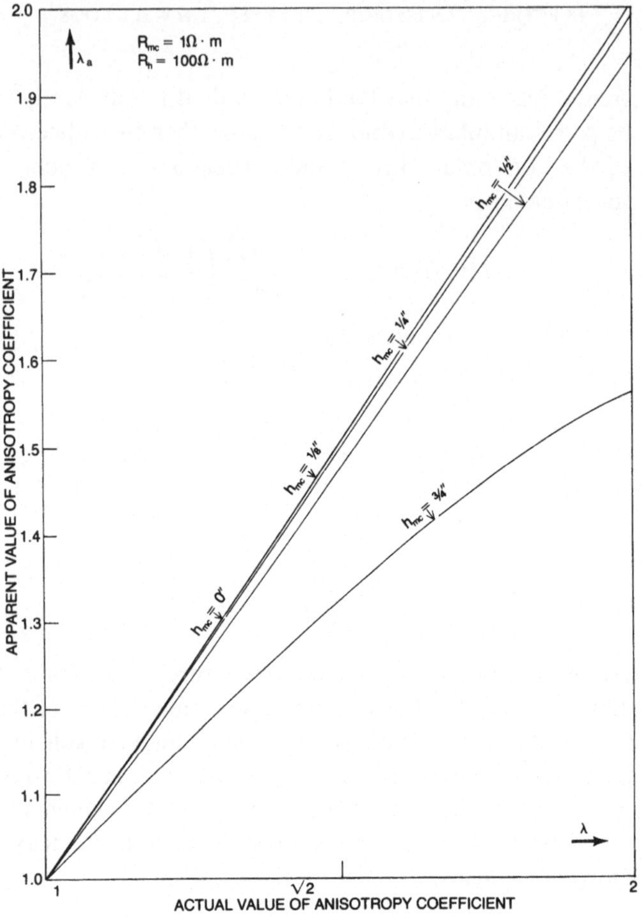

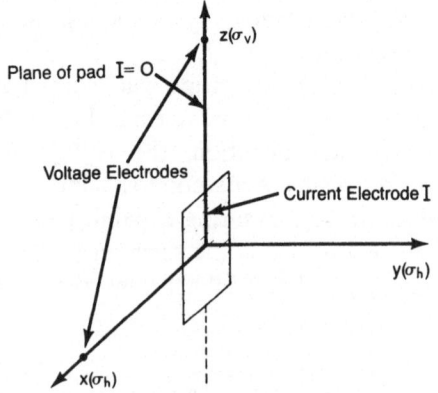

FIG. 10. Mudcake effect of conducting pad tool.

FIG. 11. Modification of the micro-normal.

where V_h and V_v are the measured voltages at the horizontal and vertical spacings respectively. Thus the modified micro-normal voltages vary directly as λ whereas for the conducting pad the currents vary as the cube of λ.

More important in the assessment of the modified micro-normal is the influence of mudcake. Figure 12 indicates an extremely large mudcake effect as opposed to that of the conducting pad as shown in Fig. 10.

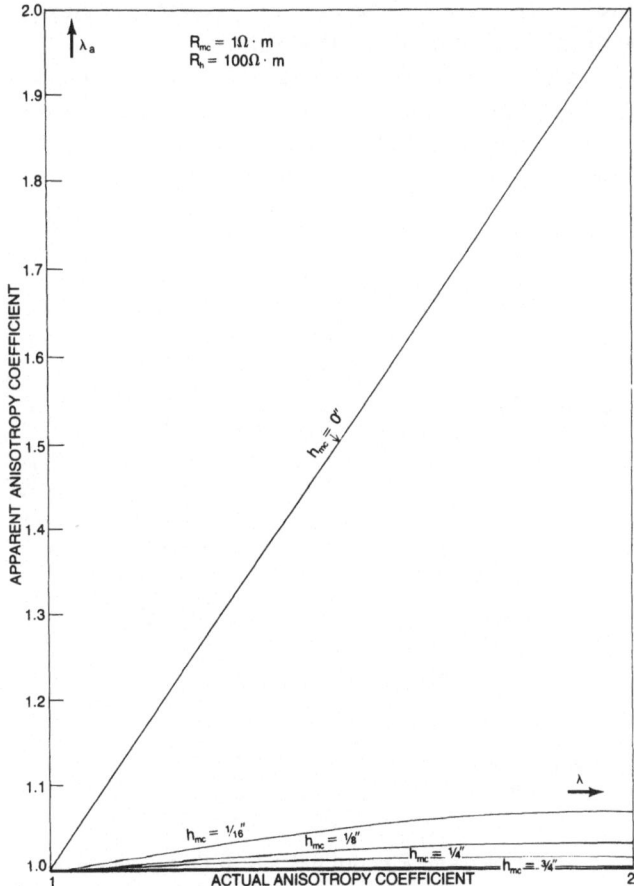

FIG. 12. Mudcake effect of modified micro-normal.

2.7. Electrode Devices in Contiguous Dipping Beds (Borehole Neglected)

The final application considered for resistivity devices is the influence of formation dip as well as anisotropy on a centralised tool. The entire discussion will be applied to the short normal. Since we have ascertained that anisotropy is a boundary effect, the most obvious generalisation to the previous analysis is to include formation dip in the case of two thick beds joined at a plane boundary.

The problem to be considered can be described as follows: Two different semi-infinite anisotropic media with conductivity tensors σ_1 and σ_2 are assumed to meet at a plane interface P. The two media may have arbitrary bedding plane dips and P may have an arbitrary orientation, so the two tensor coordinate systems are not necessarily aligned with each other or with the horizontal or vertical directions.

A short normal is considered to traverse the bed boundary as in the previous section. The solution for the potential and current density induced by the current electrode A in each medium is matched on the plane P. Consequently, it is expedient to develop an expression for the conductivity tensor as seen in the plane P coordinate system. Figure 13

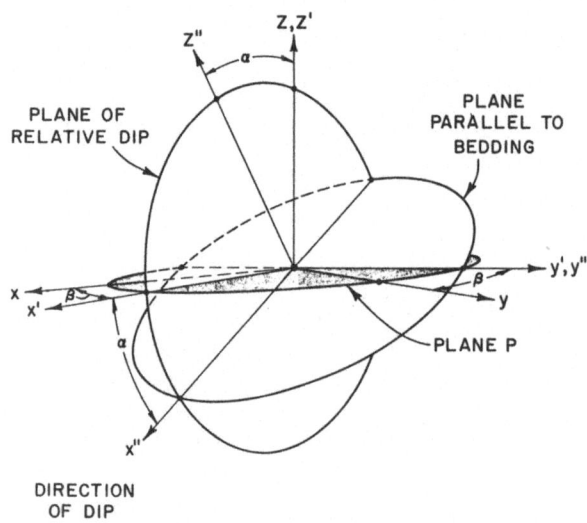

FIG. 13. Relation of plane P coordinates to bedding plane coordinates

indicates the two rotations that are necessary to align the bedding coordinate system of either medium with the coordinate system of plane P.

The rotation that brings these coordinate systems into alignment can be described by the rotation matrix **R**:

$$\mathbf{R} = \begin{bmatrix} \cos\alpha\cos\beta & \cos\alpha\sin\beta & -\sin\alpha \\ -\sin\beta & \cos\beta & 0 \\ \sin\alpha\cos\beta & \sin\alpha\sin\beta & \cos\alpha \end{bmatrix} \tag{19}$$

The overall effects of this rotation on the governing equations can be viewed as modifications upon the conductivity tensor in the bedding planes (see eqn 1). The conductivity tensor of either medium (without regard to distinguishing subscripts) in the plane P coordinate system is given by

$$\boldsymbol{\sigma} = \begin{bmatrix} \sigma_{xx} & \sigma_{xy} & \sigma_{xz} \\ \sigma_{xy} & \sigma_{yy} & \sigma_{yz} \\ \sigma_{xz} & \sigma_{yz} & \sigma_{zz} \end{bmatrix} \tag{20}$$

with

$$\begin{aligned} \sigma_{xx} &= \sigma_h + (\sigma_v - \sigma_h)\sin^2\alpha\cos^2\beta \\ \sigma_{xy} &= (\sigma_v - \sigma_h)\sin^2\alpha\sin\beta\cos\beta \\ \sigma_{xz} &= (\sigma_v - \sigma_h)\sin\alpha\cos\alpha\cos\beta \\ \sigma_{yz} &= (\sigma_v - \sigma_h)\sin\alpha\cos\alpha\sin\beta \\ \sigma_{yy} &= \sigma_h + (\sigma_v - \sigma_h)\sin^2\alpha\sin^2\beta \\ \sigma_{zz} &= \sigma_v - (\sigma_v - \sigma_h)\sin^2\alpha \end{aligned} \tag{21}$$

Note that eqn (20) reduces to eqn (1) if $\alpha = 0$, as it should.

Unfortunately, a complete solution for the potential is not possible using the theory of images as in the zero dip case. In fact, a general analytic evaluation in closed form is not possible. The solution for the potential in the respective media is given, when A is in medium 1, by

$$V_1 = V_s + \frac{I}{4\pi}\frac{1}{2\pi}\int_0^{2\pi}\frac{d\psi\,\Gamma(\psi)}{G_1}\frac{1}{\gamma_1^< z - \gamma_1^> z' - i\rho\cos(\psi - \phi)} \tag{22}$$

and

$$V_2 = \frac{I}{4\pi}\frac{1}{2\pi}\int_0^{2\pi}\frac{d\psi\,T(\psi)}{G_1}\frac{1}{\gamma_2^> z - \gamma_1^> z' - i\rho\cos(\psi - \phi)} \tag{23}$$

where

$$\Gamma(\psi)=\frac{G_1-G_2}{G_1+G_2}, \quad T(\psi)=\frac{2G_1}{G_1+G_2}$$

$$G_i=\sqrt{\sigma_{h_i}\,\sigma_{v_i}}\,\sqrt{(\lambda_i^2-1)\,\sin^2\alpha_i\,\sin^2(\psi-\beta_i)+1}$$

and

$$\gamma_i^>,\,\gamma_i^< =$$

$$\frac{-i(\lambda_i^2-1)\sin\alpha_i\cos\alpha_i\cos(\psi-\beta_i)\pm\lambda_i\sqrt{(\lambda_i^2-1)\sin^2\alpha_i\,\sin^2(\psi-\beta_i)+1}}{(\lambda_i^2-1)\sin^2\alpha_i+1}$$

Here, λ_i refers to the coefficient of anisotropy of bed i, α_i and β_i refer, respectively, to the bed's inclination (dip) and dip azimuth relation to the plane P coordinates described by the cylindrical system of coordinates ρ, ϕ, z with the z-axis normal to plane P.

In some special cases, eqns (22) and (23) can be evaluated analytically. Obviously, as mentioned earlier, if there is no dip (i.e. $\alpha_i=0$), the result reduces that obtained by Kunz and Moran (1958). Also, in the case $\sqrt{\sigma_{h2}\,\sigma_{v2}}=0$, for which $\Gamma=1$, and in the case $\sqrt{\sigma_{h2}\,\sigma_{v2}}=\infty$, for which $\Gamma=-1$, eqns (22) and (23) can be evaluated analytically.

Figures 14–23 show some typical results for the apparent resistivity R_a recorded by a normal (pole–pole) device. Here, L is the spacing between current electrode A and the measuring electrode M of the normal devices, and V is evaluated at $\rho=L\sin\alpha_{tool}$, $z-z'=L\cos\alpha_{tool}$ where α_{tool} is the angle the sonde axis makes with the normal to plane P.

Calculated apparent resistivity plots are shown in Figs. 14–19 for cases of $\alpha_{tool}=0$. Remote from the boundary, the reflection term in the V_1 solution (eqn 22) becomes small compared to the V_s term. The limiting value of the apparent resistivity R_{a_i} in bed i is given by

$$R_{a_i}(\infty)=\frac{\lambda_i\,R_{h_i}}{\sqrt{\lambda_i^2\cos^2\alpha_i+\sin^2\alpha_i}} \tag{24}$$

where R_{h_i} is the horizontal resistivity in bed i, equal to $\sigma_{h_i}^{-1}$.

Figures 20–23 present some results for various values of α_{tool}. For the computation of the limiting value of $R_{a_i}(\infty)$ in these cases, an equation

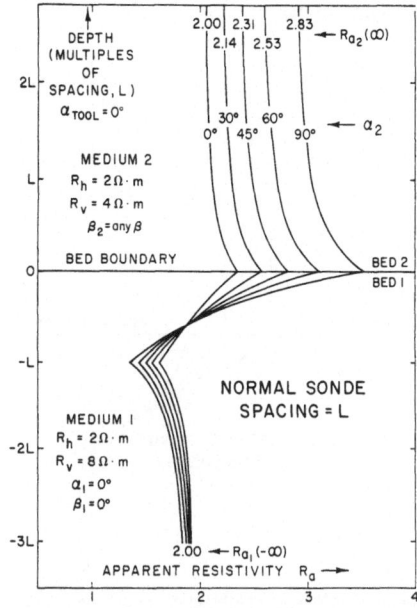

FIG. 14. Case of two adjacent, thick, anisotropic beds with no borehole. Responses of normal sonde of spacing L as a function of distance of sonde measuring point from bed boundary; sonde vertical (i.e. $\alpha_{tool}=0$), $\alpha_1=0°$, $\beta_1=0°$, α_2 as shown, β_2 any value.

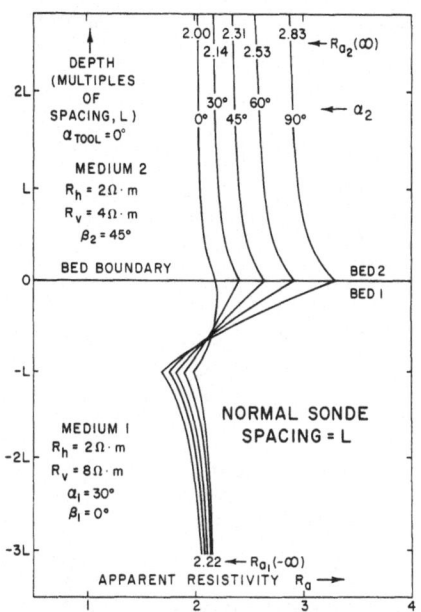

FIG. 15. Same as Fig. 14 except that $\alpha_1=30°$, $\beta_1=0°$ and $\beta^2=45°$.

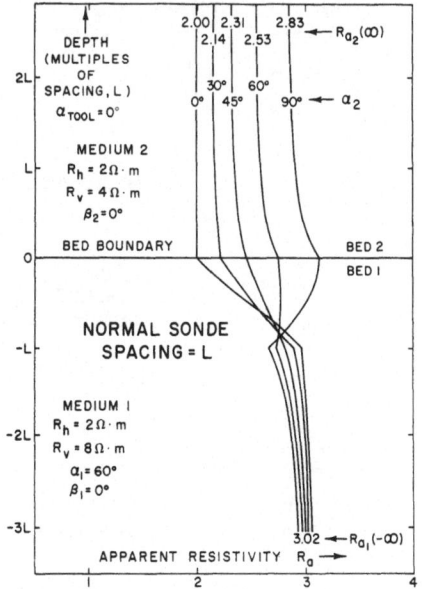

FIG. 16. Same as Fig. 14 except that $\alpha_1=60°$, $\beta_1=0°$ and $\beta_2=0°$.

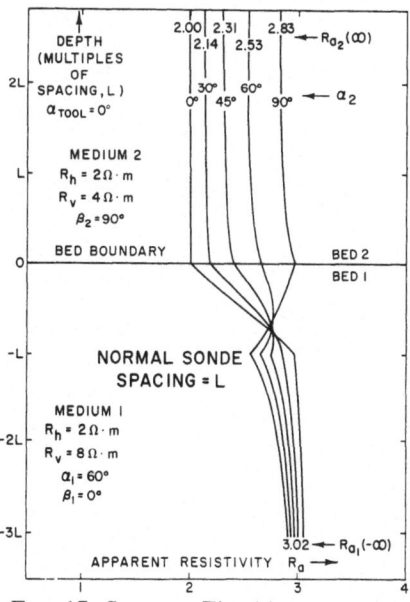

FIG. 17. Same as Fig. 14. except that $\alpha_1=60°$, $\beta_1=0°$ and $\beta_2=90°$.

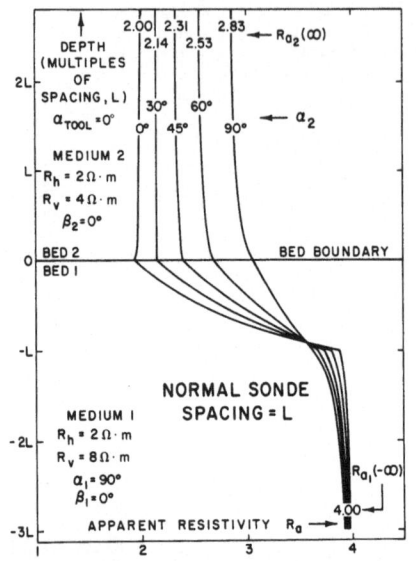

FIG. 18. Same as Fig. 14. except that $\alpha_1 = 90°$, $\beta_1 = 0°$, and $\beta_2 = 0°$.

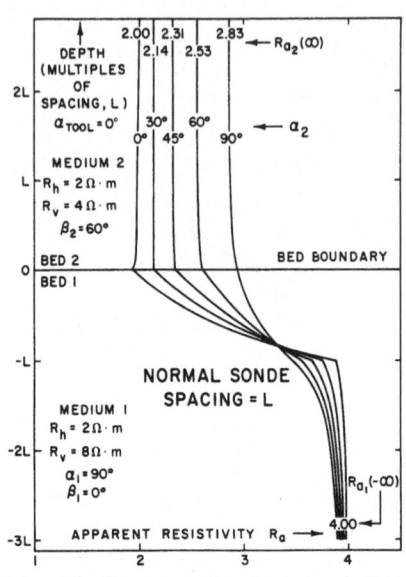

FIG. 19. Same as Fig. 14. except that $\alpha_1 = 90°$, $\beta_1 = 0°$, and $\beta_2 = 60°$.

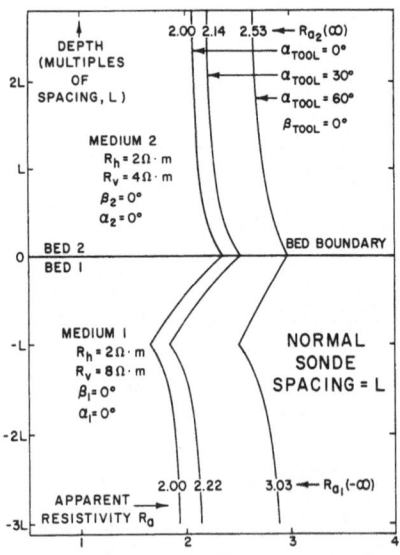

FIG. 20. Similar to Fig. 14 except that the curves are now for different values of α_{tool}, and $\alpha_1 = 0°$, $\beta_1 = 0°$, $\alpha_2 = 0°$ and $\beta_2 = 0°$.

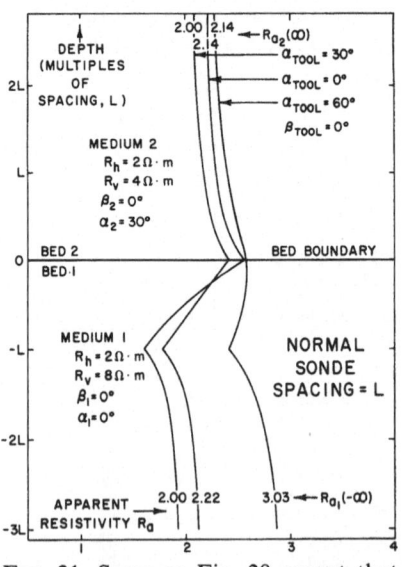

FIG. 21. Same as Fig. 20 except that $\alpha_1 = 0°$, $\beta_1 = 0°$, $\alpha_2 = 30°$, and $\beta_2 = 0°$.

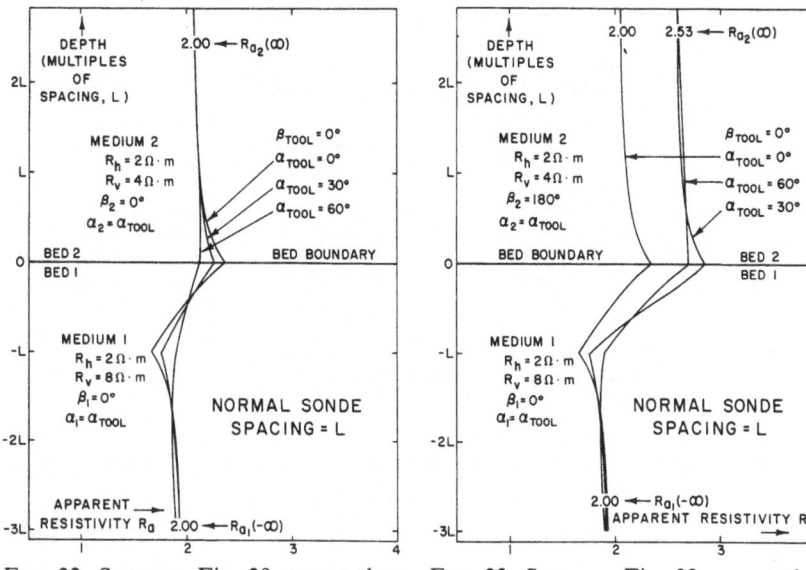

FIG. 22. Same as Fig. 20 except that $\alpha_1 = \alpha_{tool}$, $\beta_1 = 0°$, $\alpha_2 = \alpha_{tool}$ and $\beta_2 = 0°$.

FIG. 23. Same as Fig. 22 except that $\beta_2 = 180°$.

similar to eqn (24) can be derived. An inspection of Fig. 24 indicates that the essential parameter is α_i, the angle between the tool axis and the normal to the bedding plane. The angle is

$$\cos\alpha'_i = \cos\alpha_i \cos\alpha_{tool} + \sin\alpha_i \sin\alpha_{tool} \cos\beta_i \tag{25}$$

Equation (24) can be used directly if α'_i is inserted instead of α_i.

If the transitions from R_{a_1} to R_{a_2} in Figs. 14–23 are compared with similar results without dip from Kunz and Moran (1958), nothing dramatic happens because of the dipping interface. The major effect is that $R_{a_i}(\infty)$ is no longer equal to R_{h_i} when there is dip.

Equations (22) and (23) may be used for surface measurements by placing $\sqrt{\sigma_{h2}\,\sigma_{v2}} = 0$. Using the substitute $\chi = \exp(i\psi)$, the required integration becomes a contour integral over the unit circle in the χ plane. The only contribution to the integral is the pole whose residue is easily found. The result on the surface $z = 0$ is very simple and familiar:

$$V(\rho) = 2V_s(\rho) = \frac{\lambda_1 I}{2\pi\sigma_{h_i}} \frac{1}{\rho} \frac{1}{\sqrt{1 + (\lambda_1^2 - 1)\sin^2\alpha_1 \cos^2(\phi - \beta_1) + \sin^2(\phi - \beta_1)}} \tag{26}$$

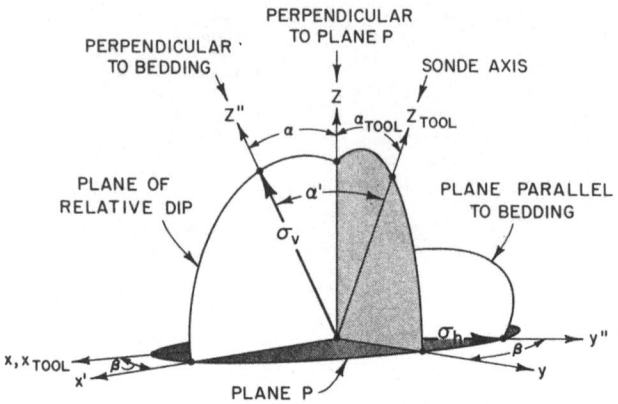

FIG. 24. Relation of sonde position to plane P coordinates and bedding plane coordinates.

This is the result expected from simple image theory and agrees with that obtained by Asten (1974). It shows that the equipotential surfaces are ellipses with axes along and at right angles to the strike. The ratio for an equipotential of the axis at right angles to the strike to that along the strike is $\sqrt{1+(\lambda_1^2-1)\sin^2\alpha_1}$ Thus, a simple surface survey could yield values for

$$\frac{\lambda_1}{\sigma_{h_1}} = \left(\frac{1}{\sqrt{\sigma_{h_1}\,\sigma_{v_1}}}\right) \text{ and } \sqrt{1+(\lambda_1^2-1)\,\sin^2\alpha_1}$$

Knowing α_1, values of σ_{h_1} and σ_{v_1} can be calculated.

2.8. Electrode Devices—Borehole Effects

The case of an electrode device centred in a borehole perpendicular to the bedding planes of an anisotropic bed was also treated by Kunz and Moran (1958). If the electrode device is eccentric, solutions can be constructed by a simple extension of the methods of Kunz and Moran (1958) and Gianzero and Rau (1977). This has been done, but the details will not be given here. The important effects of eccentricity have been covered by Gianzero and Rau (1977) and the introduction of anisotropy does not add anything interesting.

When the borehole axis is not normal to the bedding planes, the problem becomes very complicated. In spite of several efforts, no progress has been made in the analysis.

3. A.C. RESISTIVITY DEVICES

3.1. Generalities

In practice, all electrical logging devices are operated using alternating current. For devices employing electrodes, the use of alternating current is a practical necessity for a variety of reasons, such as difficulties with electrode polarisation and practical consideration of electronic instrumentation. For devices employing coils and electromagnetic induction, alternating current is also a necessity. For these reasons, any interplay between formation anisotropy and alternating currents is of definite interest.

The methods for solution of Maxwell's equations when using alternating currents will be reviewed briefly. If the sources are linear conductors terminated in electrodes, it is convenient to use the vector potential. Magnetic dipole sources are more easily treated using the Hertz potential. The basic formulae for these two methods are given and applied to specific problems later.

The starting point is Maxwell's equations with an assumed harmonic time dependence $\exp(-i\omega t)$. Using standard notation (Stratton, 1941)

$$\nabla \times \mathbf{E} = i\omega\mu_0\mathbf{H} + i\omega\mu_0\mathbf{M}_s \tag{27}$$

$$\nabla \times \mathbf{H} = \sigma\mathbf{E} + \mathbf{J}_s \tag{28}$$

where $\mathbf{M}_s$ and $\mathbf{J}_s$ represent the source distribution of magnetic dipoles and current respectively. The magnetic permeability in all the media has been taken as μ_0, the magnetic permeability of free space, neglecting the displacement currents relative to the conduction currents as is consistent with most logging applications.

For the electrode problem, it is assumed that $\mathbf{M}_s = 0$; hence from eqn (27) $\nabla\cdot\mathbf{H} = 0$. For the induction logging problem, it is assumed $\mathbf{J}_s = 0$; hence from eqn (28) $\nabla\cdot\sigma\mathbf{E} = 0$.

3.1.1. The Vector Potential

For electrode devices, since $\nabla\cdot\mathbf{H} = 0$, it is permissible to introduce the vector potential $\mathbf{A}$ such that

$$\mathbf{H} = \nabla \times \mathbf{A} \tag{29}$$

Using eqn (29) in eqn (27)

$$\nabla \times (\mathbf{E} - i\omega\mu_0\mathbf{A}) = 0 \tag{30}$$

This allows the introduction of a scalar potential Φ such that

$$\mathbf{E} = i\omega\mu_0\mathbf{A} - \nabla\Phi \tag{31}$$

$\nabla\cdot\mathbf{A}$ still has to be specified, and it is chosen to simplify the final equation. The choice found most convenient is

$$\nabla\cdot\mathbf{A} = \sigma_h\Phi \tag{32}$$

Combining eqn (32) with eqns (27) and (28), it can be shown for $\mathbf{J}$ directed along the z axis that

$$\frac{\partial^2 A_z}{\partial x^2} + \frac{\partial^2 A_z}{\partial y^2} + \frac{1}{\lambda^2}\frac{\partial^2 A_z}{\partial z^2} + k_v^2 A_z = -\mathbf{J}_s \tag{33}$$

where

$$k_v = \sqrt{i\omega\mu_0\sigma_v}$$

3.1.2. The Hertz Potential

For induction devices, we set $\mathbf{J}_s = 0$. As noted above, this implies that

$$\nabla\cdot\sigma\mathbf{E} = 0$$

This allows the introduction of the Hertz vector $\boldsymbol{\pi}$ as follows:

$$\sigma\mathbf{E} = k_h^2\nabla\times\boldsymbol{\pi} \tag{34}$$

where

$$k_h = \sqrt{i\omega\mu_0\sigma_h}$$

As for $\nabla\cdot\mathbf{A}$, the value of $\nabla\cdot\boldsymbol{\pi}$ can be chosen freely. If eqn (34) is placed in eqn (27), it follows that

$$\nabla\times(\mathbf{H} - k_h^2\boldsymbol{\pi}) = 0$$

This, in turn, allows the introduction of the scalar potential Φ, such that

$$\mathbf{H} = k_h^2\boldsymbol{\pi} + \nabla\Phi \tag{35}$$

Placing eqns (34) and (35) into eqns (27) and (28) and making

$$\nabla\cdot(\sigma\boldsymbol{\pi}) = \sigma_v\Phi \tag{36}$$

results in

$$\frac{\partial^2 \pi_x}{\partial x^2} + \frac{\partial^2 \pi_x}{\partial y^2} + \frac{1}{\lambda^2} \frac{\partial^2 \pi_x}{\partial z^2} + k_v^2 \pi_x = -\frac{M_x}{\lambda^2}$$

$$\frac{\partial^2 \pi_y}{\partial x^2} + \frac{\partial^2 \pi_y}{\partial y^2} + \frac{1}{\lambda^2} \frac{\partial^2 \pi_y}{\partial z^2} + k_v^2 \pi_y = -\frac{M_y}{\lambda^2} \tag{37}$$

$$\frac{\partial^2 \pi_z}{\partial x^2} + \frac{\partial^2 \pi_z}{\partial y^2} + \frac{\partial^2 \pi_z}{\partial z^2} + k_h^2 \pi_z = -M_z + (1-\lambda^2)\frac{\partial}{\partial z}\left(\frac{\partial \pi_x}{\partial x} + \frac{\partial \pi_y}{\partial y}\right)$$

where M_x, M_y and M_z are the components of the dipole distribution $\mathbf{M_s}$.

The virtue of the choice of $\nabla \cdot (\sigma \pi)$ in eqn (36) is now apparent, since the equations for π_x *and* π_y are completely decoupled and π_x and π_y enter into the equation for π_z only as pseudosource terms. This decoupling greatly simplifies the solution for induction sources.

3.2. Alternating Current Electrode Devices

When the z-coordinate is 'contracted' by making the substitution $z' = \lambda z$, eqn (33) is seen to reduce to the usual wave equation. The solution for a current element of strength I and length dL is thus

$$dA_z = \frac{-IdL}{4\pi} \lambda \frac{\exp(ik_v s)}{s}$$

$$d\Phi = \frac{IdL}{4\pi} \frac{\lambda}{\sigma_h} (ik_v s - 1) \frac{\exp(ik_v s)}{s^2} \frac{\lambda^2 z}{s} \tag{38}$$

where

$$s = \sqrt{\rho^2 + \lambda^2 z^2}$$

In the last expression, ρ and z are cylindrical coordinates and s is the distance contracted to account for anisotropy. For $\rho = 0$, eqn (38) reduces to

$$dA_z = \frac{-IdL}{4\pi} \frac{\exp(ik_h z)}{z}$$

$$d\Phi = \frac{IdL}{4\pi} \frac{1}{\sigma_h} (ik_h z - 1) \frac{\exp(ik_h z)}{z^2} \tag{39}$$

Equation (39) shows that along the axis $\rho = 0$, the fields do not depend on σ_v but only on σ_h. That is to say, the paradox of anisotropy holds true even though alternating currents are employed. This result will be considered in more detail after indicating how the basic solution of eqn (38) can be extended to heterogeneous media. To make this extension, we note here that eqn (38) can be written as a superposition of cylindrical waves using

$$\frac{\lambda \exp(ik_v s)}{s} = \frac{2}{\pi} \int_0^\infty K_0(\beta_h \rho/\lambda) \cos \mu z \, d\mu$$

or   (40)

$$\frac{\exp(ik_v s)}{s} = \int_0^\infty J_0(\mu \rho) \exp(-\beta_v \lambda |z|) \frac{\mu d\mu}{\beta_v}$$

where

$$\beta_v = \sqrt{\mu^2 - k_v^2} \quad \text{and} \quad \beta_h = \sqrt{\mu^2 - k_h^2}$$

and where $J_0(\mu \rho)$ and $K_0(\beta_h \rho/\lambda)$ are standard notations for Bessel functions (see McLachlan, 1961, pp. 203 and 290).

3.3. Thick Bed With Borehole Normal to Bedding Planes
The first representation in eqn (40) is useful for the problem of a borehole of radius a penetrating a thick, homogeneous, anisotropic formation. In this case, general solutions of eqn (33) can be constructed in the form

$$A_z \propto \int_0^\infty \left[K_0(\beta_m \rho) + \Gamma(\mu) I_0(\beta_m \rho) \right] \cos \mu z \, d\mu \quad \text{for } \rho < a$$

$$A_z \propto \int_0^\infty T(\mu) K_0(\beta_h \rho/\lambda) \cos \mu z \, d\mu \quad \text{for } \rho > a \qquad (41)$$

where

$$\beta_m = \sqrt{\mu^2 - k_m^2}$$

and

$$k_m = \sqrt{i\omega \mu_0 \sigma_m}$$

in which σ_m is the conductivity of the borehole fluid (mud).

The coefficients $\Gamma(\mu)$ and $T(\mu)$ are determined by requiring that H_ϕ and E_z be continuous on the boundary at $\rho = a$. This is a straightforward computation that will not be reproduced here.

3.4. Plane Bed Boundaries, No Borehole

In the same way solutions of eqn (33) appropriate for plane bed boundaries (parallel to bedding plane, perpendicular to the sonde axis) can be constructed using the second part of eqn (40). If the source is located in bed 1 and lies below bed 2 at a distance h, taking the bed boundary to be at $z=0$

$$A_z \propto \int_0^\infty J_0(\mu\rho) \left[\exp\left(-\beta_{v1}\lambda_1 |z-h|\right) + \Gamma(\mu) \exp\left(\beta_{v1}\lambda_1 z\right)\right] \frac{k \, d\mu}{\beta_{v1}} \text{ for } z<0$$

$$(42)$$

$$A_z \propto \int_0^\infty J_0(\mu\rho) \, T(\mu) \exp\left(-\beta_{v2}\lambda_2 z\right)\frac{\mu \, d\mu}{\beta_{v2}} \text{ for } z>0$$

Then using the boundary conditions that H_ϕ and E_ρ must be continuous at $z=0$, the coefficients $\Gamma(\mu)$ and $T(\mu)$ may be determined.

3.5. Frequency Effects in Homogeneous, Anisotropic Beds

Equation (38) can be applied to the practical case where a current I is emitted into the formation for an electrode A and returned via the formation to an electrode B. Placing A at the origin of coordinates and setting the distance AB equal to L, the solution of eqn (38) can be integrated over AB to yield

$$A_z = -\frac{I}{4\pi} \lambda \int_0^L \frac{\exp(ik_v s')}{s'} \, dz'$$

and

$$\Phi = \frac{I}{4\pi} \left[\frac{\exp(ik_v s)}{s} - \frac{\exp(ik_v s_L)}{s_L}\right] \qquad (43)$$

Here, z' is a variable of integration, and

$$s' = \sqrt{\rho^2 + \lambda^2(z-z')^2}$$

$$s = \sqrt{\rho^2 + \lambda^2 z^2}$$

and

$$s_L = \sqrt{\rho^2 + \lambda^2(L-z)^2}$$

where s and s_L are the contracted distances from an arbitrary point (ρ, z) to the A and B electrodes, respectively.

If some simple transformations are made, it is possible to write A_z in the form

$$A_z = -\frac{I}{4\pi} \left[\int_0^{z/\delta_h} \frac{\exp\,(-(1+i)u)}{u} dx + \int_0^{(L-z)/\delta_h} \frac{\exp(-(1+i)u)}{u} dx \right] \quad (44)$$

where

$$u = \sqrt{x^2 + (\rho/\delta_v)^2}$$

$$\delta_h = \sqrt{\frac{2}{\omega\mu_0\sigma_h}}$$

and

$$\delta_v = \sqrt{\frac{2}{\omega\mu_0\sigma_v}}$$

where δ_h and δ_v are, respectively, the skin depths parallel to and perpendicular to the bedding planes.

Then, inspecting A in the region between A and B, but remote from A and B so that $z \gg \delta_h$ and $(L-z) \gg \delta_h$, the result (McLachlan, 1961, p. 203, no. 200)

$$\int_0^\infty \frac{\exp(-(1+i)u)}{u} dx = K_0\left((1+i)\frac{\rho}{\delta_v}\right) \quad (45)$$

can be used to obtain

$$\mathbf{E} = E_z = -\frac{i\omega\mu_0 I}{2\pi} K_0\left((1+i)\frac{\rho}{\delta_v}\right) \quad (46)$$

which is independent of z. The z-component of current density, as a function of ρ, is $J_z(\rho)$ given by

$$J_z(\rho) = \sigma_v E_z = -\frac{i\omega\mu_0\sigma_v I}{2\pi} K_0\left((1+i)\frac{\rho}{\delta_v}\right) \quad (47)$$

Thus, the total current flowing within radius ρ is

$$I(\rho) = \int_0^\rho J_z(\rho)2\pi\rho d\rho = -I\left[1 - i\omega\mu_0\sigma_v\rho K_1\left(\rho\frac{(1+i)}{\delta_v}\right)\right] \quad (48)$$

Since $K_1(x) \to 0$ exponentially as $x \to \infty$, eqn (48) shows that the return current through the formation flows parallel to the current line AB and is largely confined to a radius of $\rho = \delta_v$. Thus, when dealing with alternating currents, one cannot assume that these currents are distributed in the same way as in the case of direct currents. Instead, the skin effect dominates for $L \gg \delta_h$, and the formation currents are constrained to a sheath surrounding the conductor connecting current electrodes A and B.

It can be seen from the analysis that reference electrode N cannot be placed at the surface if current return electrode B is at the surface, since that would result in intolerable errors on the measuring voltage due to the skin effect.

In the normal device for resistivity logging, current electrode A and measuring electrode M are on the sonde, and the other measuring electrode N is usually placed on the cable a distance above the sonde. Let the distances of measuring electrodes M and N from A be designated as z_M and z_N, and bear in mind that in actual logging tools the measuring potential is taken as the component of the total potential in phase with I. Then, using eqn (46) and assuming that z_N/δ_h and z_M/δ_h are small compared to unity, the following expression for the apparent resistivity can be obtained:

$$R_a \approx R_h - \frac{\pi}{2} \frac{z_N}{\delta_h} \frac{z_M}{\delta_h} \tag{49}$$

Subject to the condition stated, eqn (49) yields the skin effect error associated with the normal device at finite frequencies and also shows that σ_v has no effect on the measurement.

The above described distortion of the current pattern at finite frequencies must be considered also in order to understand the behaviour of electrode logging devices in difficult conditions. Such conditions arise, for example, when the sonde is just below a massive high resistivity bed. The current return path is also affected by skin effect phenomena in the metal of the armour or casing. These phenomena can also cause serious errors in the measured resistivity in the presence of a massive resistive bed.

4. A.C. INDUCTION DEVICES

4.1. Homogeneous Medium
Equation (37) can be solved readily in a homogeneous medium since it is essentially the well known classical wave equations. As is customary in the analysis of induction logging devices, the sources will be represented by magnetic dipoles. The solution of the first two equations of eqn (37) is

of the same form as eqn (38). The third equation for π_z is a little more subtle. Here, a solution of the homogeneous equation (found by setting the right-hand side of the equation equal to zero) must be added to ensure that π_z behaves properly at the origin, $\rho = 0$, $z = 0$, and vanishes for $\lambda = 1$ if $M_z = 0$. If $M_z \neq 0$, π_z should diverge near the origin.

The results for π are given in Table 1. From these formulae the fields **E** and **H** can be found according to the general equations in the previous section. There are many interesting results that can be derived from these formulae. Only one will be presented here.

The formulae are greatly simplified if one assumes the distance from the source to the field point is small compared to skin depth. The applications of these results lie in two different areas. In the first, the intent is to generalise the usual formula for induction logging tools to the case where the formation is anisotropic and has bedding planes not perpendicular to the coil axis system. Induction conductivity or resistivity logging tools utilise only the voltages induced in the receiver coil in phase with the transmitter current (resistive component). The out-of-phase (reactive) component is ignored. In the second area of application, the goal is to use the measurements to determine formation dip. Here, the most usable data come from the voltages induced in a receiver coil out-of-phase with the transmitter current.

Figure 25 shows schematically a sonde containing both horizontally and vertically oriented transmitter and receiver coils. Figure 26 indicates the relation of the axes of the coil system coordinates to those of the bedding planes. The transformation of field magnitudes between the bedding plane coordinates x'', y'' and z'' and the coil system coordinates x, y and z is affected by the same rotation matrix given by eqn (19).

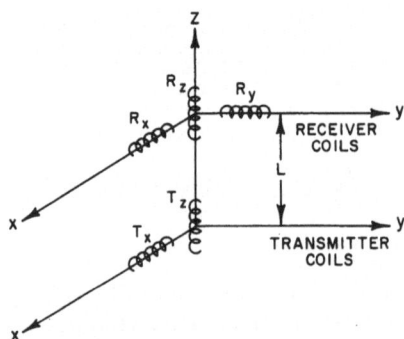

FIG. 25. Scheme of a sonde with horizontally and vertically oriented transmitter and receiver coils.

TABLE 1

HERTZ VECTOR FOR A MAGNETIC DIPOLE OF MOMENT M IN A HOMOGENEOUS ANISOTROPIC FORMATION

$$\lambda^2 = \frac{\sigma_h}{\sigma_v} \qquad r = \sqrt{x^2+y^2+z^2}$$

$$k_h^2 = i\omega\mu\sigma_h \qquad s = \sqrt{x^2+y^2+\lambda^2 z^2}$$

$$k_v^2 = i\omega\mu\sigma_v \qquad \rho = \sqrt{x^2+y^2}$$

$M_x = M, \; M_y = M_z = 0$

$$\pi_x = \frac{M}{4\pi\lambda}\frac{\exp(ik_v s)}{s}$$

$$\pi_y = 0$$

$$\pi_z = \frac{M}{4\pi}\frac{x}{\rho^2}\left(\lambda z\,\frac{\exp(ik_v s)}{s} - z\,\frac{\exp(ik_h r)}{r}\right)$$

$$\Phi = \frac{M}{4\pi}\frac{ik_h x}{\rho^2}\left[\exp(ik_v s) - \exp(ik_h r)\right.$$
$$\left. + \frac{\rho^2}{r^2}\left(1-\frac{1}{ik_h r}\right)\exp(ik_h r)\right]$$

$M_y = M, \; M_x = M_z = 0$

$$\pi_x = 0$$

$$\pi_y = \frac{M}{4\pi\lambda}\frac{\exp(ik_v s)}{s}$$

$$\pi_z = \frac{M}{4\pi}\frac{y}{\rho^2}\left(\lambda z\,\frac{\exp(ik_v s)}{s} - z\,\frac{\exp(ik_h r)}{r}\right)$$

$$\Phi = \frac{M}{4\pi}\frac{ik_h y}{\rho^2}\left[\exp(ik_v s) - \exp(ik_h r)\right.$$
$$\left. + \frac{\rho^2}{r^2}\left(1-\frac{1}{ik_h r}\right)\exp(ik_h r)\right]$$

$M_z = M, \; M_x = M_y = 0$

$$\pi_x = 0$$

$$\pi_y = 0$$

$$\pi_z = \frac{M}{4\pi}\frac{\exp(ik_h r)}{r}$$

$$\Phi = \frac{M}{4\pi}\frac{ik_h z}{r^2}\left(1-\frac{1}{ik_h r}\right)\exp(ik_h r)$$

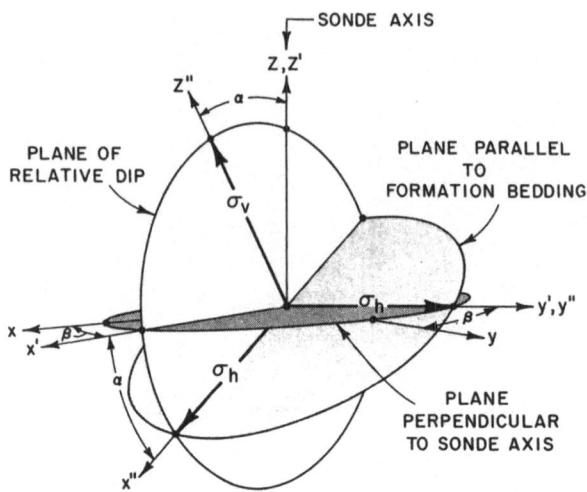

FIG. 26. Relation of coil system (sonde) coordinates to bedding plane coordinates. α is the relative dip of the bedding planes.

In a basic two-coil induction logging sonde, the source has only a
single component M_z directed along the z-axis, the receiver coil R_z is
sensitive only to H_z, and the in-phase (resistive) component of the
voltage induced in the receiver coil is used for the measurement of
conductivity. Using the general formulae of Table 1, the following result
can be deduced for the apparent resistivity obtained from such an
induction device in a dipping anisotropic bed, where the ·dip angle
relative to the tool axis is α:

$$R_a = \frac{\lambda R_h}{\sqrt{\sin^2\alpha + \lambda^2 \cos^2\alpha}} \tag{50}$$

This, perhaps surprisingly, is exactly the same result as given by eqn (24)
for electrode tools.

The second application, to an induction dipmeter, requires the use of
two transmitter coils T_z and T_x and all three receiver coils R_x, R_y and
R_z. Here it is assumed that the transmitter–receiver spacing L is small
compared to δ_h and that only the out-of-phase voltage in the receiver
coils is considered. Using the general formulae of Table 1, it is possible to
determine the relative couplings between the various coils. The relative
couplings for zero spacing between transmitter–receiver coil pairs are
designated in Table 2.

TABLE 2
RELATIVE REACTIVE FORMATION COUPLING OF A ZERO SPACING SONDE

$$T_z R_z = 1 - \frac{3}{4}\left(1 - \frac{1}{\lambda^2}\right)\sin^2\alpha$$

$$T_z R_x = T_x R_z = 3\left(1 - \frac{1}{\lambda^2}\right)\sin\alpha\cos\alpha\cos\beta$$

$$T_z R_y = 3\left(1 - \frac{1}{\lambda^2}\right)\sin\alpha\cos\alpha\sin\beta$$

$$T_x R_x = \left(\sin^2\alpha + \frac{3+\lambda^2}{4\lambda^2}\cos^2\alpha\right)\cos^2\beta + \frac{3+\lambda^2}{4\lambda^2}\sin^2\beta$$

$$T_x R_y = \frac{3}{2}\left(1 - \frac{1}{\lambda^2}\right)\sin^2\alpha\sin 2\beta$$

To convert the relative couplings in Table 2 to voltage levels, each of them must be multiplied by the instrument coefficient K:

$$K = \frac{2}{3} \times \frac{M}{4\pi} \times N_R A_R \times \omega^2 \mu_0^2 \times \frac{\sigma_h}{\delta_h} \tag{51}$$

where M is the source magnetic dipole moment, N_R is the number of receiver coil turns, and A_R is the receiver coil area.

From the results of Table 2 it is seen that the reactive components of voltage couplings $T_z R_x$ and $T_z R_y$ will be, respectively, proportional to $\cos\beta$ and $\sin\beta$. Thus the dip azimuth β can be determined directly from the relative phases of the two signals. Then, using only the amplitudes of $T_z R_x$, $T_z R_y$ and $T_x R_y$, one finds for the dip angle α

$$\tan^2\alpha = \frac{|T_x R_y|^2}{|T_z R_x| \times |T_z R_y|} \tag{52}$$

A dipmeter based on these results would be feasible if it were not for the perturbation caused by the borehole. Because of the magnitude of the borehole effect, it is unlikely that such an induction method can succeed except perhaps in shales using non-conductive muds.

4.2. Heterogeneous Configurations
Solutions of eqn (37) can be constructed for the usual cases of heterogeneous configurations of media if one restricts the geometry so that bedding planes are parallel to interfaces or so that the borehole is

perpendicular to the bedding planes. Again, only an outline of the method of solution will be offered since the details are extremely tedious.

As in the preceding section, the initial step is to represent the solutions for a heterogeneous medium in terms of cylindrical waves adapted to either the case of plane boundaries or the case of cylindrical boundaries. For π_x and π_y, terms then are introduced to account for reflection and transmission. For example, for plane boundaries, if M_x is the only non-zero component of $\mathbf{M}$, one assumes a term, in addition to the homogeneous solution, of the form

$$\int_0^\infty \frac{d\mu \mu J_0(\mu\rho)}{\beta_{v_1}} \exp(\beta_{v_1}\lambda_1 z)\Gamma_1(\mu) \text{ for } z<0$$

and   (53)

$$\int_0^\infty \frac{d\mu\ \mu J_0(\mu\rho)}{\beta_{v_2}} \exp(-\beta_{v_2}\lambda_2 z)T_1(\mu) \text{ for } z>0$$

and also adds particular solutions of eqn (37) without the cross terms in π_x and π_y.

For the case of no borehole, particular solutions for π_z are then generated by placing the expressions of eqn (53) into the third part of eqn (37) and also by adding particular solutions of this same equation without the cross terms in π_x and π_y. This latter step results in the introduction of two further functions, $\Gamma_2(\mu)$ for $z<0$ and $T_2(\mu)$ for $z>0$. Thus, for a single plane interface, four functions must be determined. The necessary conditions come from the continuity of E_ϕ, E_ρ, H_ϕ and H_ρ on the plane interface at $z=0$. If there are two plane boundaries, there are eight functions to be determined.

A procedure similar to that described above also works for the case of a borehole penetrating a thick bed, using in this case solutions and expressions analogous to eqn (53) but in forms appropriate for cylindrical geometry (i.e. employing modified Bessel functions). Again, four coefficients must be determined for a single boundary from the continuity requirements for E_ϕ, E_z, H_ϕ and H_z at the borehole wall $\rho=a$.

Proceeding this way, solutions of eqn (37) have been constructed for plane bed boundaries with no borehole and for cylindrical boundaries in a thick bed. In the application of these results, only one case is of much interest; that is when the dipoles are oriented parallel to the bedding planes. (If they are perpendicular to the bedding planes, the effect of anisotropy is nil for both the plane boundary case and the case of the borehole. Also, the dipoles (receiver and transmitter) must be oriented in the same direction for any coupling.) Some results will be obtained now for horizontally oriented dipoles in a two-coil sonde of spacing L.

First, it is straightforward, using Table 1, to obtain the coupling between the two coils in a homogeneous medium. The voltage V induced into the receiver coils is given by

$$\frac{V}{K} = \frac{2\sigma_h}{(k_h L)^2} \left[\frac{(k_h L)^2}{2} + (ik_h L - 1) + \frac{(k_v L)^2}{2} \right] \exp(ik_h z) \qquad (54)$$

where for non-zero transmitter–receiver spacing L

$$K = N_R A_R (\omega \mu_0)^2 \frac{M}{4\pi} \times \frac{1}{2L}$$

and N_R is the number of receiver turns on the coil of area A_R. For $k_h \to 0$, the real part of eqn (54) reduces to σ_v. Since V/K is to be interpreted as the apparent conductivity σ_a produced by the horizontally coupled dipoles, an expression equivalent to eqn (54) is

$$\sigma_a = \frac{V}{K} = \sigma_r + \sigma_x = \sigma_h \left[\frac{\left(\frac{L}{\delta_h} + 1 \right) \left(\frac{L}{\delta_h} + i \right)}{\left(\frac{L}{\delta_h} \right)^2} \right] \exp\left[(i-1)L/\delta_h \right]$$

$$+ \sigma_v \exp\left[(i-1)L/\delta_h \right] \qquad (55)$$

The dominant term in eqn (54) for small values of L/δ_h is

$$i \frac{\sigma_h}{(L/\delta_h)^2} \left(= \frac{2}{\omega \mu_0 L} = (\sigma_x)_0 \right) \qquad (56)$$

This term $(\sigma_x)_0$, which is independent of σ_h, represents the free-space mutual coupling of the coils. In the following discussion, this term will be subtracted from eqn (55) as it carries no sensitivity to the conductivity. This procedure is consistent with the usual practice in induction logging of cancelling the direct mutual voltage.

4.3. Determination of σ_h and λ

Figure 27 is a plot derived from eqn (55) after the correction (eqn 56) is applied. This plot allows the determination of values of σ_h and λ from measured values of σ_r (resistive induction logging signal) and $(\sigma_x)_f$

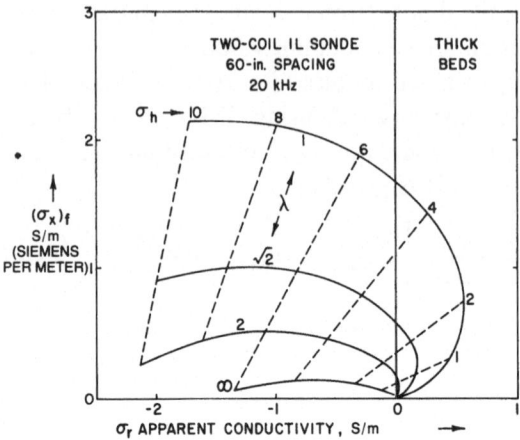

FIG. 27. Chart derived from eqns (55) and (56) which is entered with resistive and reactive components of induction logging signal to find σ_h, σ_v and λ.

(reactive signal from the formation). A device of this type could be the basis for the measurement of formation anisotropy.

This possibility of measurement of formation anisotropy has been considered further with regard to the influence of formation heterogeneity, due either to bed boundaries or the influence of the borehole. The calculations have been performed as outlined above. The presentation, however, will be given in terms of the interpretation suggested by Fig. 27. That is, after computing σ_r and σ_x, Fig. 27 is used to arrive at apparent values σ_{ha} and σ_{va} or their reciprocals R_{ha} and R_{va}. Of course, in a homogeneous medium $R_{ha} = R_h = \sigma_h^{-1}$ and $R_{va} = R_v = \sigma_v^{-1}$. The influence of heterogeneity is now to be considered.

An example of the effect of heterogeneity is shown in Fig. 28. It is seen that the device reproduces rather well the desired results. However, in another example (Fig. 29), strange artefacts are seen on the R_{va} curve reminiscent of the response of lateral electrode devices. A further more discouraging result is shown in Fig. 30. Results for finite beds also have been computed with even more discouraging results.

Another example of the effect of heterogeneity is that due to a borehole. Figure 31 shows some results obtained in this case where the ratio σ_m/σ_h (borehole conductivity divided by formation horizontal conductivity) is moderate. The results are rather encouraging. However, another example (Fig. 32) shows that the results deteriorate rapidly when σ_m/σ_h reaches larger values often encountered in practice.

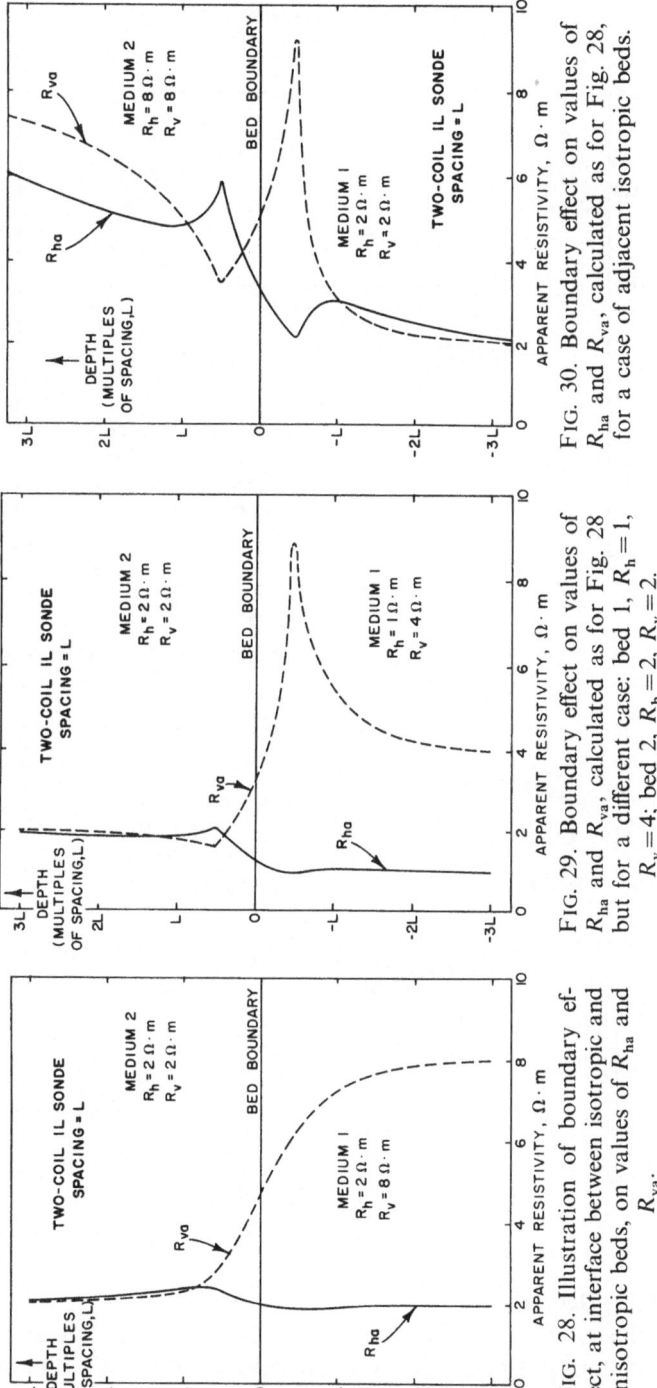

FIG. 28. Illustration of boundary effect, at interface between isotropic and anisotropic beds, on values of R_{ha} and R_{va}.

FIG. 29. Boundary effect on values of R_{ha} and R_{va}, calculated as for Fig. 28 but for a different case: bed 1, $R_h = 1$, $R_v = 4$; bed 2, $R_h = 2$, $R_v = 2$.

FIG. 30. Boundary effect on values of R_{ha} and R_{va}, calculated as for Fig. 28, for a case of adjacent isotropic beds.

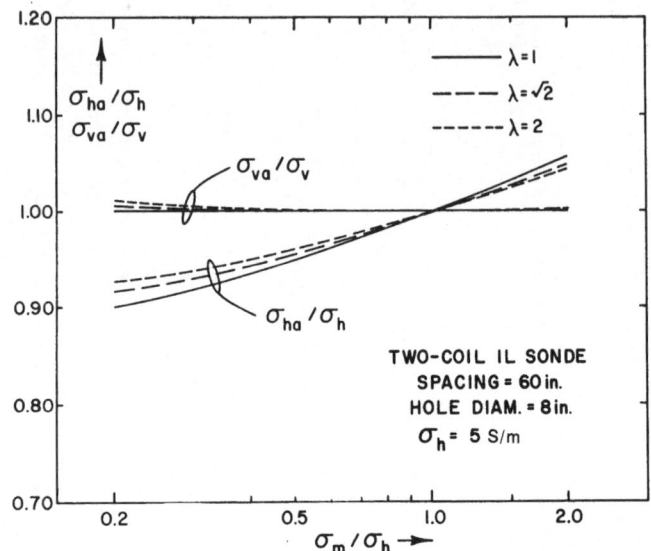

FIG. 31. Plots of σ_{va}/σ_v and σ_{ha}/σ_h versus σ_m/σ_h for a thick bed.

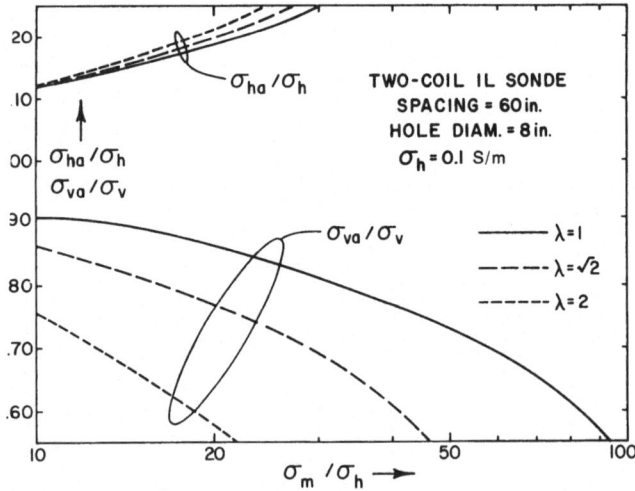

FIG. 32. Chart similar to Fig. 31.

4.4. Closing Remark

The conclusion of the studies described in this section is that, although in ideal conditions one could measure formation anisotropy with horizontal magnetic dipoles, in practice formation and borehole heterogeneities pose significant difficulties. Because of these inherent limitations, the only suitable candidate considered here for measuring formation anisotropy is the conducting pad device described in Section 2.5.

REFERENCES

AL'TGAUZEN (Althausen), N. M. (1969). Electromagnetic wave propagation in layered anisotropic media, *Izvestiya, Earth Physics*, Acad. Sci. USSR. (English edition, *Physics of the Solid Earth*, published by AGU, No. 8, 510–14.)

ASTEN, M. W. (1974). The influence of electrical anisotropy on mise a la masse surveys, *Geophys. Prop.*, **22**, 238–45.

BROWN, J. (Ed.) (1967). Electromagnetic wave theory part 1, *Proc. Symp. Delft, The Netherlands*, September 1965. Pergamon Press, Elmsford, N.Y.

CAMBELL, D. L. (1977). Electrical sounding near Yellow Creek, Rio Blanco County, Colorado, *J. Res., US Geol. Surv.*, **5**(2), 193–205.

CHETAYEV, D. N. (1962). On the field of low-frequency electric dipole situated on the surface of a uniform anisotropic conducting half-space, *Soviet Physics—Technical Physics*, **7**, 991–5. (Translated from *Zhumal Tekhnicheskoi Fizika*, **32**, 1962, 1342–8.)

CHETAYEV, D. N. (1966a). A new method for the solutions of problems in the electrodynamics of anisotropic media, *Izvestiya, Earth Physics*, Acad. Sci. USSR. (English edition, *Physics of the Solid Earth*, published by AGU, No. 4, 233–6.)

CHETAYEV, D. N. (1966b). On electromagnetic potentials in anisotropic layered media, *Izvestiya, Earth Physics*, Acad. Sci. USSR. (English edition, *Physics of the Solid Earth*, published by AGU, No. 10, p. 651–7.)

CHETAYEV, D. and Belen'kaya, B. N. (1971). The electromagnetic field of an inhomogeneous plane wave in a horizontally inhomogeneous half-space, *Izvestiya Earth Physics*, Acad. Sci. USSR. (English edition, *Physics of the Solid Earth*, published by AGU, No. 3, 212–13.)

DAVYDOV, V. M. (1970). The electromagnetic field of low-frequency ionospheric currents, *Izvestiya, Earth Physics*, Acad. Sci. USSR (English edition, *Physics of the Solid Earth*, published by AGU, No. 6, 366–73.)

FELSEN, L. B. and MARCUWITZ, N. (1973). *Radiation and Scattering of Waves*. Prentice-Hall, Englewood Cliffs, N. J.

GIANZERO, S. and RAU, R. (1977). The effect of sonde position in the hole on responses of resistivity logging tools, *Geophysics*, **42**, 642–54.

HABBERJAM, G. M. (1975). Apparent resistivity, anisotropy and strike measurements, *Geophys. Prospecting*, **23**, 211–47.

HILL, D. G. (1972). A laboratory investigation of electrical anisotropy in Precambrian rocks, *Geophysics*, **37**(6).

KELLER, G. V. and FRISCHKNECHT, F. C. (1966). *International Series of Monographics in Electromagnetic Waves, Vol. 10, Electrical Methods of Geophysical Prospecting*. Pergamon Press, Elmsford, N. Y.

KUNZ, K. and MORAN, J. H. (1958). Some effects of formation anisotropy on resistivity measurements in boreholes, *Geophysics*, **23**, 770–94.

238   J. H. MORAN AND S. GIANZERO

MCLACHLAN, N. W. (1961). *Bessel Functions for Engineers*. Oxford University Press, Oxford.

MAILLET, R. and DOLL, H. G. (1932). Sur un théorème relatif aux milieux électriquement anisotropes et ses applications à la prospection électrique en courant continue, *Ergänzungshefte für angewandte Geophysik*, 3, 109.

MITCHELL, F. R., ROBINSON, J. D., VOGIATZIS, J. P., PEHOUSHEK, Fr., MORAN, J. H., AUSBURN, B. E. and BERRY, L. N. (1972). Using resistivity measurements to determine distance between wells. *J. Petroleum Technol.*, 24, 723–40.

MORAN, J. H. and GIANZERO, S. (1979). Effects of formation anisotropy of resistivity-logging measurements, *Geophysics*, 44, 1266–86.

O'BRIEN, D. P. and MORRISON, H. F. (1967). Electromagnetic fields in an n-Layer anisotropic half-space, *Geophysics*, 32, 668–77.

RUNGE, R. J. and HILL, D. G. (1971). The role of anisotropy in ULSEL, *SPWLA 12th Annual Logging Symp.*

SCHLUMBERGER, C. (1920). *Etude sur la Prospection Électrique du Sous-Sol.* Gautheir-villars, Paris, p. 40.

SCHLUMBERGER, D. and SCHLUMBERGER, M. (1930). Sur la détermination électromagnetique du pendage des couches sédimentaires, *Comptes Rendus.*, Academie des Sciences, Paris.

SCHLUMBERGER, C., SCHLUMBERGER, M. and DOLL, H. G. (1933). *The Electromagnetic Teleclinometer and Dipmeter*. World Petroleum Congress, London.

SCHLUMBERGER, C., SCHLUMBERGER, M. and LEONARDON, E. G. (1934). Some observations concerning electrical measurements in anisotropic media and their interpretation, *Trans. Am. Inst. Mining Engrs*, 110, 159.

SINHA, A. K. (1968). Electromagnetic fields of an oscillating dipole over an anisotropic earth, *Geophysics*, 33, 346–53.

SINHA, A. K. and BHATTACHARYA, P. K. (1967). Electric dipole over an anisotropic and inhomogeneous earth, *Geophysics*, 32, 652–67.

SMYTHE, W. R. (1950). *Static and Dynamic Electricity*, McGraw-Hill, New York and London.

STRATTON, J. A. (1941). *Electromagnetic Theory*. McGraw-Hill, New York and London.

TSANG, L., BROWN, R., KONG, J. A. and SIMMONS, G. (1974). Numerical evaluation of electromagnetic fields due to dipole antennas in the presence of stratified media, *J. Geophys. Res.*, 79, 2077–80.

Chapter 7

BOREHOLE GEOPHYSICS IN GEOTHERMAL EXPLORATION

W. Scott Keys

US Geological Survey, Denver, Colorado, USA

SUMMARY

The geophysical logging of geothermal wells is still in its infancy, but advantage can be taken of the great technological advances in the field of borehole geophysics as applied to petroleum exploration. The chief differences between these applications are in the temperatures and in the unique lithology of geothermal reservoir rocks. Equipment that will make logs dependably at temperatures greater than 200°C is not widely available, and the interpretation of data obtained at those temperatures in hydrothermally altered igneous and metamorphic rocks is still more an art than a science. Another unique characteristic of most geothermal reservoirs is the importance of fractures in controlling the production of hot water and steam. The log analyst faces a real challenge if borehole geophysics is to play a major role in the exploration, evaluation and development of the world's geothermal resources in hydrothermally altered, fractured rocks.

1. INTRODUCTION

Borehole geophysics is broadly defined as the science of making measurements in wells related to their construction and to the chemical and physical characteristics of the surrounding rocks and their contained fluids. For purposes of this brief chapter, a knowledge of the principles and terminology of borehole geophysics is assumed. Few geothermal reservoirs are known well enough that drilling and geophysical logging are not considered to be exploratory in nature. Geophysical well logs have been widely utilised in the petroleum field for decades; and, more recently, in mineral and groundwater exploration and development. Geothermal well logging is, however, still in its infancy because neither the instrumentation nor interpretive techniques are highly developed. Wells with temperatures higher than 200°C cannot yet be logged dependably, and because the borehole conditions and geothermal aquifers are poorly understood, the interpretation of logs of most geothermal wells is subject to many errors. Although there are many geothermal wells at temperatures less than 200°C, where logging may be more dependable, these have not yet proven economic for power generation.

Only the very deepest oil wells attain temperatures in excess of 200°C so the economic incentive to develop probes, cableheads and cables for this environment has been lacking. The US Geological Survey began to develop instrumentation for such conditions in 1973, and a workshop sponsored by the US Department of Energy in 1975 marked the beginning of a major and more widespread effort (Baker et al., 1975). Recommendations of this workshop included a list of parameters that should be measured in a geothermal well. That list, in order of priority, is still valid although subsequent groups have made minor modifications:

1. temperature;
2. formation pressure;
3. flow rate;
4. fracture system (location, orientation, permeability, etc.);
5. fluid composition (pH, dissolved solids and gases, redox potential);
6. permeability;
7. porosity (interconnected and isolated);
8. formation depth and thickness.

Other parameters of lower priority were thermal conductivity or resistivity, heat capacity, lithology and mineralogy, P and S wave velocities and hole size. This chapter describes which of these measurements can be made in geothermal wells and briefly explains log interpretation methods for these less well known environments.

Table 1 is a compilation of geothermal reservoir parameters that may be obtained from well logging, the logs that may provide the required

TABLE 1

Required parameter	Logs that might provide data	Operational status of 250°C probes	Log interpretation
Temperature	Temperature	Relatively reliable	Calibration errors common
Pressure	Pressure transducer	Relatively reliable	Multiphase problems in flowing wells
Flow rate	Impeller flowmeter	Unreliable	Requires hole diameter information
Fractures	Acoustic televiewer	Unreliable	Interpretation straightforward, except for aperture
Fractures	Dipmeter	Not available?	Interpretation questionable
Fractures	Caliper	Fairly reliable	Interpretation of fractures difficult
Fractures	Resistivity and porosity	Reliable?	Fracture information very questionable
Fluid composition	pH, specific conductance, redox, SP	Only SP available?	Classic assumptions of rock and fluid properties for SP may not apply
Permeability	No direct measurement	——	Flowmeter and temperature logs provide relevant data
Porosity	Acoustic velocity	Reliable?	Calibration errors, no data in steam
Porosity	Gamma–gamma	Reliable?	Major hole diameter effect, calibration errors
Porosity	Neutron	Reliable?	Hole diameter effect; calibration errors and hydro-thermal alteration
Lithology, depth and thickness	Gamma (also resistivity and porosity logs)	Reliable?	Redistribution of radioisotopes, some anomalies exceed normal scales

data, the present status of the equipment and the interpretation techniques. All of the probes are dependent on the use of high temperature cable and cableheads. New multiconductor cables that can be used at 275°C and single conductor cables with insulation that will not break down at higher temperatures are available. Cableheads are, however, much less dependable at these temperatures. The reliability factor in Table 1 is relative, differs between companies, and is improving rapidly owing to much research aimed at improving high temperature logging equipment (Veneruso and Coquat, 1979).

It is beyond the scope of this report to discuss the availability of high temperature equipment, but much of it can be obtained on a service basis from international well logging companies. A few instruments of this type are in use only by US Government groups at the present time. Note, however, that moderate temperature geothermal wells, between 150°C and 200°C, can be logged with considerable reliability although log interpretation problems still exist. Glenn et al. (1980b) summarise the results of logging a number of geothermal exploration wells in the Basin and Range Province of the western US. Some of these wells were intentionally cooled by circulation; as a result, the average maximum temperature recorded during the logging of 23 wells was only 142°C. In contrast, the US Geological Survey recorded a temperature of 261°C while logging a flowing well in the Roosevelt Hot Springs field.

The lack of reliable high temperature logging equipment does not necessarily eliminate the possibility of geophysical logging. In some wells, it is possible and economically feasible to circulate cold water to lower the temperature in the hole. The cooling technique has several limitations: a drill rig must be on the hole and large quantities of cold water must be available; it is not possible to cool some dry steam wells or those with high pressure at the surface; the cold water may cause spalling, chemical reactions, or other undesirable changes in the reservoir rocks, and these unknown changes may affect the response of logs.

2. PURPOSES OF GEOTHERMAL WELL LOGGING

The most important objective of geothermal well logging is the evaluation of the well as a potential producer and of the reservoir for additional exploration by drilling. Wells drilled in a high temperature environment cost two to three times as much as oil wells of comparable depth at lower temperatures. Often the decision to abandon or continue development of a thermal anomaly is based on the evaluation of a single well. An understanding of the geohydrology of a potential reservoir is also essential to help make decisions on drilling additional exploration wells, as aids in their location, and to increase the likelihood of success of each well. An example would be the use of log information on the location and orientation of a producing fracture zone to directionally drill another well to intercept that zone.

Reservoir evaluation generally requires information on the location of producing intervals, their relative contributions, the type of porosity and the degree of permeability. Porosity derived from logs has been used to determine whether or not the size of the resource warrants further exploration or construction of a power plant. Economic decisions of this importance must be based on accurately calibrated logs. Logs are also used to make important decisions on the construction of geothermal production and injection wells. These decisions include where to set casing, cement and pumps, and where to stimulate production by hydraulic fracturing.

Almost all oil wells drilled anywhere in the world are logged geophysically. This is not true of all geothermal wells, and the decision whether to log or not is usually based on a cost-benefit analysis. Rigby and Reardon (1979) describe a study of geothermal well logging that identifies many millions of dollars of savings that might be realised through improved application of borehole geophysics. Their data show that wells ranging from 1·5 to 2·5 km deep cost on average $1–2·5 million each. The potential benefits of improved applications of logging are reduced costs of well drilling, flow testing and reinjection, improved drilling success, and reduced expenditures on unsuccessful exploration wells. Certainly, savings of this magnitude will not be realised until improved equipment for logging high temperature wells becomes readily available at lower cost and log interpretation techniques approach the reliability found in the petroleum industry. Until these problems are solved, geothermal wells will not be universally logged.

3. INSTRUMENTATION AND LOGGING OPERATIONS

High temperature is the chief cause of failure of geothermal logging equipment. Reliability seems to decrease as an exponential function of temperature increase. Failure of materials such as insulation and seals is the most important cause, but operator error due to lack of experience in the logging of high temperature wells is also a factor. Many geothermal wells in the temperature range 150–200°C have been logged, but probe failures were frequent and some of the data obtained are questionable. A few logs have been made at temperatures above 250°C. Probes, electronic components and materials frequently fail at temperatures much lower than their rating. It has been found that laboratory testing of high temperature probes is not sufficient to ensure success in a well. The combined effects of high temperature and pressure, along with exposure to steam, mechanical shock and oil or grease in some probes may only be attained while logging. Thus, the user of geothermal logs should be forewarned that he may not obtain the data needed.

Under some conditions, particularly in a flowing well, the fluid column may contain unknown proportions of steam and water. The response of most logging tools is based on the premise of a constant borehole fluid. In a multiphase system, deflections on the logs may be due to changes in the borehole fluid. Furthermore, most acoustic and resistivity logging devices require a coupling or conductive liquid in the well, and therefore will not work in steam. The effect of interstitial steam on porosity measurements will be discussed later. In some geothermal reservoirs, such as the Imperial Valley in California, and Cerro Prieto in Mexico, reservoir fluids may be highly corrosive to most materials used in logging equipment.

The introduction of logging probes to flowing geothermal wells under pressure is difficult. A lubricator and riser pipe assembly capable of withstanding the temperature and pressure must be used. In addition, a line cooler is required if high temperatures exist at or near the well head. Logging cable that is hot cannot be spooled on a drum because contraction during cooling may crush the drum. It may be necessary to use a vertical line cooler above the lubricator to cool the cable before it goes over the upper sheave. If hot cable is bent over a sheave while under considerable tension, the insulation may be extruded. Line coolers can be made of a piece of pipe with rubber packers at each and through which cold water is circulated. It is also advisable to use sheaves of the order of 1 m in diameter to further reduce the possibility of damage to the cable.

4. PLANNING THE LOGGING PROGRAMME

Proper planning of geothermal log suites, log calibrations, and correlation of geologic, drilling and logging information is required (Benoit *et al.*, 1980). As for any other logging programmes, those for geothermal wells should be planned before drilling is started. If results from borehole geophysics are important, the well should not be drilled too large or it should be logged before reaming. Excessive deviation from the vertical makes logs requiring centralised probes more difficult to obtain. Logs should be run before casing is set if the depth interval to be cased is likely to assume any importance in the future. It is necessary to budget for more rig standby time for logging a high temperature well than a comparable oil well.

The necessity to contact the logging service company very early about well and environmental conditions is usually overlooked. Probes designed to overcome some of the corrosion problems are in limited supply and sometimes modifications to improve chances of success can be made if sufficient notice is given. Unusual well conditions in this category include high pressure, deviation, large well diameter or washouts, junk or obstructions .in the hole and corrosive fluids. Prior notice of unusual lithologic conditions may also produce better results. Many geothermal wells are drilled in rock types unfamiliar to oil well logging engineers. Such conditions as very high gamma radiation in acidic igneous rocks, bulk density greater than 3 g/cm^3 in metamorphic rocks and very high porosity may produce off-scale logs unless planned for ahead of time. It is also highly desirable to pay the extra cost of field digitising the raw data. In some commercial logging trucks, both the analogue and digital records may be processed by the use of a function that does not apply to geothermal reservoirs. An example is the erroneous use of a grain density of 2·65 g/cm^3 to calculate a field porosity curve from a gamma–gamma log in basic igneous rocks. Often the customer must specify either a sandstone or limestone matrix for these real-time calculations, and neither may apply.

If reasonably accurate quantitative results are desired from any geophysical well log, some core is essential for log calibration. This is particularly true of geothermal reservoir rocks that commonly present unusual characteristics such as hydrothermal alteration or intense fracturing. If there are other wells in the area to guide the selection of coring intervals, then core should be taken of each different and important lithologic unit. In a wildcat well, a core run should be made each time the cuttings indicate a significant change in lithology. Interpretation of the logs will then provide the basis for selecting samples from the core for laboratory analysis. Only cores that are representative of lithologic units thick enough to provide true log response should be analysed. One reason for this is to minimise the effect of depth errors that are common in both core and logs. Due to space limitations, it is not possible to go into all of the factors affecting the use of core for the calibration of logs, but the log analyst must be aware of these. Several analyses should be made of what is apparently the same lithology in order to establish the variability in such parameters as porosity or bulk density. Furthermore, changes in porosity and permeability may occur when the core is analysed at temperatures or pressures that are lower than those in the reservoir, when permeability is measured with non-reservoir fluids, or if the core is allowed to dehydrate. Even though some of these factors may be corrected in the laboratory, it is almost impossible to return to true reservoir conditions. Unfortunately, representative cores are almost never recovered from open fracture zones which are one of the most important sources of geothermal fluids. Drill cuttings are a poor substitute for core because of return lag, depletion or concentration of some mineral components, and grinding of particles to a size smaller than the average grains or crystals in the rock (Glenn *et al.*, 1980*a*). In addition, cuttings may not be recovered because of zones of lost circulation which are common in geothermal wells.

5. LOG INTERPRETATION PROBLEMS

5.1. The Downhole Environment

The interpretation of geophysical logs is usually divided into two overlapping categories: qualitative and quantitative. Qualitative interpretation provides information on lithology, depth and thickness, and permits correlation of rock units. Quantitative interpretation may provide data on the amount and quality of fluid in place and, possibly, the relative magnitude of hydraulic conductivity. Although quantitative log interpretation is widely used to evaluate petroleum reservoirs, accuracy is questionable in most geothermal reservoirs. The values calculated from logs for such important parameters as porosity, bulk density, 'permeability index' (relative magnitude of hydraulic conductivity) and water quality seldom compare well with laboratory analyses or well test data. The chief reason for this is the wide and unknown variation in hole conditions and rock types. Oil field log interpretation techniques have been successfully applied to the Cerro Prieto geothermal field in Mexico where a sand–shale lithology is predominant (Ershagi *et al.*, 1979). Even in this fairly well known reservoir matrix the lack of accurate calibration data, including temperature, and the need to better understand the effects of hydrothermal alteration have limited the quantitative results.

Even the qualitative interpretation of logs of geothermal wells is subject to considerable error. Log response to lithology is not unique so that information from core and cuttings is usually necessary to develop a functional interpretation scheme in each area. The so-called porosity logs, neutron, gamma–gamma and acoustic velocity, respond to the type of rock matrix as well as to porosity. Matrix response is well known only for limestone, dolomite, sandstone and shale because of many years of research in the petroleum industry on the most common reservoir rocks. In contrast, the response of logs to such rock types as quartz monzonite, biotite schist and hydrothermally altered tuffaceous sediments is relatively unknown. For this reason, it is a questionable practice to accept, at face value, porosity scales on logs of geothermal wells that are based on calibration carried out in sedimentary rocks. In low porosity igneous rocks, response of the three 'porosity' logs may be more related to matrix mineralogy than to percent pore space. If the pore spaces are filled with steam or a mixture of steam and water then porosity scales on the neutron, gamma–gamma and acoustic velocity logs will not be correct unless calibrated for steam. The neutron log will provide low porosity values because of a decrease in hydrogen content. The gamma–gamma log will give high porosity values because of a decrease in bulk density. Therefore, the divergence of these logs may be evidence of the presence of steam. It is unlikely that any useful data will be obtained from an acoustic velocity log under these conditions.

Hole diameter effects introduce a serious error in the quantitative interpretation of many porosity logs. Caliper logs that might be used to correct such errors are often of poor quality in high temperature wells. The construction of calibration pits using blocks of igneous rocks and research on log interpretation in igneous and metamorphic rocks will aid efforts to locate sites for the storage of radioactive waste as well as geothermal exploration. The state of the art in geothermal well log interpretation and recommendations for further research are described by Sanyal et al. (1980).

The most significant effect of high temperatures on geothermal well logs is thermal drift of instrument response. Instrument drift can usually be recognised by a consistent slope to a log, such as a caliper, on which deflections in either direction are superimposed. Drift may sometimes be identified by comparison with a temperature log of the well, but since it may be caused by combined effects of changing temperature on the cable and on the electronics, correction may be impossible. Dewar flasks are used in some probes to thermally insulate the electronics, and drift in these instruments will be related to time rather than temperature alone. Recording probe response for a period of time while stationary in the hole may be helpful in correcting for thermal drift. The difference between temperature in the hole and laboratory temperature at which core analyses are made can also have an effect on log interpretation guided by core. The basic measuring principles of such logging devices as resistivity, neutron and acoustic velocity probes are temperature dependent, but the magnitude of this error on actual log response is unknown.

5.2. Well Construction

Every geophysical log is affected to some degree by the character of the well in which it is made. In some cases, corrections can be made for these extraneous effects if they are known. Well diameter is the most significant source of error; consequently an accurate caliper log is absolutely essential to the correct interpretation of other logs. Large diameter washouts and hole diameter increases due to fractures or alteration are common in geothermal wells. Where the diameter exceeds known log correction factors, it is best to eliminate these depth intervals from quantitative consideration.

Many geothermal wells are intentionally deviated. Hole deviation in excess of 10° from the vertical is quite common and can cause errors in data interpretation if not recognised. The US Geological Survey has successfully made sophisticated logs in holes in igneous rock deviated as much as 43° from the vertical, but a number of problems were encountered. Sometimes it is not possible to properly centralise probes when this is essential to correct response. Furthermore, probes are likely to hang up on the way down deviated holes. This produces major depth errors in logs made while the probe is going down the hole. For this reason, it is preferable to log on the way up except for temperature and fluid conductivity.

Logs are frequently made on the trip down geothermal wells because of the risk of probe failure as temperature increases. Tool hangups can often be recognised by straight line intervals on the log and can usually be detected early by close attention to the line weight indicator. Hangups are most likely to occur in irregular sections of the hole shown on the caliper log.

5.3. Calibration and Quality Control

Geophysical logging of geothermal wells is extremely expensive and quality control by the geoscientist in charge of the operation is essential if useful logs are to be obtained. As previously mentioned in Section 4, early contact with the logging equipment operators is necessary. Not only should all needs and anticipated hole conditions be described, but information should be obtained on calibration procedures carried out before the equipment arrives at the well. All log calibration should be done in models or pits that represent in-hole conditions as closely as possible. Although several sets of logging calibration pits are available to the public in the United States, and all major logging service companies as well as many oil companies maintain private calibration pits, none of these is designed for geothermal conditions. Mathews (1980) describes three pits presently under construction for this purpose that will be available for public use at the US Geological Survey log calibration facility in Denver, Colorado. Each of the pits will be constructed of blocks of igneous rocks 2·4 m in diameter by 6 m deep. The rock types are a fine-grained and a coarse-grained granite and an altered diorite; artificial fractures are included. A large variety of laboratory measurements are being made on the cores.

After proper log calibration is assured, it is essential to insist on the use of field standards or calibrators to check probe response before and after each log. The organisation which will make the logs must also be informed of the horizontal and vertical log scales and types of recording that are expected. Although it adds to the cost, real-time field digitisation of logs is highly desirable. Digital data recorded in the field are more accurate and less costly than data digitised later from analogue records. Field digitising also permits the correction of some types of operator errors. The data to be recorded in both analogue and digital format should always include the raw, unprocessed signal(s) from the probes. Many logging companies process data as it comes up the cable to provide corrected or compensated logs. Such logs may be useful; however, they result from calculations in a black box known as a function former. Because of the present lack of knowledge about log response in geothermal reservoir rocks, records of the raw data are essential for further processing. Commercial gamma–gamma, gamma, caliper and resistivity logs made in geothermal wells have been observed to go off scale without back-up curves. Prior warning of the expected conditions and field digitising of the data may prevent the loss of important information. Some logging trucks have an on-board computer which may provide on-site processing, redisplay, recalibration and a logging job history not available with analogue systems.

The geoscientist in charge, who should be a log analyst, must observe and question all logging operations in order to obtain the best data possible. The time to request a rerun because of possible malfunction or improperly selected scale is when the probe is on the end of the cable. A charge will be made for later reruns. A short rerun of each log will not increase accuracy but may increase confidence that the probe was responding properly. It is also essential that all in-hole conditions, probe parameters and logging scales be properly recorded on each log.

Depth errors seem to be more common and of greater magnitude on geothermal logs than logs of other wells. Errors up to 7–10 m in a 2000 m hole have been reported. Depth discrepancies occur between logging companies and between logs run by the same company. Such errors make correlation between logs very difficult, and log calibration with core subject to great error. Most equipment operators will reference the depth on logs to the kelly bushing on the drill rig unless told otherwise.

Often it is more useful to reference to a permanent datum that will be used for testing and logging after the rig is gone. The depth reference used by the driller is important when comparing core to logs. Depth inconsistencies are often due to operator error which must be recognised at the time. Consistent errors such as cable stretch can often be corrected by use of the computer. Oil well logging companies use magnetic cable markers as a reference to manually enter periodic corrections on the log. This sudden correction at unknown depth may cause more problems in using the logs than a regularly distributed error. Depth errors can often be recognised on logs and partially corrected using identifiable anomalies on several logs as markers. Significant hole diameter increases may be detectable on most logs in a suite and provide such markers. In Fig. 1 the hole enlargement at a depth of approximately 225 m is clearly seen on the neutron, resistivity, caliper and gamma logs.

6. RESERVOIR PARAMETERS

6.1. Lithology

The first information usually needed from an exploration or development well is the rock types and the depths at which they have been penetrated. Unfortunately it is not possible to make accurate interpretations from borehole geophysical data in a new area without some background information on lithologies to be expected. This can come from other wells in the same geologic environment, information extrapolated from the outcrop or samples, or core from the exploration well. A log analyst with knowledge of the principles of log response and petrophysics can

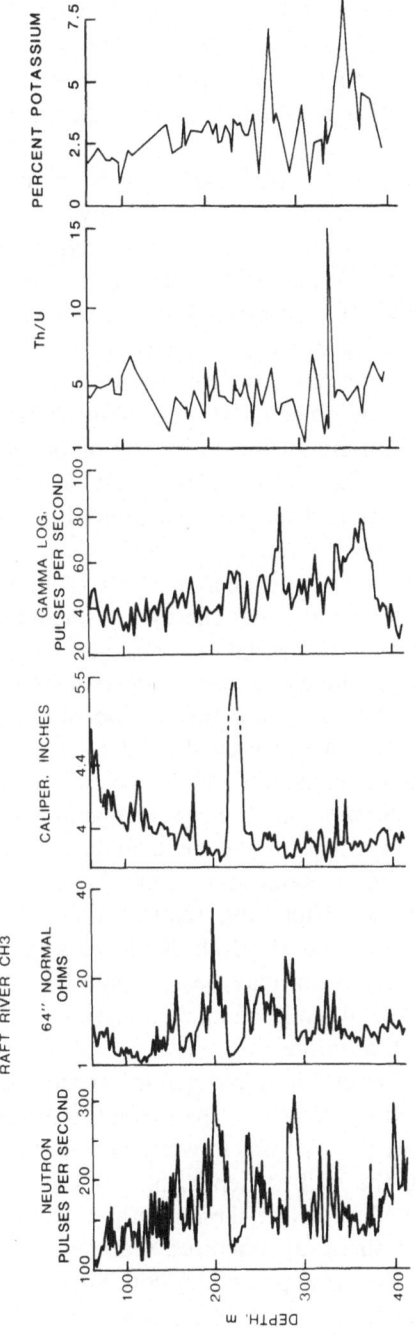

FIG. 1: A composite of logs and core data showing a hydrothermally altered zone containing several hot water producing fractures, Raft River, Idaho.

make preliminary interpretations of lithology by studying the depth-correlated logs as a group; this is synergistic log interpretation in its crudest form. The combined response of several logs is likely to provide a more positive identification of a rock type than any of the individual logs.

A more sophisticated synergistic study of logs is reported for a geothermal well in a complex volcanic sequence (Sanyal et al., 1979). Using information from drill cuttings, frequency distribution plots and cross plots, three basic rock types were distinguished: rhyolite, volcanic ash and basalt. By combining the interpretation of the three porosity logs and a gamma log, the well was divided into 36 zones with different lithology or pore geometry. Using porosity, temperature and caliper logs, several permeable zones were identified in basalt at a depth greater than 2700 m. The common rule of thumb for the interpretation of gamma logs in igneous rocks is that the acidic rocks are more radioactive than the basic rocks. In contrast, density and magnetic susceptibility are usually greater in basic rocks than in acidic rocks.

The identification of hydrothermally altered zones is an important problem in the exploration for geothermal energy and metallic mineral deposits. The Raft River, Idaho, geothermal field provides an example of synergistic log interpretation of altered zones. Use of several logs permits the identification of biotite schist or gneiss and the location of intervals where the biotite has been hydrothermally altered to chlorite. Concentrations of the mica minerals can usually be recognised by an indication of high bulk density on the gamma–gamma log and contrasting high apparent porosity on the neutron log. Resistivity is likely to be less than in other metamorphic rocks or granite, and a smooth hole indicated by the caliper log reduces the possibility of extraneous effects on other logs. In the Raft River reservoir, chlorite derived from the alteration of biotite may be recognised by a reduced gamma response, probably due to the leaching of potassium and possibly uranium and thorium from the biotite.

Figure 1 is a composite of several logs and gamma spectral data from a core hole in the Raft River reservoir (Keys and Sullivan, 1979). Temperature logs made while the well was flowing indicate that most of the hot water was entering the well at depths of 337 and 347 m. The caliper log has two sharp deflections at these depths and the acoustic televiewer substantiates that these are fractures. Gamma spectral data indicate an interval of relatively low potassium concentration bounded by high concentrations at approximate depths of 285 and 370 m. X-ray analyses of core show this to be a hydrothermally altered zone that is enriched in quartz and analcime, and low in potash feldspar and clay. Alteration was probably produced by hot water moving along fractures. Conventional neutron and 2 m normal resistivity curves are included in Fig. 1 to demonstrate that they do not indicate the location of either the altered zone or the producing fractures.

Borehole gamma spectrometry has also been used in New Hampshire to identify altered zones that had a similar migration of potassium (Keys, 1979). West and Laughlin (1976) describe the use of spectral gamma logs to identify felsite dikes in a geothermal well by concentrations of natural radioisotopes near their contacts. They also report that the upper 50 ft of the pre-Cambrian at this location is distinguished by a leaching of potassium. The evidence for self-sealing of permeable zones in geothermal systems is very strong. Keith *et al.* (1978) describe self-sealing in Yellowstone Park, and point out that potassium feldspar of hydrothermal origin may indicate the upflow channels of the hottest water. Therefore, gamma spectral data on the distribution of potassium may shed light on the genesis and plumbing of a geothermal system. Thus borehole gamma spectrometry seems to be a useful geothermal logging technique for lithologic purposes and, as discussed later, for fracture identification.

After visual correlation, computer cross plotting is the next logical step in determining lithology because it makes the process of comparing the responses of many logs somewhat easier. Figure 2 is an example of a

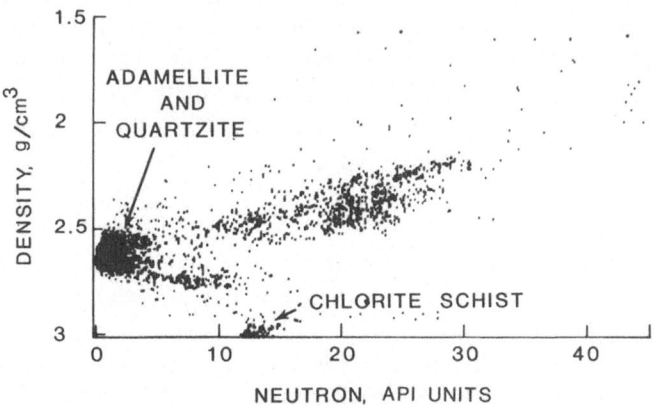

FIG. 2. A computer generated cross plot of gamma–gamma density and neutron logs for the depth interval 1286–1825 m, well RRGE-2, Raft River, Idaho.

cross plot of density and neutron logs from a producing geothermal well in the Raft River reservoir. A wide variety of igneous, metamorphic and sedimentary rocks were first identified from cuttings and a few pieces of core. The adamellite and quartzite are clearly distinguished from the chlorite schist and overlying altered tuffaceous sedimentary rocks. The linear trend of data points represents a gradual decrease in porosity and increase in bulk density with depth. Note that part of the cluster of data points representing chlorite schist is off-scale at 3 g/cm^3. The gamma–gamma log from which the data were obtained was off-scale because it was recorded at a span commonly used in oil wells. The off-scale data were lost because the logs were not digitised in the field.

One of the most useful cross plots in igneous rocks is gamma versus neutron, which is not widely used for interpreting oil well logs. In many intrusive igneous rocks, with very low porosity, the neutron probe responds mostly to chemically bound water in mica minerals. The range of gamma radiation in such intrusives is very wide and can be related to rock type as well as to subsequent alteration. An example would be a change in the content of orthoclase which contains significant quantities of the natural radioisotope potassium-40. The content of orthoclase decreases from acidic to basic igneous rocks and gamma logs can be used to distinguish granite from diorite, for example. Biotite is also radioactive due to potassium-40 and trace amounts of uranium and thorium. Although gamma logs are useful, it is clear that gamma spectral data will provide more definitive answers to log interpretation problems.

Benoit *et al.* (1980) describe the interpretation of a suite of logs of a geothermal well at Desert Peak, Nevada. In spite of some equipment malfunctions and significant errors due to a greatly enlarged hole, they were able to obtain useful lithologic information. The most useful logs were the gamma and the induction, which is usually not recommended in high resistivity rocks. In this case however, the induction log provides positive identification of the phyllites.

6.2. Porosity

Porosity is the reservoir characteristic that determines the amount of fluid in storage and it is related to thermal conductivity. The reservoir engineer wants to know effective porosity—the volume fraction from which fluids can be drained. Unfortunately, logs such as neutron and gamma–gamma respond to total porosity, which may include pore spaces that are not interconnected. In some igneous rocks, isolated pores are common, e.g. gas vesicles in basalt. Even more serious log interpretation problems are caused by variations in mineralogy and by chemically bound water. None of the so-called porosity logs measures porosity directly. Response of the neutron log is primarily due to hydrogen with lesser effects from rare elements with a high capture cross-section. Chemically bound hydrogen in clays and other hydrous minerals, such as zeolites, is common in hydrothermally altered rocks. The presence of these minerals will produce values for porosity that are too high, and the error becomes much more significant at the low porosities common in igneous rocks.

A further error may be introduced by the possible non-linearity of neutron log response in rocks with low porosity. The presence of steam in pore spaces or the wellbore can produce errors in the interpretation of neutron and gamma–gamma logs, but this effect may be useful. From well logs made at the Geysers, the world's largest dry steam field, Ehring et al. (1978) reported the identification of intervals of steam entry or exit. Identification was based on a computer plot of the difference between apparent porosity from both neutron and gamma–gamma logs, and porosity from a water normalised gamma–gamma log. If the pore space or fracture is filled with water, the two porosities will be equal. If some of the porosity is steam filled, then the neutron porosity will be lower. A study of logs at the Cerro Prieto field suggests that hydrothermally altered zones with lower porosity may be identified by the response of the neutron log and by higher resistivity (Ershagi, 1980).

The response of the gamma–gamma log is largely due to changes of electron density which should correlate with changes in bulk density if the atomic number and atomic mass are approximately equal. The corrected bulk density from the log is then entered in the following equation:

$$\phi = \frac{\rho_m - \rho_b}{\rho_m - \rho_f} \tag{1}$$

where ϕ = porosity, ρ_m = matrix or grain density, ρ_b = bulk density from gamma–gamma log and ρ_f = fluid density. It can be seen that the calculated porosity in geothermal wells may be subject to three sets of errors. The correct ρ_m may be unknown, although 2·65 gives reasonable results in some acidic igneous rocks. The plot in Fig. 2 shows that adamellite and quartzite would have a ρ_m of approximately 2·65 at zero porosity. ρ_b is subject to hole diameter and Z/A ratio errors, and ρ_f may be less than $1\,g/cm^3$ if steam is present, or greater than $1·1\,g/cm^3$ in brine. Considering these errors and possible hole diameter effects it is not surprising that porosities calculated in an unknown environment may not be close to true values. In many commercial logging trucks, this calculation is made automatically and a 'porosity' curve is plotted. This approach is acceptable only if the raw bulk density data are recorded simultaneously for later correction. The most accurate bulk density logs are made with probes employing two detectors at different spacings from the source.

The empirically derived time-average equation is the basis for calculation of porosity from acoustic velocity logs:

$$\phi = \frac{\Delta T_1 - \Delta T_m}{\Delta T_f - \Delta T_m} \tag{2}$$

where ΔT_1 = acoustic velocity from logs, ΔT_m = acoustic velocity of matrix and ΔT_f = acoustic velocity of fluid. The equation is useful only for rocks with uniformly distributed intergranular pore spaces, and it is considered to provide reliable values for effective porosity in such rocks. Many references in the literature describe the derivation of secondary porosity from the numerical difference between acoustic porosity and neutron or gamma–gamma porosity. Although this interpretation technique has apparently been successful in a few areas, it cannot be used with any confidence in an unknown environment. Acoustic logs of acidic igneous rocks at a number of localities in the United States and Canada have demonstrated surprisingly consistent transit times of 5·9–6·1 km/s. Porosities in these rocks are reported to average less than 1%, so these values may approximate matrix velocity. Slower velocities in such rocks would suggest the possibility of changes that would favour the production of geothermal fluids; however, evidence from other logs would be essential to corroborate such an interpretation.

The similarity between the equations for calculations of porosity from gamma–gamma and acoustic logs is obvious. The general equation for the response of a number of logs takes the form

$$L = L_f(\phi) + L_m(1 - \phi) \tag{3}$$

where L denotes a measured physical value from a geophysical log, such as acoustic velocity. The expression merely states that the log response value in a well is the sum of that value for the interstitial fluid times the fractional porosity, plus the value for the rock matrix times the fraction of rock matrix present. If the rock matrix has more than one mineral component (N), or the fluid has several components, such as water, steam or carbon dioxide, then N equations can be solved for N

unknown matrix values. In practice, the three porosity equations for neutron, gamma–gamma and acoustic velocity logs permit the solution for porosity and two matrix components. The typical cross plot represented by Fig. 2 permits the solution for porosity and one type of lithology. Obviously, if nothing is known about the rock types intersected by a geothermal well, then the matrix response cannot be identified and the porosity will probably be in error. Physical characteristics such as resistivity, acoustic velocity, neutron capture cross-section, etc., have been measured in the laboratory for a number of minerals, but these data are difficult to extrapolate to borehole conditions and to rocks with unknown concentrations of minerals.

7. PERMEABILITY

7.1. Fractures

Permeability, or hydraulic conductivity, is the most needed result from borehole geophysics, yet no log measures permeability directly. The often quoted relationship between porosity and permeability is valid only in a few sedimentary rocks having primary intergranular porosity and containing no clay or shale in the pores. Although intergranular porosity and permeability are thought to be important in a few geothermal reservoirs, such as the Imperial Valley in California, fractures are probably the most important type of conduit for geothermal fluid on a worldwide basis. In some reservoirs, such as the Geysers in California, a single fracture may provide sufficient production to make a well economic. This is why information on the location, orientation and aperture of fractures is so important in geothermal exploration. If the location, strike and dip of a producing fracture or set is known, then a well can be directionally drilled to increase the possibility of intersecting that fracture set. Furthermore, if the state of stress is known, it may be possible to create a hydraulic fracture to intersect a producing system.

The interpretation of fracture location, orientation and width from geophysical logs is still quite ambiguous in spite of much study on the subject. Sometimes the porosity logs indicate the presence of fractures if the porosity change is large enough, but substantiating evidence is necessary. Fractures partly filled with quartz crystals may cause only a small porosity deflection. If a fracture is large enough, it will produce a deflection on all logs. High resolution caliper and resistivity devices can provide fracture data, but results are inconsistent. Figure 3 shows a high

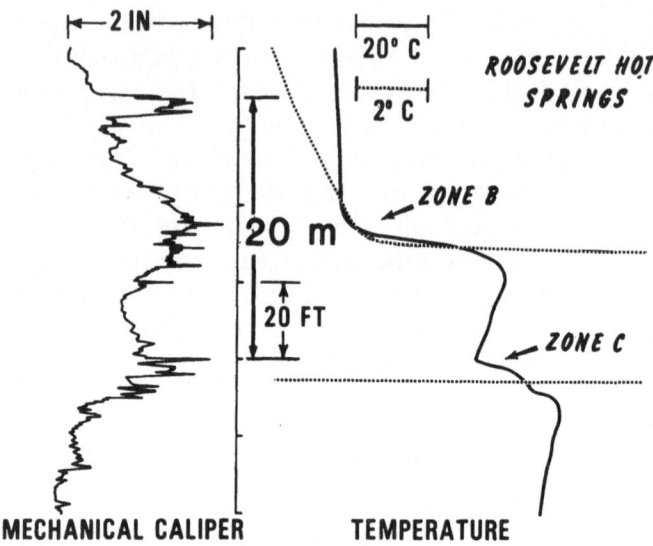

FIG. 3. Caliper and temperature logs of a well in the Roosevelt Hot Springs reservoir, Utah. The temperature log indicates that zones B and C produce hot water.

resolution caliper log and a temperature log of a geothermal well. The temperature log indicates that two of the three rugose intervals on the caliper log produce hot water. The dipmeter is advertised as a fracture finder, and the raw analogue data can provide excellent fracture information. The so-called fracture finder computer programs for analysis of dipmeter data do not, however, compare with the accuracy of programs for the calculation of strike and dip of beds in sedimentary rocks. It is very difficult to write a program that will correlate the resistivity deflections due to a complex system of intersecting fractures. Streaming potential, displayed as noisy intervals on spontaneous potential logs, has been used to locate producing intervals in water wells for years. Rigby (1980) reports the use of streaming potential to provide fracture location in a geothermal well. Figure 1 is an example of redistribution of uranium along fractures which can be detected by gamma spectral logging.

The geophysical log that provides the most consistent data on fractures is the acoustic televiewer. However, it also has several shortcomings (Zemanek *et al.*, 1969; Keys, 1979; Keys and Sullivan, 1979). A high temperature version is not yet available commercially, although it has been utilised by the US Geological Survey at borehole temperatures of

261°C. This probe is not yet reliable at high temperatures and resolution may be affected by heavy mud in the well or mudcake on the wall. Problems with mud are not common in geothermal wells, however, Figure 4 shows the details of a complex system of intersecting fractures

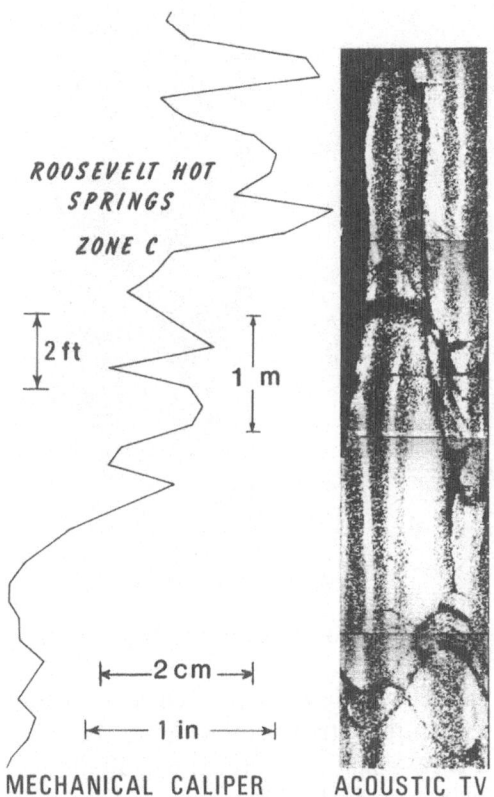

FIG. 4. Mechanical caliper and acoustic televiewer logs of production zone C, Roosevelt Hot Springs, Utah.

which constitutes production zone C in Fig. 3. The caliper log provides no clear understanding of this fracture system which is shown by the televiewer log to be the intersection of a near-vertical open fracture dipping northwest with a set of fractures with shallower dip in the same direction. Near the bottom of this log, there is also a set of tight fractures with shallow dip to the northeast. The direction of dip with respect to magnetic north can be read directly from the televiewer log, and the angle of dip calculated from the caliper and televiewer logs. The acoustic caliper log is produced by a time- rather than amplitude-dependent output of the televiewer probe and provides much higher resolution-oriented traces than the mechanical caliper. Figure 5 is a televiewer log

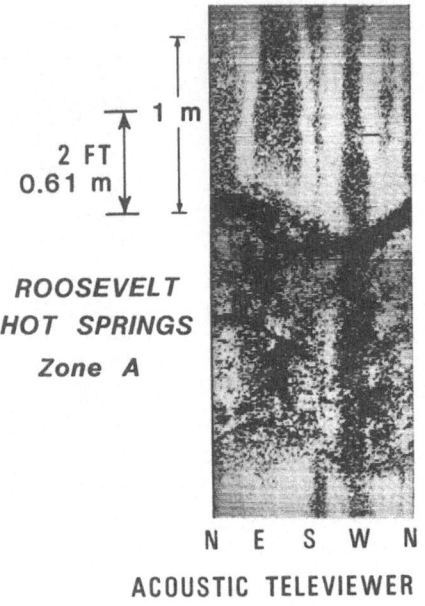

ACOUSTIC TELEVIEWER

FIG. 5. Televiewer log of production zone A, Roosevelt Hot Springs, Utah.

of another production zone in the same well. The zone is produced by a series of parallel fractures with a shallow dip to the west. The televiewer log suggests that there might be hydrothermal alteration in this interval. The caliper logs in Fig. 6 show that the well is somewhat larger in diameter in this interval, possibly due to softer, fractured rock. The oriented acoustic caliper traces have higher resolution than any mechanical device and clearly substantiate the direction of dip of the uppermost open fracture.

Although the televiewer log does provide a way of measuring the apparent aperture of a fracture at the wall of the borehole, this information may indicate a greater than true width due to fragmentation or breaking of the rock at the acute angle of intersection of fracture and borehole wall. The 1·3 MHz frequency does not permit the signal to penetrate the rock.

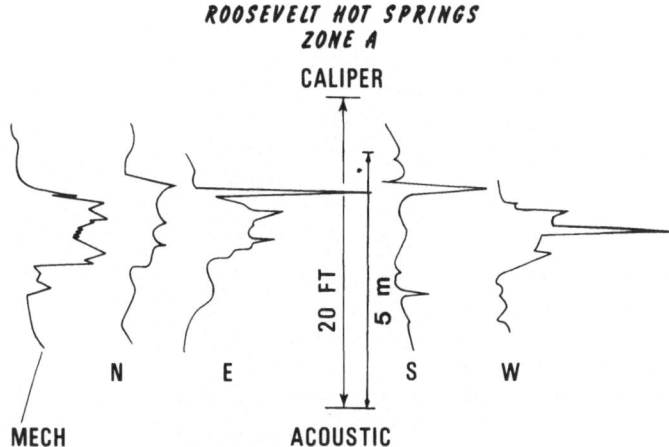

FIG. 6. Mechanical and acoustic caliper logs, Roosevelt Hot Springs, Utah.

Acoustic waveform studies utilising the digitised signal from an acoustic velocity probe have yielded information related to the hydraulic conductivity of fractures in igneous rocks (Paillet, 1980). The output of acoustic velocity probes, 15–20 kHz, has a lower frequency than the televiewer which permits some penetration of the rocks surrounding the borehole with fairly high resolution. Both shear and tube wave amplitudes are clearly affected by open, fluid-filled fractures. Furthermore, a relationship between the tube wave amplitude and the effective fracture permeability was demonstrated. Fracture permeability was determined by packers tests (Davison, 1980).

A further use of the televiewer or other high resolution fracture logging devices that might be developed is to obtain data on hydraulic fractures. Hydraulic fractures are sometimes produced accidentally during drilling or testing. More importantly, the technique is now being used experimentally to stimulate production from geothermal wells that are otherwise uneconomic. Figure 7 is a comparison of two televiewer logs made before and after fracturing a well at Raft River, Idaho, (Keys, 1980). The east dipping natural fracture system at a depth of 1430 m apparently terminated the downward propagation of the open portion of the vertical hydraulically induced fracture.

Artificially induced fractures are also used to hydraulically connect two wells in the hot dry rock approach to extracting geothermal energy. The televiewer has been used to identify these artificial fractures at the well but provides no information on their propagation away from the

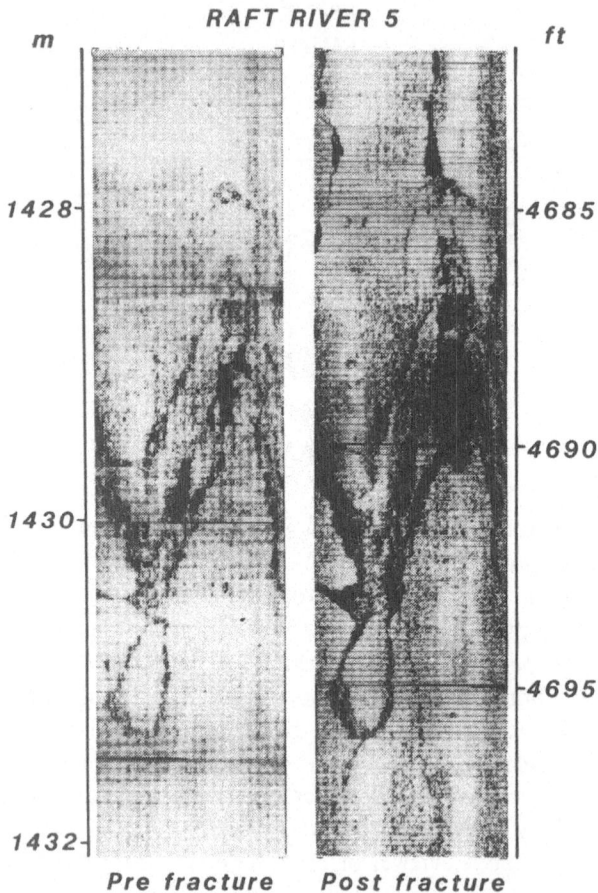

FIG. 7. Natural fracture system in Raft River well 5 which terminates the open
interval of a hydraulically induced fracture.

well. Albright *et al.* (1980) describe the transmission of compression (P) and shear (S) waves between wells as much as 46 m apart in fractured granite rocks at Los Alamos, New Mexico. The general character of the waveforms reportedly depends on the presence and number of fractures along the acoustic paths and their pressure states. The authors have also demonstrated that the growth of hydraulic fractures can be mapped acoustically. Existing fractures were studied by simultaneously operating two standard acoustic velocity logging probes in adjacent wells. The waveform data were digitised from a stationary receiver in one well while the transmitter was moved in another well. These data were collected at hydrostatic pressures and at pressures elevated sufficiently to separate fracture faces. When fractures are pressurised, S waves are not transmitted and P waves are attenuated significantly as the transmitter frequency increases. The attenuation of the seismic signals at Las Alamos was probably due to an increase in microfractures and large open fractures through which hot water can move. The approximate location of a large fracture between the wells was also derived from the attenuation data.

7.2. Borehole Flow

Measurement of flow in a borehole under either natural head conditions or while pumping or injecting can provide data on the location of permeable zones and the magnitude of hydraulic conductivity. In most uncased wells, head differences between aquifers or fracture zones produce enough vertical movement in the borehole to be detected by an impeller flowmeter, temperature log, or possibly a fluid conductivity log. Under natural conditions, the head differences are probably not known, so producing zones may be located, but permeability cannot be quantified. In contrast, if injection or pumping rates are known, hydraulic conductivity can be estimated. Schimschal (1981) describes a technique utilised to estimate field hydraulic conductivity during an injection test in the Raft River geothermal reservoir, Idaho. Water at a temperature of 129°C was injected at a pressure of 1020 kPa (kilopascals), and a constant rate of 1700 litres/min. During injection, logs were made with a centralised impeller flowmeter operating through a pressure lubricator. Both static and dynamic (trolling) calibration were carried out in the casing and open hole. An accurate caliper log is essential for the quantitative interpretation of flowmeter logs in an uncased hole. Logging speed, impeller rotation rate and hole diameter information were digitised, data analysis was carried out in a computer, and a log of apparent hydraulic conductivity was plotted. Schimschal used the relationship

$$K = \frac{Q_r \ln 2H/d}{2\pi H \Delta S} \text{ where } H \geq 5d \tag{4}$$

where K = hydraulic conductivity, Q_r = rate of radial fluid flow, H = thickness of layer, d = borehole diameter and ΔS = differential head of water in the interval tested.

The difference in vertical flow rates measured at the top and bottom of a layer is considered equal to Q_r. Because the conditions used in deriving the above relationship are not exactly met in a well, the results are considered to represent an apparent or field hydraulic conductivity at in-hole fluid temperature, viscosity and density. A shortcoming in this very useful technique is the lack of vertical resolution sufficient to define very thin producing intervals caused by fractures. In the example cited, a moving interval average of 12 m was chosen.

The use of both flowmeter and temperature logs to analyse the permeability distribution of a producing geothermal well at Raft River is described by Keys and Sullivan (1979). Although this well had a complex history of production and injection of large amounts of cold water, temperature logs and flowmeter logs did enable the identification of producing fracture zones. Because large variations in hole diameter caused turbulence and affected the accuracy of flowmeter logs, an attempt was made to estimate the volumes produced from various intervals using differences between temperature logs made while the well was shut in and flowing. If one makes the simplifying assumptions of no change in heat energy from the wall rock and adiabatic mixing of fluid in the well with fluid entering, then the ratio of the masses of water will be a function of the change in temperature caused by the producing zone. The simple energy balance equation would be

$$T_1 M_1 + T_2 M_2 = T_3 M_3 \tag{5}$$

where T_1, T_2 and T_3 are temperatures below the fracture, from the fracture and above the fracture, respectively, and M_1, M_2 and M_3 are the respective water masses.

In the well at Raft River, it was possible to obtain temperature logs of the well both while shut in and flowing for initial and mixed values. This method permits the approximation of the percentage of total flow contributed by each producing fracture. Because of the assumptions that must be made, the temperature method should only be used if flowmeter logs are not available.

8. TEMPERATURE

Although temperature logs can be used as described above to locate and quantify producing zones, an important economic question in determining the value of the well is the actual temperature of the reservoir. The values taken from temperature logs are subject to several errors, yet an accurate estimate of the long-term fluid temperature to be expected from the reservoir is essential to economic calculations. The errors stem from incorrect equipment calibration, thermal drift and temperature differences between the well and the surrounding rock. Differences of 10°C or more are common between logs made with different equipment in a well that has reached equilibrium. Such errors are usually due to inadequate calibration, possibly compounded by long-term or short-term drift. Copies of recent calibration data should be requested, and at least a couple of temperatures should be checked with secondary or field standards. The selection of the proper recording scales for temperature logs is essential if maximum benefit is to be obtained from the data. Too many logs of geothermal wells are run on an insensitive, compressed scale because that is the practice in oil wells, and because a wide range of temperature is expected. In a well under static conditions, changes of several tenths of a degree Celsius may be significant in locating a permeable fracture. On some analogue recorders, it is possible to record on two scales; one sufficiently compressed to display the full thermal range encountered and a second to resolve changes of 0·1°C or better. One advantage of simultaneous digital recording of raw data in the field is the ability to replot at the needed sensitivity and the ability to plot differential or gradient temperature logs. The differential presentation of temperature data simplifies the location of intervals of water entry and changes in gradient that might be due to rocks of different thermal conductivity.

If water is moving vertically in a well, it is likely to have temperatures that differ from the surrounding rocks. Temperature reversals are very common in hot wells. Consequently, the practice of straightline extrapolation of geothermal gradient can be greatly misleading. Because convective fluid movement is so common in geothermal systems and vertical flow is so prevalent in uncased hot wells, the interpretation of changes of thermal conductivity from temperature profiles is difficult. These problems do not apply to shallow, small diameter holes drilled for the specific purpose of making temperature gradient measurements. Benoit et al. (1980) describe the analysis of logs at the Desert Peak geothermal field in Nevada, which was discovered solely on the basis of temperature logs of gradient holes. Large numbers of such heat flow holes have been drilled and logged by the US Geological Survey and private companies in the western US. They are usually relatively shallow, and small diameter casing is cemented in. It may be necessary to wait a year or more for thermal equilibrium so that the equipment accuracy of greater than 0·01°C will provide meaningful data.

Predicting the return of thermal equilibrium between well and rock after drilling or injection of fluids or under conditions of natural vertical circulation is extremely difficult. Equations designed to calculate thermal equilibrium may be totally inaccurate under the unstable conditions created by a well penetrating an anistropic geothermal system. In Raft River, production well RRGE 2, 4·2 million litres of 43°C water was injected prior to deepening the well in March 1976. Temperature logs made more than 4 months after deepening still exhibited temperature inversions due to the injected cold water. These inversions were located at the most permeable zones, and their persistence was almost impossible to predict because the amount of injected water accepted by these zones was unknown.

A theoretical method for the calculation of true formation temperature from data obtained after circulation or fluid injection is described by Sanyal et al. (1980).

$$T_i - T_{ws} = \frac{q}{2\pi Kh} [T_D(T_p + \Delta t) - T_D(\Delta t)] \qquad (6)$$

where T_i = true rock temperature (°C), T_{ws} = shut-in temperature (°C), q = rate of heat removal from rocks during circulation (cal/s), K = thermal conductivity of rocks (cal/scm°C), h = thickness of rock unit (cm), T_p = circulation time (s), Δt = shut-in time (s) and T_D = a dimensionless heat flow function. Edwardson et al. (1962) define T_D, the function that relates the heat flow to spatial changes in temperature, as follows:

$$T_D = \frac{Kt}{C_p \rho r_w^2} \qquad (7)$$

where t = real time (s), C_p = specific heat of rocks (cal/g °C), ρ = rock density (g/cm³) and r_w = well bore radius (cm).

Theoretically, at infinite shut-in time compared to circulation or injection time, the measured shut-in temperature will equal the true reservoir or rock temperature. Obviously, this solution ignores the problems of non-uniform distribution of permeability and vertical flow in the borehole under shut-in conditions.

Fluid conductivity logging presents some of the same problems as temperature logging. It also provides the potential for locating permeable intervals and determining how water quality varies among the production zones. To date, fluid conductivity logging has not been widely used in geothermal wells.

REFERENCES

ALBRIGHT, J. N., PEARSON, C. F. and FEHLER, M. C. (1980). Transmission of acoustic signals through hydraulic fractures, *21st Annual Logging Symposium, SPWLA*, 806 Main St., Suite B-1, Houston, Texas, 77002, USA, p. R1.

BAKER, L. E., BAKER, R. P. and HUGHEN, R. L. (1975). Report of the geophysical measurements in geothermal wells workshop, Sandia Laboratories Report 75-0608, Albuquerque, New Mexico, USA.

BENOIT, W. R., SETHI, D. K., FERTL, W. H. and MATHEWS, M. (1980). Geothermal well log analysis at Desert Peak, Nevada, *21st Annual Logging Symposium, SPWLA*, 806 Main St., Suite B-1, Houston, Texas, 77002, USA, p. AA1.

DAVISON, C. C. (1980). Physical hydrogeology measurements conducted in boreholes WN1, WN2, and WN4 to assess the local hydraulic conductivity and hydraulic potential of a granitic rock mass—report of 1978–1979 activities. National Hydrology Research Institute, Environment Canada, 70 p.

EDWARDSON, M. S., GUNER, H. M., PARKINSON, H. R., WILLIAMS, C. D. and MATHEWS, C. S. (1962). Calculation of formation temperature disturbances caused by mud circulation, *J. Petroleum Technol.*, April.

EHRING, T. W., LUSK, L. A., GRUBB, J. M., JOHNSON, R. B., DEVRIES, M. R. and FERTL, W. H. (1978). Formation evaluation concepts for geothermal resources, *19th Annual Logging Symposium, SPWLA*, 806 Main St., Suite B-1, Houston, Texas, 77002, USA, p. FF1.

ERSHAGI, I., PHILLIPS, L. B., DOUGHERTY, E. L., HANDY, L. L. and MATHEWS, M. (1979). Application of oilfield well log interpretation techniques to the Cerro Prieto Geothermal Field, *20th Annual Logging Symposium, SPWLA*, 806 Main St., Suite B1, Houston, Texas, 77002, USA, p. PP1.

ERSHAGI, I. (1980). Detection of hydrothermal alteration in a sedimentary type geothermal system using well logs, paper presented at the *55th Annual Fall Technical Conference and Exhibition of the Society of Petroleum Engineers of AIME*, 6200 N. Central Expressway, Dallas, Texas, 75206, USA.

GLENN, W. E., HULEN, J. B. and NIELSON, D. L. (1980a). A comprehensive study of LASL Well C/T-2, Roosevelt Hot Springs, KGRA, Utah and applications to geothermal well logging, Los Alamos Scientific Laboratories Preliminary Report, Los Alamos, New Mexico.

GLENN, W. E., ROSS, H. P. and ATWOOD, J. W. (1980b). Review of well logging in the basin and range known geothermal resource areas, Society of Petroleum Engineers Publication No. 9496, AIME, 6200 N. Central Expressway, Dallas, Texas, 75206, USA.

KEITH, T. E. C., WHITE, D. E. and BEESON, M. H. (1978). Hydrothermal alteration and self-sealing in Y-7 and Y-8 drill holes in northern part of Upper Geyser Basin, Yellowstone National Park, Wyoming, US Geological Survey Professional Paper 1054–A.

KEYS, W. S. (1979). Borehole geophysics in igneous and metamorphic rocks, *20th Annual Logging Symposium, SPWLA*, 806 Main St., Suite B-1, Houston, Texas, 77002, USA.

KEYS, W. S. (1980) The application of the acoustic televiewer to the characterization of hydraulic fractures in geothermal wells, *Proceedings of the Geothermal*

268 W. SCOTT KEYS

Reservoir Well Stimulation Symposium, San Francisco, California, USA, p. 176.

KEYS, W. S. and SULLIVAN, J. K. (1979). Role of borehole geophysics in defining the physical characteristics of the Raft River Geothermal Reservoir, Idaho, *Geophysics*, **44**, 1116.

MATHEWS, M. (1980). Calibration models for fractured igneous rock environments, *21st Annual Logging Symposium, SPWLA*, 806 Main St., Suite B-1, Houston, Texas, 77002, USA, p. L1.

PAILLET, F. L. (1980). Acoustic propagation in the vicinity of fractures which intersect a fluid-filled borehole, *21st Annual Logging Symposium, SPWLA*, 806 Main St., Suite B-1, Houston, Texas, 77002, USA, p. DD1.

RIGBY, F. A. (1980). Fracture identification in an igneous geothermal reservoir, Surprise Valley, California, *21st Annual Logging Symposium, SPWLA*, 806 Main St., Suite B-1, Houston, Texas, 77002, USA, p. 21.

RIGBY, F. A. and REARDON, P. (1979). Well logs for geothermal development, benefit analysis, *20th Annual Logging Symposium, SPWLA*, 806 Main St., Suite B-1, Houston, Texas, 77002, USA, p. CC1.

SANYAL, S. K., JUPRASERT, S. and JUSBACHE, M. (1979). An evaluation of a rhyolite–basalt–volcanic ash sequence from well logs, *20th Annual Logging Symposium, SPWLA*, 806 Main St., Suite B-1, Houston, Texas, 77002, USA, p. TT1.

SANYAL, S. L., WELLS, L. E. and BICKHAM, R. E. (1980). Geothermal well log interpretation—state of the art, Los Alamos Scientific Laboratory, Informal Report No. LA-8211-MS, Los Alamos, New Mexico, USA.

SCHIMSCHAL, U. (1981). Flowmeter analysis at Raft River, Idaho, *Ground Water*, January–February.

VENERUSO, A. F. and COQUAT, J. A. (1979). Technology development for high temperature logging tools, *20th Annual Logging Symposium, SPWLA*, 806 Main St., Suite B-1, Houston, Texas, 77002, USA, p. KK1.

WEST, F. G. and LAUGHLIN, A. W. (1976). Spectral gamma logging in crystalline basement rocks, *Geology*, **4**, 617.

ZEMANEK, J., CALDWELL, R. L., GLENN, E. E., Jr., HOLCOMB, S. V., NORTON, L. J. and STRAUS, A. J. D. (1969). The borehole televiewer—a new logging concept for fracture location and other types of borehole inspection, *J. Petroleum Technol.*, **21**, 762.

Chapter 8

MEASUREMENT AND ANALYSIS OF GRAVITY IN BOREHOLES

Joseph R. Hearst and Richard C. Carlson
Lawrence Livermore National Laboratory,
Livermore, California, USA

SUMMARY

Small variations in the vertical component of gravity (g_v) *have been found to be of considerable use in geophysical exploration. The phenomenon of interest to conventional exploration geophysics is the variation of* g_v *with position on the surface. Two-dimensional contour maps of this variation are usually prepared. The change in* g_v *caused by a buried mass will, of course, be greatest directly above the mass, but will vary with the horizontal distance between the mass and the measuring point. This variation can be used to attempt to locate and describe the mass. Surface gravimetry is typically used to locate buried faults, domes and other structures of interest to the exploration geophysicist, and a vast literature exists describing the method.*

Borehole gravimetry is a fairly recent extension of surface gravimetry to the third dimension. The phenomenon of interest is the variation of g_v *with depth in the hole. This variation, like that on the surface, is caused by both the vertical and lateral position of the buried mass, but while with the surface gravimeter the lateral distance between the mass of interest and the instrument is changed, with the borehole gravimeter the vertical distance is changed instead. This permits a different look at the mass and a somewhat different method of analysis.*

This paper discusses the history of the development of borehole gravimetry and the borehole gravimeter. Important aspects of field operations are described and the corrections necessary for data reduction are explained. The several methods of analysing the data to attempt to infer subsurface structure from borehole gravity and density log data are outlined. Finally, a number of applications of the method are listed.

1. INTRODUCTION

The acceleration due to gravity, often called the force of gravity, or just 'the gravity' is perceived by all of us. However, we cannot notice the small variations in this fairly large force. But these small variations have been found to be of considerable use in geophysical exploration.

Because the force is both large and pervasive, only variations of the force parallel to its original direction can be measured easily. These are the variations in the vertical component of gravity. In this paper, we shall discuss only the vertical component and shall often refer to it merely as the 'gravity'. It can be represented as

$$g_v = km_1 m_2 \cos\theta / r^2$$

where g_v is the vertical component of gravity, k the gravitational constant, m_1 and m_2 the two masses that are mutually attracting (in our case, usually the mass of interest and a 'test mass' in a gravity meter), r the distance between them and θ the angle between the line connecting the two masses and the vertical. Usually g_v is taken as positive downward. The common unit for measurement of gravity is the milligal (mgal or mg) defined as $10^{-5}\,\text{m/s}^2$. Thus the normal value of g at the surface of the earth is about a kilogal.

The phenomenon of interest to exploration geophysics is the variation of the vertical component of gravity with position on the surface. Two-dimensional contour maps of this variation are usually prepared, and the contour interval is generally 1–5 mgal. The change in g_v caused by a buried mass will, of course, be greatest directly above the mass, but will vary with the horizontal distance between the mass and the measuring point. This variation can be used to attempt to locate and describe the mass. Surface gravimetry is typically used to locate buried faults, domes and other structures of interest to the exploration geophysicist, and a vast literature (for example, Nettleton, 1976) exists describing the method.

Borehole gravimetry is a fairly recent extension of surface gravimetry to the third dimension. The precision required for borehole gravity work is generally one or two orders of magnitude greater than that for surface gravimetry, primarily because the masses of interest are often considerably smaller than those investigated in routine surface gravimetry.

The phenomenon of interest is the variation of g_v with depth in the hole. This variation, like that on the surface, is caused both by the vertical and lateral position of the buried mass, but while with the surface gravimeter the lateral distance between the mass of interest and the instrument is changed, with the borehole gravimeter the vertical distance is changed instead. This permits a different look at the mass and a somewhat different method of analysis.

Borehole gravimetry, so far, has typically been used to locate variations in subsurface bulk density when these variations cannot be defined from conventional borehole logging methods. For example, density logs cannot be used in cased wells, and their short range does not permit them to locate density variations that occur more than about 10 cm from the hole. More recently, the borehole gravimeter has been used to investigate buried structures such as dipping beds, faults or domes, in a manner similar to the conventional use of surface gravimetry. However, because depth rather than lateral position is varied, the borehole gravimeter is more suitable than the surface gravimeter to locate the depth to a buried structure.

Borehole gravimetry is still a new technique, and far less experience is available than for surface gravimetry. This report is intended to summarise our knowledge of the use of borehole gravimetry to date..

2. HISTORY

While some work was done in the 19th and early 20th centuries (Airy, 1856; Jung, 1939), the practical use of subsurface gravity measurements can really be considered to start with the seminal papers of Smith (1950) and Hammer (1950) in the same issue of *Geophysics*.

Smith (1950) discussed many possible uses of borehole gravimetry. In particular he considered the subsurface gravity gradient caused by buried anomalous masses (now called the borehole Bouguer anomaly (Snyder, 1976), and its use in interpretation of surface gravity surveys, the use of layer density measured by borehole gravity in interpretation of seismic surveys, and the effects of hole diameter on borehole gravity readings. Hammer (1950) reported on a subsurface gravity survey and compared the results with densities from samples from the same mine shaft. He discussed the likely sources of error, and showed the basic method of data reduction. He derived the basic equation for density from subsurface gravity:

$$\rho = \frac{1}{4\pi k}\left[F - \frac{\Delta g + c}{\Delta z} \right] \qquad (1)$$

where ρ is the density, Δg the difference in gravity between two measuring points a vertical distance Δz apart, c the sum of the necessary corrections (Hammer discussed only the terrain correction), F the free air gradient (or $\Delta g/\Delta z$ above the surface) and k the gravitational constant.

For a number of years work on the use of borehole gravity proceeded quite slowly. Rogers (1952) used gravity measurements in a mine shaft to locate an anomalous mass and computed the size of a sphere that would be equivalent to that mass.

Miller and Innes (1953) measured layer densities in shafts and also computed the mean density of the earth. Facsinay and Haaz (1953) also measured mean layer densities in mines.

Domzalski (1954, 1955) measured effects of ore bodies on gravity measurements in a mine shaft. He obtained apparent average densities, calculated the effects of the ore bodies on the averages and compared the results to the calculated values. He also discussed some of the corrections needed.

Algermissen (1961) and Plouff (1961) combined surface and subsurface surveys over long profiles to determine subsurface density. Many authors (Kumegai *et al.*, 1960; Bodemuller, 1963; Thyssen-Bornemisza, 1963, 1964) considered uses of the vertical gradient of gravity either above or below the surface.

In 1963 the US Geological Survey began work on subsurface gravimetry. Beyer (1980) describes the history in some detail. It began with a remeasurement by McCulloh (1965) of Hammer's (1950) original data and a discussion of the need for accuracy as well as precision in such data. McCulloh then made strong arguments (McCulloh, 1966a,b, 1967a,b) for the development and use of a borehole gravimeter. Several organisations (Gilbert, 1952; Goodell and Fay, 1964; Howell *et al.*, 1966; Anon., 1966) had been working on the development of a borehole gravimeter. The first commercially successful meter was completed by LaCoste and Romberg, under contract to the US Geological Survey, in 1966 (McCulloh *et al.*, 1967a,b).

The early tests were quite successful and the instrument was used both in oil fields (McCulloh *et al.*, 1968) and at the Nevada Test Site (Healey, 1970.)

The US Geological Survey then began extensive use of the new gravimeter to study oil and gas fields in California (Beyer, 1971, 1977*a,b*) and elsewhere (Beyer and Clutsom, 1978*a,b*, 1980). In particular Beyer (1971) made a detailed study of the sources of error and methods of data reduction for the borehole gravimeter. Two additional borehole gravimeters were built for industry in the late 1960s (Jones, 1972), and then became useful to some oil companies (Rasmussen, 1973.) In 1977 a new, smaller-diameter higher-temperature version of the LaCoste and Romberg meter became available, and eight of these are now in use in the United States (LaFehr *et al.*, 1979).

3. THE INSTRUMENT

In principle, the direct measurement of gravity is not difficult. The difficulty of constructing a useful gravimeter is due entirely to the requirement for very high precision. Although the gravity field is quite strong and pervasive, it is quite homogeneous and the perturbations, usually the objective of the measurement, are quite small compared to the total. Thus, to be useful, gravity measurements must be precise to at least one part in 10^6, and often to one part in 10^8.

For some purposes, high accuracy is required as well as high precision. In such situations, it is best to use a method that does not depend on the accurate knowledge of the mass of an object, because this can only be obtained with knowledge of the gravity. Such methods use, for example, the angular inertia of a pendulum, or the linear inertia of a falling body. The period T of a simple mass point depends only on its length and g. In fact, the pendulum is not a mass point but a physical body of moment of inertia I and radius of gyration i about a knife-edge of some sort. Then

$$T^2 = I/mgh = i^2/gh$$

where m is the mass of the pendulum and h the distance from the knife-edge to the centre of gravity of the pendulum. The pendulum apparatus, with the pendulum thermostatted in a vacuum, was used in the field for many years until the development of simpler gravity meters.

In the case of the falling body, the expression, from Newton's first law, is

$$g = 2s/t^2$$

where s is the distance the body falls in time t. Until recently it was quite difficult to time the falling body accurately. Nettleton (1976) discusses modern instruments using falling corner reflectors (which reflect light parallel to the direction of incidence) as part of a laser interferometer to measure absolute gravity to one part in 10^8. He also describes simpler systems.

The standard surface gravimeter now in use incorporates a 'zero-length spring' (LaCoste, 1934, 1935) primarily to measure relative gravity. The usual expression for force on a spring is

$$F = K(L - L_0)$$

where K is the spring constant and L and L_0 the lengths of the spring with loads F and 0, respectively. A zero-length spring, on the other hand, is prestressed, so that a force is required to separate the coils and so wound and terminated that $F = KL$ (LaCoste and Romberg, 1942, describes the process in detail.) The spring is used to support a weight at the end of a beam (Fig. 1.) The gravitational torque on the beam, caused

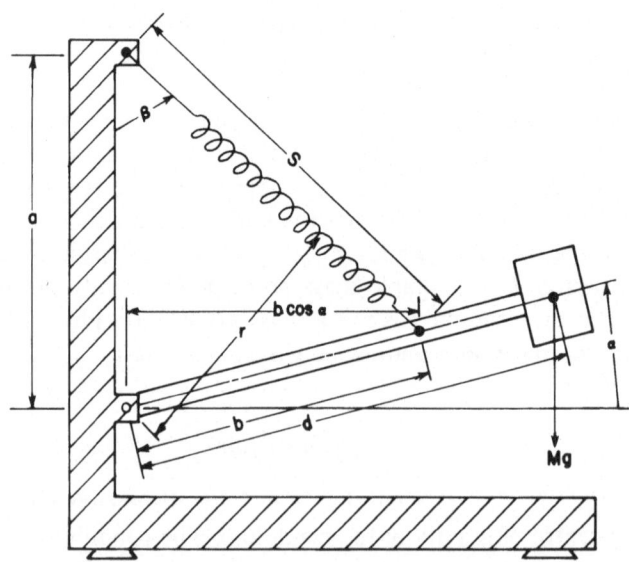

FIG. 1. Schematic of the zero-length spring (LaCoste and Romberg) gravimeter.

by the weight, is

$$T_g = -mgd\cos\alpha$$

where m is the mass of the weight, d the length of the beam and α is the angular displacement of the beam. The torque exerted by the spring is

$$T_s = Ksr$$

where

$$s\sin\beta = b\cos\alpha \quad \text{and} \quad r = a\sin\beta$$

Thus

$$T_s = Kba\,\cos\alpha$$

and the net torque is

$$T = T_s + T_g = (Kba - mgd)\cos\alpha$$

In operation, Kba is set equal to mgd so that the net torque is zero. Then the torque is insensitive to small angular displacements of the beam. At each measuring station, a is adjusted through a complex system of levers and springs (LaCoste and Romberg, 1945), driven by a screw, to set $Kba = mgd$. The amount of turning of the screw is a measure of a and thus of the relative value of g and of the absolute value of g if the system has been calibrated against an absolute standard. The instrument is carefully temperature-compensated and also thermostatted.

The first attempt at a borehole gravimeter was the vibrating-string type. The fundamental frequency F of a vertical vibrating string of mass m per unit length and length L, held under tension Mg by the weight of a mass M, is

$$F = \frac{1}{2L}\left[\frac{Mg}{m}\right]^{1/2}$$

and then the change in frequency ΔF caused by a change in gravity Δg is

$$\Delta F/F = 0\cdot5\ \Delta g/g$$

Two borehole meters have been built embodying this principle, one by Shell (Goodell and Fay, 1964) and the other by Esso (Howell $et\ al$, 1966). The first is said to have a precision of 1 mgal, the second 0·01 mgal.

The LaCoste and Romberg instrument that has been the most successful borehole gravimeter to date is a modification of the company's standard geodetic gravity meter. The meter is thermostatted at 101°C and is gimbal-mounted for remote levelling (LaCoste, 1977). The nulling screw of the meter is rotated by a motor-driven gear train and the train also turns a step switch that is followed by a synchro motor at the surface (McCulloh *et al.*, 1967*a*). The synchro motor in turn drives a counter that is calibrated so that g can be obtained from the counter units. The position of the beam near the null position is determined by observing the reading of a galvanometer which is driven by a pair of photo-sensors illuminated by a light reflected from a mirror attached to the gravity meter beam. A strip chart recorder is also driven by the output of the photo-sensors. When the meter is in exact null position, the trace is, except for minor fluctuations, parallel to the direction of chart motion; when it is not, the trace is at an angle to the chart direction. In practice, the readings are taken with the trace on either side of the null and the angle between the two traces on the strip chart recorded. The gravity can be calculated from the counter reading and the slope of the traces on either side of the null. This is a much faster technique than obtaining a perfect null.

Schmoker (1978*a*) has studied the precision of this instrument and determined that the gravity difference between two stations in a borehole can be measured to $10\,\mu$gals, or one part in 10^8.

This prototype borehole gravimeter was a great improvement over previous methods of measuring gravity at depth, but had some limitations. It was 14 cm in diameter and a deviation of only 6·5° from the vertical was permitted. In addition, the maximum temperature at which it could be used for any length of time was 101°C, somewhat low for many boreholes. Consequently second-generation instruments have been built. The nulling data are transmitted electronically. The meter has an outside diameter of 10·5 cm and is thermostatted at 124°C. (The sensing element must be kept within about 0·1 °C to avoid changes in calibration, so care must be taken to keep it connected to power at all times.) The meter can be tilted as much as 14°. LaFehr *et al.* (1979) have found this new meter to be precise to $3\,\mu$gal in the laboratory and $7\,\mu$gal in the well.

A special case to permit use of this gravimeter at temperatures up to 350 °C has been designed (Baker, 1977) but never built.

A servo accelerometer, designed for use as a lunar gravimeter, (Henderson and Iverson, 1968) has been considered for borehole use but never tried for the purpose.

Gravimeter development is still proceeding (Buck, 1973; Veneruso, 1978) but the authors are not aware of any new additions to the list of successful borehole meters. A more promising avenue is the revival of the gravity gradiometer, which has been modified for airborne and space-borne use (Bell *et al.*, 1969; Trageser, 1970, 1975; Forward, 1965). A problem with gradiometers is their great sensitivity, especially to objects close to them. Hole size changes could be a severe problem. They have been designed for continuous operation in spacecraft and might be much faster to use than standard gravimeters, which usually take about 10 min for a reading.

4. FIELD OPERATIONS

Rasmussen (1973) gives a good overview of the standard field operation of the first-generation borehole gravimeter. The original instrument required a special 13-conductor cable. Later versions can be operated on a standard 7-conductor logging cable. It is important to give serious consideration to the method of lowering the meter into the well. First, error in measurement of depth is often the largest error in a borehole gravity survey. Standard logging trucks can measure depth to 1% or better, but this is often not good enough. If gravity stations are to be spaced less than 15 or 30 m apart (see below) it is desirable to flag the cable and measure the distance between flags directly. It is possible to measure relative depth between stations to less than 1 cm with equipment developed by EDCON (1977).

In some wells it is necessary to clamp or spring-load the meter to the borehole wall. Even slight vibration of the meter will make the readings useless. It is sometimes necessary to clamp the logging cable at the top of the well casing to prevent wind-induced noise being transmitted to the meter, especially for readings at depths less than a few hundred metres.

The US Geological Survey has developed a special logging truck for its borehole gravimeters (Robbins, 1979). At the Nevada Test Site, where a modified undersea gravimeter is used as a borehole meter in large (1·5–3 m) diameter holes, a large boom and a hydraulic clamping mechanism are used.

Because the gravity meter is expensive and few are available, the well should be in excellent condition, with no possibility of caving. Many meter owners require a cased or lined well. Often a caliper log of the well will be required, and perhaps a dummy run. Some individual instruments will not read properly during the first readings after they have entered the hole, or after direction of travel has been changed (Schmoker, 1978*b*). Surveys should be planned accordingly.

The choice of which gravity stations to use in a borehole is an important operational decision. Hearst (1977a) discusses this in some detail and concludes that the finer the spacing between stations, the better an object will be resolved, regardless of distance from the well. One must certainly select a station spacing equal to or less than the depth resolution desired. Beyond that criterion, additional data points at least allow one to average out local perturbations in the gravity field that tend to degrade the precision of the measurement. If it is known that the structure in the vicinity of the borehole consists of flat beds of homogeneous density, it is advantageous to locate stations at bed boundaries. The usual interpretation method produces the average density between gravity stations, and the choice of stations according to lithology produces a truer curve of density versus depth than if arbitrary stations were chosen.

The new slim-hole LaCoste and Romberg gravimeter is said to be very temperature-sensitive (Caton, 1981) although the manufacturer will evidently supply a correction circuit upon request. Caton (1981) has developed a method for correcting for this sensitivity by least-squares fitting a drift curve to data from three or more surveys at the same depths. However, Jageler (1981) states that the drift as a function of temperatures can be determined for each individual meter, and the problem of temperature compensation can be avoided by simply waiting until the meter has reached a temperature at which the drift is known to be acceptable.

Caton (1981) has also developed a convenient method of recognising tares (or sudden changes in gravity readings caused by small changes in a gravimeter spring, support wire, hinge point, or other bit of the machinery), skips in the counter, or reading errors. He plots a 'reduced gravity', or the difference between the measured gravity and an average gradient, and then compares this reduced gravity for at least three repeat surveys. If one reduced gravity point disagrees with the other two, it is likely wrong.

The difference in gravity readings is usually reduced to density with some form of eqn (1). If the precision of the gravity readings is known, the smallest measurable density contrast for any station spacing can be calculated:

$$\delta\rho = 4\pi k \delta g / \Delta h$$

where δg is the smallest measurable gravity difference and Δh is the station spacing. This is illustrated in Fig. 2.

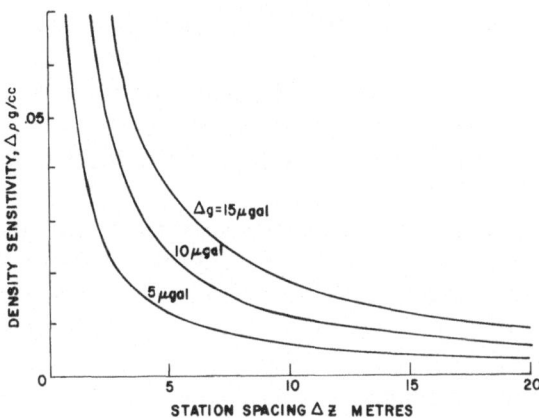

FIG. 2. Minimum observable average density difference $\delta\rho$ for a minimum resolvable gravity difference δg as a function of station spacing.

The resolution in apparent density is inversely proportional to station spacing. This does *not* mean that better resolution is obtained with wider spacing, since $\delta\rho$ refers to average density between stations. But it does mean that one can learn about a large volume of material with more accuracy than one can learn about a small volume.

There are many corrections (see below) that must be applied to the gravity data, and each correction is obtained from field data. For example, the tide correction requires a knowledge of time to within about 5 min and position to 24 km. Terrain correction requires a knowledge of the elevation of nearby terrain to an accuracy depending on the distance to the terrain feature (see below). Often a survey of nearby terrain is needed for the best results. The readings must be corrected for hole deviation. The measured depth difference in a hole is of course equal to the true vertical difference only for a vertical hole. A directional survey is often useful.

Changes in the borehole diameter can affect the data. These are particularly important at the top and bottom of very large (2–3 m) holes, or in shafts. Nearby mudpits or subsurface workings can cause errors as well. It is important to record all information about known values of the density of nearby material.

The first term in eqn (1) involves F, the free air gradient. Most workers calculate this term from the latitude and elevation of the top of the well (Heiskanen and Vening Meinesz, 1958), or by assuming a layer of known-density material at some depth interval, but we have found that it is better to measure it. This can be done either by calculation ('upward continuation') from surface gravity data (Grant and West, 1965) or by direct measurement on a tower. In the latter case, it is important that all measurements be made well above the surface (Fajklewicz, 1976), as otherwise effects of nearby terrain will invalidate the data (Hearst *et al.*, 1980.)

Drake (1967) has also found that the measured vertical gradient above the surface is inconsistent with the gradient calculated from latitude and elevation, and several authors (Thyssen-Bornemisza, 1965*a* and *b*; Kuo *et al.*, 1969; Arzi, 1975; Fajklewicz, 1976) have used such measurements to infer the position of bodies that cause the anomalies.

Two other useful descriptions of fielding methods have been published by Jones (1972) and Brown *et al.* (1975), and EDCON's (1977) borehole gravity manual is a fine presentation of current field techniques.

5. CORRECTIONS

The major corrections to the borehole gravity data are for terrain, tide, drift, free air gradient and non-verticality of the borehole.

The terrain correction is the most complex. It is the most time-consuming and is the most important for depths less than a few hundred metres. Vaschilov (1964) appears to have been the first to develop terrain corrections for subsurface gravity, but his method was cumbersome. Hearst (1968) produced a simple method for terrain corrections to borehole gravimetry for use with a computer, and Beyer and Corbato (1972) improved the formulation.

Hearst *et al.* (1980) have studied the precision in the terrain elevation needed to obtain a given accuracy in the gravitational effect g_t of a given terrain element. If we consider that an error in the apparent density calculated from eqn (1) caused by imprecision in terrain elevation should be no more than $0.01 \, g/cm^3$, and we use the 13 Hammer (1939) zones for dividing the terrain, then Fig. 3 shows the permissible elevation error in each zone. The zones are usually divided into n compartments, where n differs for each zone. The permissible error in compartment elevation is that shown in Fig. 3 multiplied by $\sqrt{n}$.

Inclined zones have a different terrain effect from flat zones of the same elevation. Figure 4 shows the inclination of each zone that will not cause an unacceptable density error. If the inclination of a zone is greater than that shown in Fig. 4, the zone should be subdivided radially for best results.

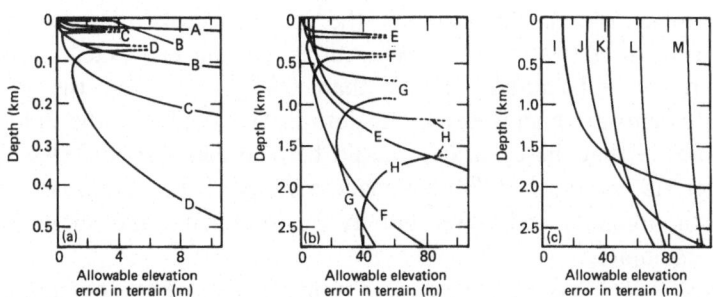

FIG. 3. Allowable elevation error of each Hammer (1939) zone vs depth in the borehole (for a RMS density error of no more than 0.01 g/cm^3 with all zones having allowable error). (a) Zones A–D; (b) Zones E–G; (c) Zones I–M. Terrain density is 2.67 g/cm^3.

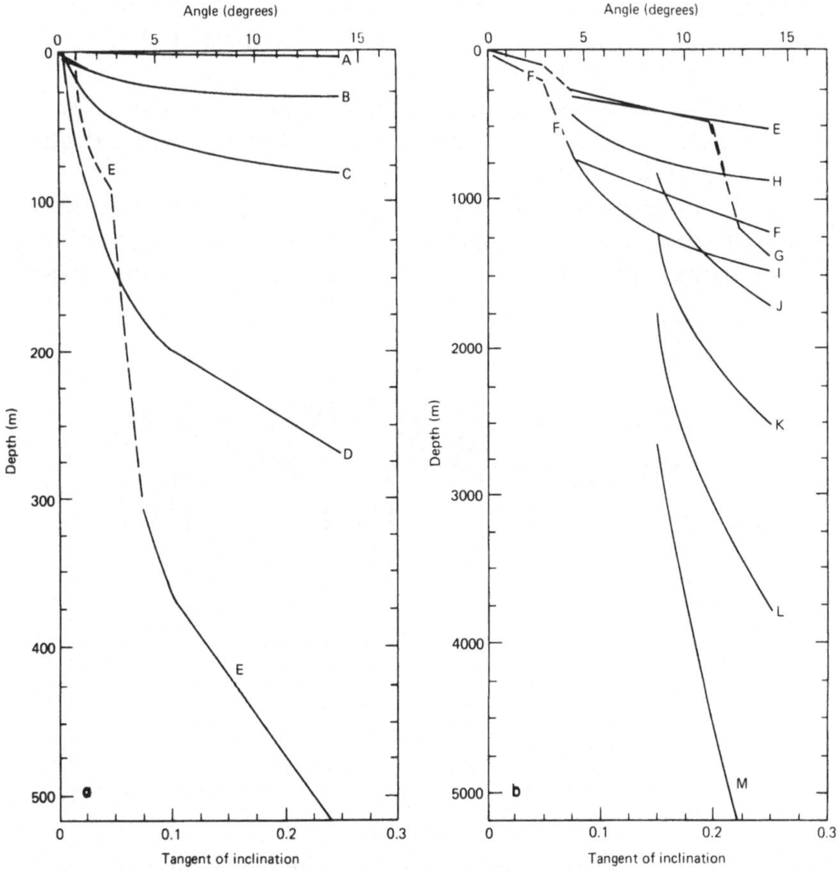

FIG. 4. Allowable inclination of each Hammer (1939) zone versus depth in the borehole. Assumptions as in Fig. 3. (a) Zones A–E; (b) Zones E–M.

Schmoker (1980a) has studied the terrain effects of very nearby cultural features (water tanks, excavations, drilling pads, etc.) and finds them to be substantially larger than $0.01\,g/cm^3$ when the gravity measurements are made close to the datum. This reinforces the statement made above that these features should be recorded and corrected for.

Beyer (1979a and b, 1980) has made a study of the terrain correction for both borehole and tower gravity measurements and shows many useful examples.

Plouff (1966) has developed a convenient method for obtaining the terrain correction from elevations digitised on a latitude–longitude grid. Tapes of elevations digitised from 1/250 000 topographic maps are available (National Cartographic Information Center, no date) but the accuracy is not adequate for shallow holes. Healey (1977) has tabulated elevations on 1 min squares for a number of 15 min quadrangles in Nevada, and these are quite satisfactory for distances of about 2·5 km and more from the wellhead.

Hearst (1978b) has written a program to adapt Plouff's method to terrain corrections for borehole gravity at the Nevada Test Site, where Healey's digitised data are available.

All gravity measurements must be corrected for earth tide effects. These effects are functions of time (Fig. 5) and are large enough to cause serious errors if not used. Computer programs for calculation of earth

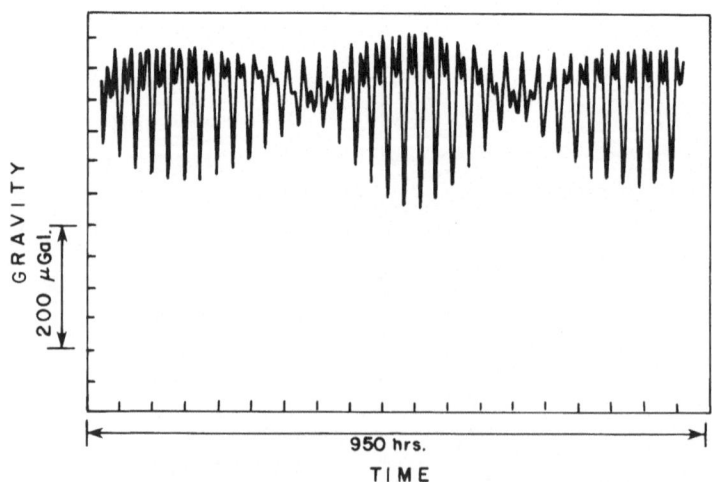

FIG. 5. Typical gravimetric effect of earth tides vs time.

tides are available, or can be written from the original papers (Longman, 1959). The input required is Greenwich mean time, date and the geographic coordinates of the measuring points. Hearst (1978b) incorporated tide corrections into the terrain correction program, and this is a convenient method for routine operations.

Correction for drift of the gravity meter is obtained by repeated readings at the same gravity station. It is sometimes important that the gravity meter should approach this station from the same direction every time the reading is made, because some gravimeters are affected by the direction of last motion (Schmoker, 1978b). The gravimeter readings at the same station are plotted as a function of time, and the change appropriate to the time of any station reading is applied to that reading as a drift correction.

If free air gradient, required for eqn (1), is measured, those measurements too must be corrected for terrain, tide and drift. Also if subsurface structure is found to have a substantial effect on the data (see below) the free air gradient data should be structure-corrected.

Corrections for the diameter of the borehole are ordinarily unnecessary, except for large diameters with readings within 1 m or so of the top or bottom of the hole, the water level, or major diameter changes. In that case the borehole can be treated as a terrain zone.

A non-vertical hole, if its angle is not greater than that permitted for the gravimeter, does not cause errors in the gravity data. But the angle does affect Δz used in eqn (1). If the deviation from the vertical θ of the well is known, then $\Delta z'$ (the corrected value) $= \Delta z \cos \theta$. If θ is not known, it must be determined from a directional survey. At $14°$, a $2°$ error in the angle can give a 1% difference in the cosine, and therefore a 1% error in the calculated density. At smaller angles the effect is considerably less. If the beds dip, the effect is quite different, but we consider that part of the data analysis rather than a correction.

6. DATA ANALYSIS

Before density can be calculated from eqn (1), values for the constants must be inserted. Robbins (1981) has recently pointed out that some authors have used an obsolete value of the gravitational constant k. The currently accepted value of k is $6·672 \times 10^{-8} \, cm^3/g \, s^2$, or $6·672 \times 10^{-6}$ for length in metres, mass in kilograms and gravity in milligals.

Equation (1) represents the density of an infinite homogeneous, horizontal slab. The vertical component of gravity caused by such a slab below a measuring station is (Heiskanen and Vening Meinesz, 1958)

$$g_s = 2\pi k \rho \Delta z$$

where k is the gravitational constant, ρ the density of the slab and z its thickness. If the slab is above the station the sign of g_s is reversed, and so the difference in gravity between two stations above and below the slab is

$$\Delta g_s = 4\pi k \rho \Delta z$$

Note that the distance from the point to the slab does not affect g_s provided that the point is not within the slab. Within the slab, g_s varies as a linear function of depth as shown in Fig. 6.

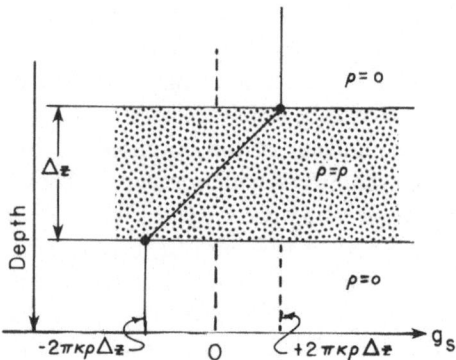

FIG. 6. Vertical component of gravity as a function of depth caused by an infinite homogeneous horizontal slab.

Of course, the earth is not made up of infinite homogeneous horizontal slabs, but is approximately spherical. However, Hearst and Carlson (1977) show from Gauss' theorem that eqn (1) is equally valid for homogeneous spherical shells. But the world is not made up of homogeneous spherical shells either. A major value of borehole gravimetry is the possibility of inferring the subsurface structure from the apparent density of the shells or slabs. There are two ways in which the subsurface structure can cause the apparent density between two stations to be different from the true density of the material close to the borehole and between the stations. One, of course, is non-uniformity of density of the material in the infinite horizontal but now inhomogeneous slab between the stations, i.e. for density between stations to vary as a function of distance from the well. The other is the effect on the apparent density of material that is not between the stations.

It was stated above that the distance from the measuring point to the slab does not affect g_s. This is true only for the infinite horizontal homogeneous slab. For a mass point, the vertical component of gravity is

$$g_p = kmz(r^2 + z^2)^{-3/2}$$

where r is the horizontal distance and z the vertical distance to the point of mass m. Figure 7 shows the gravitational effect of a point mass at

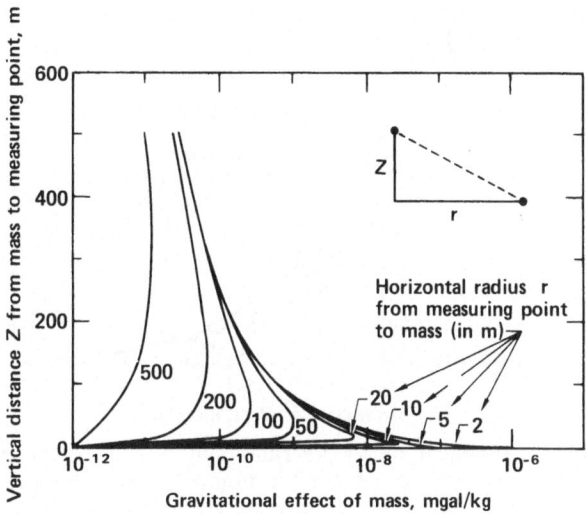

FIG. 7. Vertical component of gravity caused by a point mass at a horizontal radius r from a borehole as a function of vertical distance z from the measuring point to the mass.

different horizontal radii as a function of vertical distance z. The maximum effect is found at $z = r/\sqrt{2}$ and therefore the maximum sensitivity is at an angle from the vertical of $\tan^{-1}\sqrt{2}$ or about 55°.

Since any gravimeter has a minimum resolvable δg, the minimum visible mass for any r must be that which gives a vertical component of gravity equal to δg at $z = r/\sqrt{2}$. This minimum mass is

$$M_m = 3·86 \times 10^5 \, r^2 \delta g$$

where M_m is in kilograms and r is in metres. A reasonable value for δg is 0·01 mgal ·(Schmoker, 1978a) and so for $r = 1$ m, M_m is about 4000 kg or the mass of an elephant, and for $r = 100$ m, about 4×10^7 kg, or that of a battleship.

Note that Fig. 7 shows the effect of a point *at constant horizontal distance* as a function of depth. Clearly the smaller the radius, the greater the effect on the vertical component (the greatest effect is of course a point directly under or above the meter). Figure 8 shows the effect of a point at constant vertical distance as a function of horizontal distance.

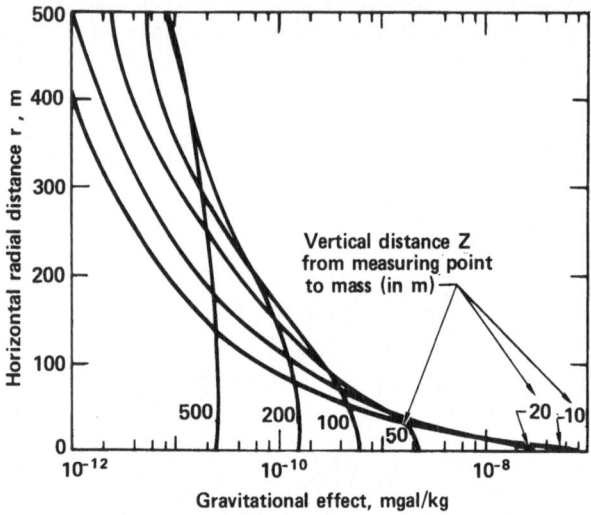

FIG. 8. Vertical component of gravity caused by a point mass at a vertical distance z from a measuring point as a function of the horizontal radius r from the measuring point to the mass.

Thus far we have been considering the effect of a mass anomaly on the vertical component of gravity. When the apparent density, or difference in vertical component between two measuring stations, is used, a mass anomaly whose density is *greater* than that of the surrounding medium (a positive mass anomaly) can give rise to a *decrease* in the apparent density in some parts of the borehole. Figure 9 shows contours of equal effect of an infinite (into the paper) line source of linear density

$$\sigma = 4\pi \Delta Z$$

on the apparent density calculated from gravity measured at two stations a vertical distance $2\Delta z$ apart. The position of the line source is given in dimensionless units $x/\Delta z$ and $z/\Delta z$. The surfaces defined by

$$(z/\Delta z)^2 = (x/\Delta z)^2 + 1$$

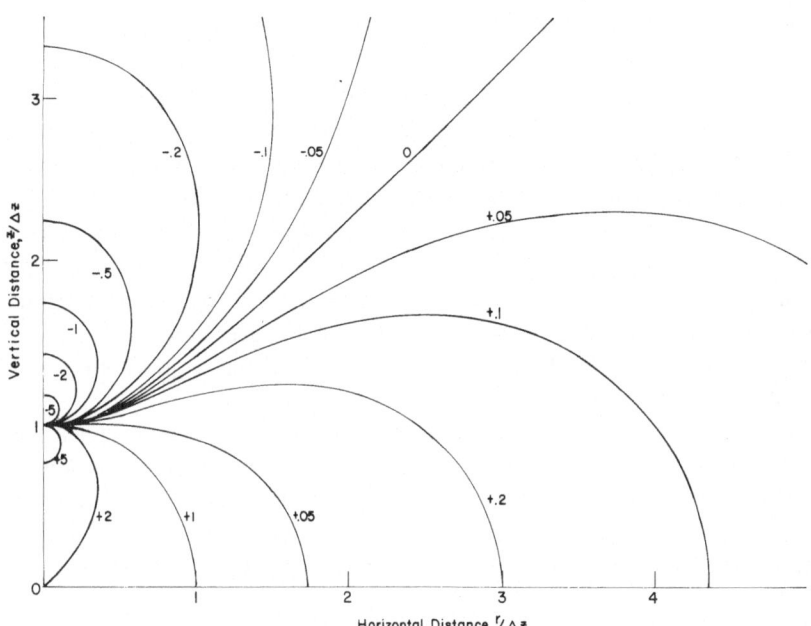

FIG. 9. Contours of equal value of apparent density caused by an infinite line
source of linear density $\sigma = 4\pi\Delta z$ with measuring stations $2\Delta z$ apart.

(only one is shown in this quarter-space) are the boundaries between
regions of positive and negative effect of the anomaly on the apparent
density. In other words, the contours are values of the function

$$\frac{4\pi\Delta z\Delta\rho_0}{\sigma}$$

where σ is the mass per unit length of the anomaly and ρ_0 the apparent
density caused by the anomaly. For example, if the measurement interval
is 20 m, $\Delta z = 10$ m, and if we take a line source of linear density
$4\pi\Delta z = 40\pi\ Mg$ per metre of length, then the values on the contours
correspond to the apparent densities caused by the source. Then, if the
anomaly is 10 m from the axis ($r/\Delta z = 1$) and centred on the interval
($z/\Delta z = 0$), we are on the $+1$ contour, and the apparent density caused by
the anomaly is 1 g/cm^3. On the other hand, if the anomaly is 3 m from
the axis ($r/\Delta z = 0\cdot3$) and 15·5 m above or below the centre ($z/\Delta z = 1\cdot55$),
we are on the -1 contour, and the apparent density caused by the
anomaly is -1 g/cm^3.

Borehole gravity data are usually interpreted by comparing the apparent density between stations to the density measured between stations by sampling or, more commonly, by another logging method such as a gamma–gamma density log. Alternatively, one can calculate the gravity as a function of depth that would be caused by the density measured near the hole if it were in infinite homogeneous horizontal beds and then compare the measured gravity to that calculated gravity.

In either case, the difference in density or gravity is compared to that which would be caused by some hypothetical model, either a simple geometrical shape such as a sphere or a fault, or a complex subsurface structure, and the model is varied until the agreement between the observed and calculated density or gravity difference is good enough to satisfy the interpreter.

Such interpretations are not unique. It is always possible (at least in principle) to find infinitely many models that can cause the same density or gravity difference within the error of the measurement. Consequently, the interpreter must have some idea of what he is looking for, and some knowledge of the range of plausible models, before sensible models can be made.

The borehole-gravity user should be familiar with the principles of the gamma–gamma log and with the pitfalls in its operation and the corrections that must be applied to its output before he attempts to use it for comparison with gravity data. The difficulties with the density log are unlikely to be severe unless the hole is both rough-walled and dry. In such circumstances the density log data should be regarded with suspicion (Hearst, 1976.) A caliper log would be useful to find those parts of the hole where density log data are likely to be poor, and repeat density logs may help to build confidence.

The range of investigation of a density log is about 20 cm (Sherman and Locke, 1975). This short range explains the virtue of a gravity log even in the case of a seemingly infinite homogeneous bedding. It often occurs that the rock very close to the borehole is invaded by drilling fluid, in the case of a liquid-filled hole, or dried out in the case of a dry hole. Then the density observed by the log is not representative of the true density of the material near the hole. If the user thinks the beds are infinite and homogeneous, or knows the structure and can correct the gravimetric density for that structure, then the gravity data can be used to infer the true density. However, if one wants to use the difference between gravimetric and log density to infer the structure, then fluid invasion or drying can be a severe problem. Sometimes the matter can be resolved by waiting (for weeks or months) for the perturbed zone to reequilibrate with the surroundings. One can never be sure that the problem is solved, however, and in fact fluid invasion is one of the limiting factors in the accuracy of gravity interpretation.

For many situations, the well is liquid-filled and the pores of the medium are liquid-saturated before drilling, and the only effect of invasion is a small change in the density of the saturating fluid. If the natural fluid is oil or gas and the invading fluid water or mud, this change can be significant, and in fact has been used to locate oil or gas (Jones, 1972; Jageler, 1976) in formations of high porosity. If not, the change can probably be neglected.

Another possible source of difference between the density logs and borehole gravity data is depth correlation. Depth discrepancies between logs run on different passes in a hole are not uncommon. If the density log and gravimeter are run with different cables, or even by different companies, the problem is compounded. In any case where depth accuracy is important, the use of permanent reference points in the hole, e.g. casing collars, hole bottom, or any distinctive lithologic bed, is recommended.

When the log densities have been computed, they are averaged over the gravity station intervals, and then the differences between log and gravimetric densities can be computed and one can begin the interpretation.

Rasmussen (1973) calculated the effect of a number of buried structures on borehole gravity data. The work of Nettleton (1942) and Hammer (1974) can sometimes be extrapolated for use with borehole gravity. Rasmussen (1975) calculated the general effects of buried reefs and salt domes. Coyle (1976) calculated the effects of a large variety of buried objects and Beyer (1977a) calculated the effect of several interesting structures.

The most comprehensive work to date involving postulated simple structures is that of Snyder (1976). Rather than the density, Snyder used the 'borehole Bouguer anomaly' or the difference between measured gravity and the gravity that would be observed if the world were indeed infinite homogeneous horizontal slabs with density equal to the log density. Hammer (1974) showed, for example, that the anomaly in the vertical component of gravity caused by an infinite horizontal line source is

$$\Delta g_\mathrm{a} = \frac{2k\sigma z}{r^2 + z^2}$$

where σ is the mass per unit length, r the horizontal distance from the hole to the line and z the vertical distance from the gravity station to the line. Figure 10 shows the anomaly for a line source with a mass of π Mg

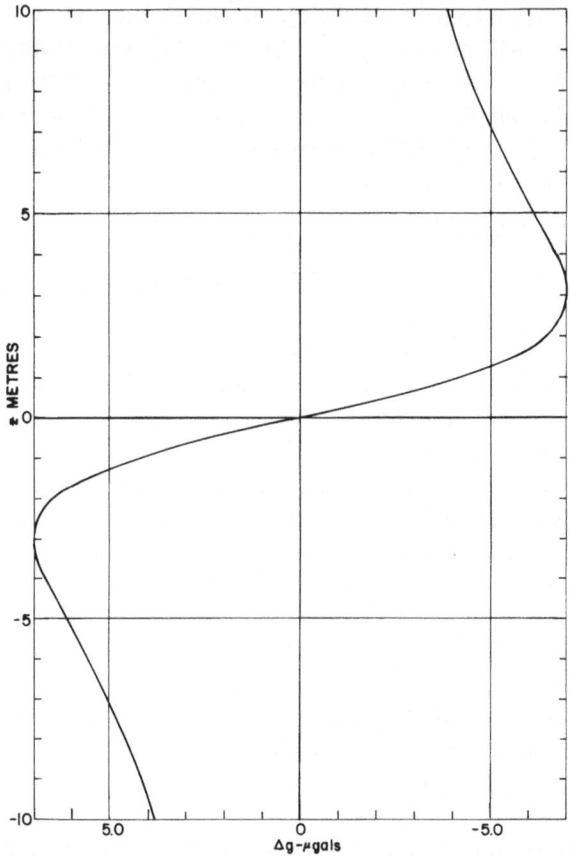

FIG. 10. Vertical component of gravity (borehole Bouguer anomaly) caused by an infinite horizontal linear mass of linear density π Mg/m, 2m from the borehole.

per metre of length (this corresponds to an infinite cylinder of density 1 g/cm^3, 2 m in diameter). The anomaly reaches a maximum at a distance $z = r$ above or below the line, where r is the horizontal distance to the line. The maximum value of Δg_a is

$$\Delta g_{max} = k\sigma/r$$

so the diameter of a cylinder of density contrast ρ at a distance r is

$$d = \left[\frac{4\Delta g_{max} r}{\pi k \rho}\right]^{1/2}$$

Snyder showed that one can distinguish between, say, a cylinder and a truncated bed by the shape of the anomaly, and can then estimate the parameters of the perturbing shape from the size of the anomaly.

Beds are rarely horizontal or infinite. Hearst (1977b) has shown that the apparent difference in density between a dipping bed and its surroundings ρ_{ca} is related to the true difference (measured with a density log) ρ_{ct} by

$$\rho_{ca} = \rho_{ct} \cos \theta$$

where θ is the dip angle.

For a horizontal bed of finite extent

$$\rho_{ca} = \rho_{ct}\left[1 - \frac{2 \tan^{-1}(t/w)}{\pi}\right]$$

where t/w is the ratio of bed thickness to its horizontal extent. The effect is more complicated for dipping beds of finite lateral extent and is shown in Fig. 11. Snyder and Merkel (1977) did a similar calculation and included the effect of the deviation of the borehole through the dipping bed.

A more complicated structure can be represented by a two-dimensional cross-section. Hearst and McKague (1976) used equations developed by Elkins (1974) to calculate the effect of any cross-section that could be represented by straight lines. This calculation is useful in deciding which of two possible cross-sections is the most plausible. For example, Figs. 12 and 13 show two postulated cross-sections for a region near a proposed nuclear test at the Nevada Test Site. Measured and calculated gravity in two holes shown on the figures are compared in Figs. 14 and 15. The sum of squares of the differences between measured and calculated gravimetric density for the data in Fig. 14 is 0·67, whereas that sum of squares for the data in Fig. 15 (covering the same depths in the same holes) is 0·40. Therefore it appears that the cross-section in Fig. 13 is more plausible than that in Fig. 12.

This method has been modified (Hearst, 1978a) to use Talwani and Ewing's (1960) method for bodies with polygonal cross-section to provide more versatility and simpler input.

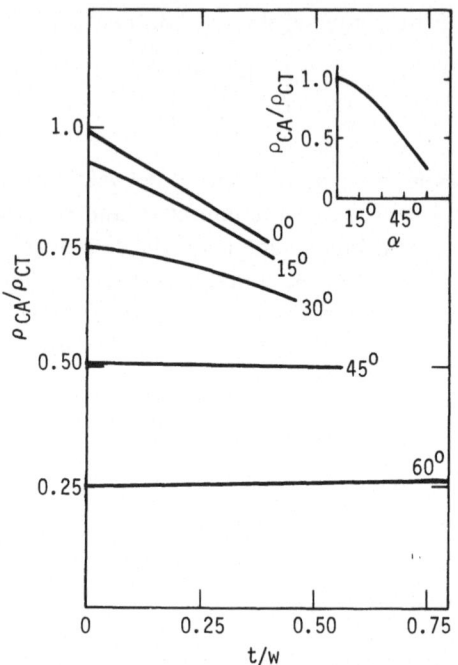

FIG. 11. Ratio of apparent density contrast ρ_{ca} to true density contrast ρ_{ct} as a function of bed thickness t divided by horizontal width of bed w for various values of dip. Inset: ratio as a function of dip for an infinite bed.

7. APPLICATIONS

The most common use of borehole gravimetry to date has been in the location of oil and gas. McCulloh (1966a and b) suggested this method before a borehole gravimeter was available and later (McCulloh et al., 1968) demonstrated its feasibility. The basic principle of the technique is that oil is less dense than water, and gas less dense than oil. The formation very close to the wellbore is usually water-saturated and therefore higher density than the natural formation if oil or gas is present. The detection of oil, of course, requires much more accuracy than the detection of gas.

The problem of the masking of nearby oil or gas by fluid invasion, however, is not the most important, since conventional well logging can usually detect oil or gas just outside the invaded zone. It is more important to use the range of the gravimeter to detect oil or gas that is not present near the wellbore, for example in reefs that are tight at the well and porous farther away (Rasmussen, 1975). Gravimetry can also be useful if the well has been cased, when most of the conventional methods cannot be used.

Jageler (1976) describes some of the applications used by Amoco. This company appears to use borehole gravity more than most other oil companies and in fact has recently (SpaN, 1980) published a short paper on borehole gravity in its quarterly report to stockholders. Bradley (1976) also discusses this work. He mentions, for example, a well in a limestone reef in the Michigan Basin. Conventional density logging showed no indication of useful porosity. However, the borehole gravimeter indicated densities significantly more than $0.1 \, g/cm^3$ below those indicated by the density log in several zones in the well. These zones were perforated and were found to yield commercially valuable gas flows, in a well that had appeared to be useless.

The US Geological Survey (Beyer, 1971, 1977a,b, 1980; Beyer and Clutsom, 1978a,b, 1980) has done extensive borehole gravity work in oil and gas fields. Often they have located oil or gas. Always, they have elucidated the lithology of the formations, thereby increasing the understanding of the geology.

A consortium of oil companies working with a geophysical company (LaFehr, 1979) has acquired several of the new LaCoste and Romberg gravimeters, and the gravimeter is now becoming almost a routine method of analysis of oil and gas wells.

Many other applications of borehole gravimetry have either been developed or proposed. Head and Kososki (1979) have used gravimetry to evaluate aquifers. Heintz and Alexander (1979) attempted to locate sulphur deposits with borehole gravity, and Schmoker (1979) compared amounts of sulphur estimated by borehole gravity to those observed in core and cuttings.

The borehole gravimeter is beginning to become a useful method of studying lithology and geology without attempting to locate minerals. Hinze et al. (1978) used a borehole gravity survey to obtain bulk densities in a deep exploratory well. Schmoker (1977a,b) has used the gravimeter to investigate density in granite and Devonian shales.

Comparision of borehole gravity to log density to infer structure, discussed in the previous section, has been extensively applied at the Nevada Test Site, and has now been used by Schmoker (1980b) to attempt to define salt structures near a well.

294

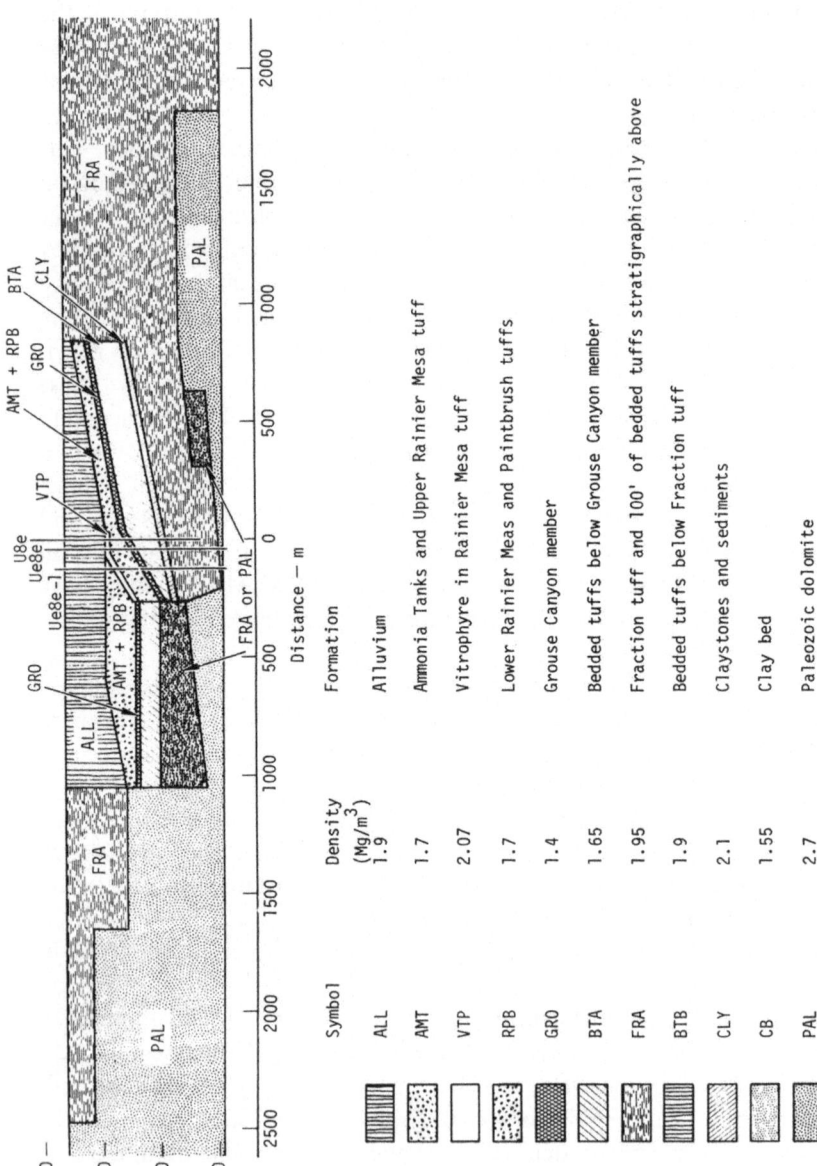

Symbol	Formation	Density (Mg/m³)
ALL	Alluvium	1.9
AMT	Ammonia Tanks and Upper Rainier Mesa tuff	1.7
VTP	Vitrophyre in Rainier Mesa tuff	2.07
RPB	Lower Rainier Meas and Paintbrush tuffs	1.7
GRO	Grouse Canyon member	1.4
BTA	Bedded tuffs below Grouse Canyon member	1.65
FRA	Fraction tuff and 100' of bedded tuffs stratigraphically above	1.95
BTB	Bedded tuffs below Fraction tuff	1.9
CLY	Claystones and sediments	2.1
CB	Clay bed	1.55
PAL	Paleozoic dolomite	2.7

FIG. 12(a). Single-scarp cross-section for comparison with cross-section in Fig. 13. Gravity measured and calculated in wells marked U8e and Ue8e.

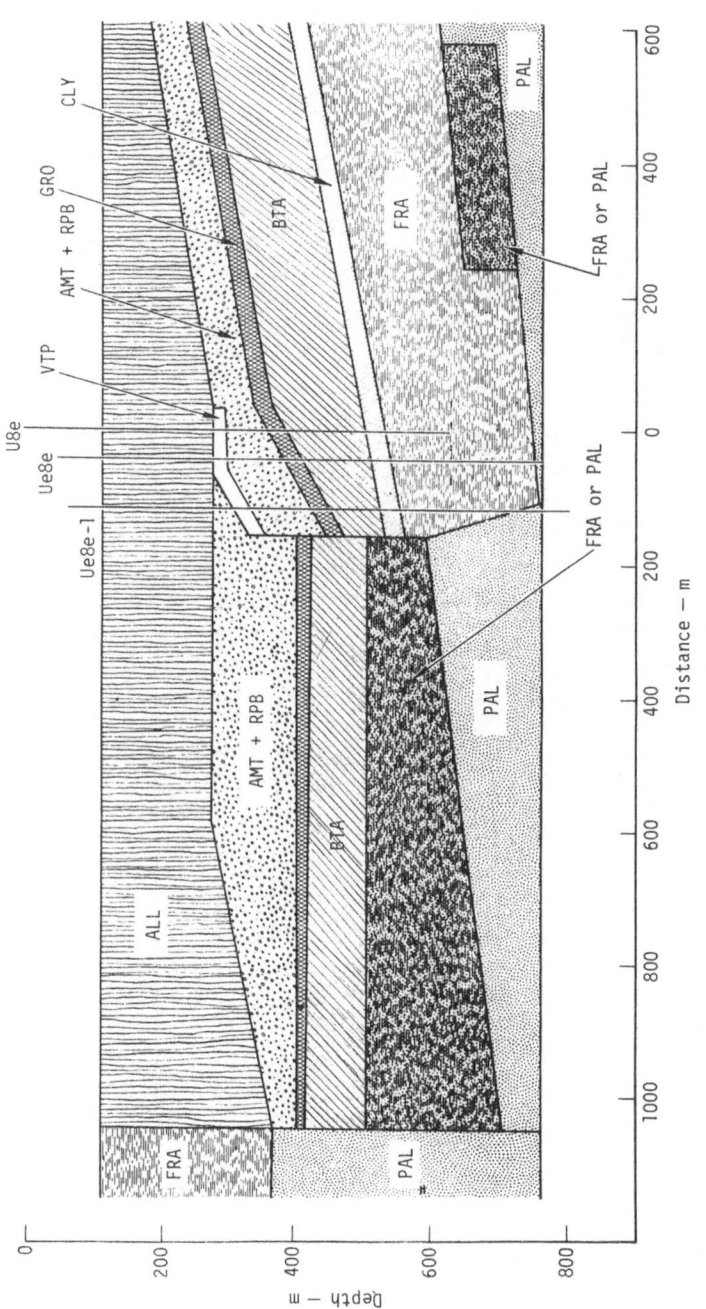

F<small>IG</small>. 12(b). Enlargement of centre part of Fig. 12(a).

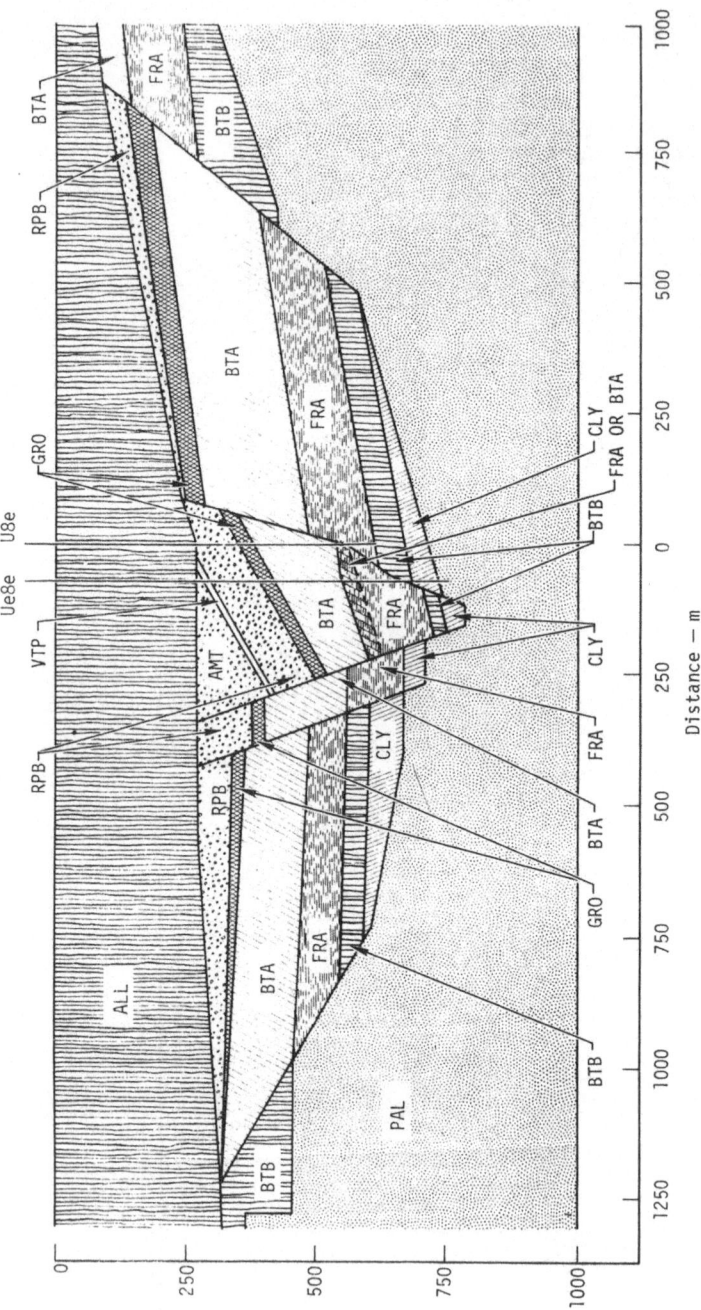

FIG. 13. Two-scarp cross-section for comparison with cross-section in Fig. 12.

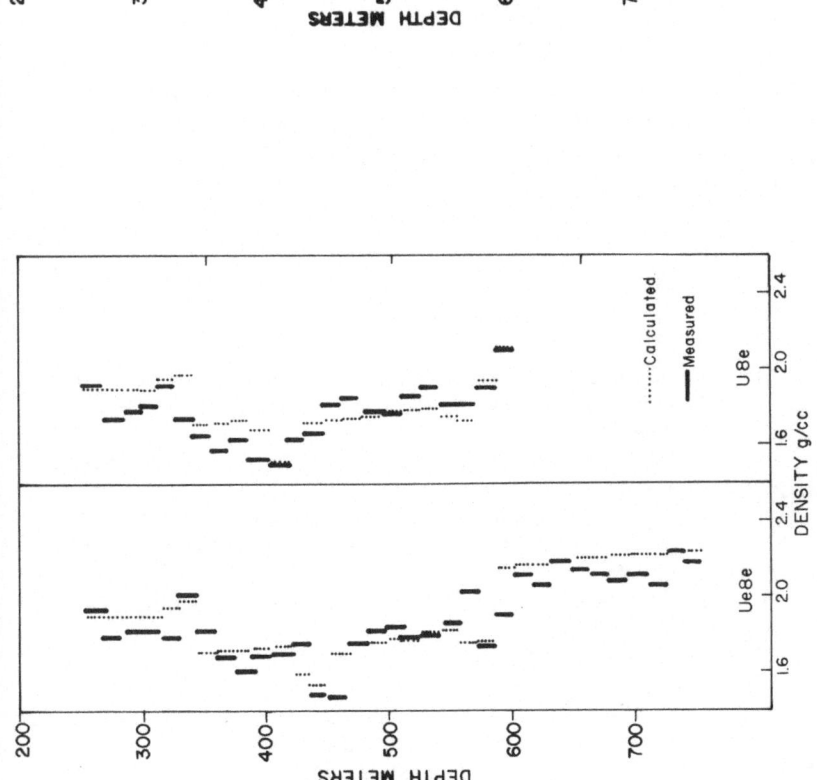

FIG. 14. Calculated gravimetric density vs depth for cross-section in Fig. 12 compared with measured gravimetric density in two wells, U8e and Ue8e.

FIG. 15. Calculated gravimetric density vs depth for cross-section in Fig. 13 compared with measured gravimetric density in two wells. U8e and Ue8e.

Hearst *et al.* (1978) have proposed use of a borehole gravimeter to detect subsidence upon withdrawal of formation fluids, specifically for geothermal applications, and Baker (1977) has proposed a method of adapting the new LaCoste and Romberg instrument to geothermal wells.

Robbins (1980) gives an almost complete bibliography of work to date in practical applications of borehole gravity.

As applications of the borehole gravimeter become more common, and new uses are found, we expect that more and more geophysicists will find the borehole gravimeter a familiar, useful and perhaps ultimately indispensable tool.

ACKNOWLEDGEMENTS

This work would have been much more difficult without the annotated bibliography prepared by Robbins (1980). We have paraphrased it in several places. We are indebted to the publications of L. A. Beyer and J. W. Schmoker. And we thank P. W. Kasameyer for a valuable review. Finally we should like to thank David Gilmore for his help in the preparation of the manuscript.

REFERENCES

AIRY, G. B. (1856). Account of pendulum experiments undertaken in the Harton Colliery for the purpose of determining the mean density of the earth, *Royal Soc. Phil. Trans.*, **146**, 297–355.

ALGERMISSEN, S. T. (1961). Underground and surface gravity survey, Leadwood, Missouri, *Geophysics*, **26**, 158–68.

ANON. (1966). Esso licenses down-hole gravity meter. *Oil and Gas J.*, **64** (26), 101–2.

ARZI, A. A. (1975). Microgravity for engineering applications, *Geophys. Prospecting*, **23**, 408–25.

BAKER, G. E. (1977). Gravity instrument cannister feasibility study, EG&G Report No. GEB77–57, 63 p.

BELL, C. C., FORWARD, R. L., MILLER, L. R., BEARD, T. D. and BARAN, T. M. (1969). Static gravitational gradient field detection with a rotating torsional gradiometer, *Proc. 1969 Symp. on Unconventional Inertial Sensors*, Naval Applied Science Laboratory, Brooklyn NY, 29–30 January, p. 143–68.

BEYER, L. A. (1971). The vertical gradient of gravity in vertical and near-vertical boreholes, US Geological Survey Open-file Report 71–42, 229 p.

BEYER, L. A. (1977a). The interpretation of borehole gravity surveys (abstract only), *Geophysics*, **42**, 141.

BEYER, L. A. (1977b). Interpretation of borehole gravity in the southern San Joaquin Basin (abstract only), Geophysics, 42, 1100.

BEYER, L. A. (1979a). Terrain corrections for borehole and tower gravity measurements, US Geological Survey Open-file Report 79–121, 17 p.

BEYER, L. A. (1979b). Terrain corrections for borehole and gravity measurements, Geophysics, 44, 1584–7.

BEYER, L. A. (1980). Borehole gravity program of the US Geological Survey (1963–1975)—Brief history and basic data, US Geological Survey Open-file Report 80–903, 76 p.

BEYER, L. A. and CLUTSOM, F. G. (1978a). Borehole gravity survey in the Dry Piney oil and gas field, Big Piney–LaBarge area, Sublette County, Wyoming, US Geological Survey Oil and Gas Investigations Chart OC-84 2 pl, 12 p.

BEYER, L. A. and CLUTSOM, F. G. (1978b). Density and porosity of oil reservoirs and overlying formations from borehole gravity measurements, Gebo Oil Field, Hot Springs County, Wyoming, US Geological Survey Oil and Gas Investigations Chart OC-88 3 pl, 16 p.

BEYER, L. A. and CLUTSOM, F. G. (1980). Density and porosity of Upper Cretaceous through Permian formations from borehole gravity measurements, Big Polecat oil and gas field, Park County, Wyoming, US Geological Survey Oil and Gas Investigations Chart OC-103 3 pl, 12 p.

BEYER, L. A. and CORBATO, C. E. (1972). A FORTRAN IV computer program for calculating borehole gravity terrain corrections, US Geological Survey Open-file Report (Available from NTIS as PB-208–679).

BODEMULLER, H. (1963). Measurement and geodetic evaluation of vertical gradients of gravity, Boll. Geol., 69, 261–79.

BRADLEY, J. W. (1976). The commercial application and interpretation of the borehole gravimeter, in Tomorrow's Oil from Today's Provinces, Ed. R. E. JANTZEN. American Association of Petroleum Geologists, Pacific Section, Misc. Pub. 24, p. 98–109.

BROWN, A. R., RASMUSSEN, N. F., GARNER, C. O. and CLEMENT, W. G. (1975). Borehole gravity logging fundamentals, Preprint, SEG 45th Annual Meeting, Denver, CO, 12–16 October, 20 p.

BUCK, S. W. (1973). Traverse gravimeter experiment final report, Charles Stark Draper Laboratory Report No. R-739, 75 p.

CATON, P. W. (1981). Improved methods for reducing borehole-gravity data—Applications and analyses of reduced gravity plots, Soc. Prof. Well Log Analysts 22nd Annual Logging Symp. Trans., 22–26 June.

COYLE, L. A. (1976). The application of borehole gravimetry to remote sensing of anomalous masses, Purdue Univ. M.S. Thesis. (Available from University Microfilms).

DOMZALSKI, W. (1954). Gravity Measurements in a vertical shaft, Trans. Inst. Mining Metall., 63(9), 429–45.

DOMZALSKI, W. (1955). Three dimensional gravity survey, Geophys. Prospecting, 3, 15–55.

DRAKE, R. E. (1967). A surface–subsurface measurement of an anomaly in the vertical gradient of gravity near Loveland Pass, Colorado, Univ. of California Riverside M.S. Thesis, 41 p.

EDCON (1977). Borehole Gravity Meter Manual. EDCON, Denver, CO.

ELKINS, T. A. (1974). Personal Communication.

FACSINAY, L. and HAAZ, H. (1953). Density determinations of rocks, based on subsurface gravimeter measurements at different depths, *Magyar Allami Eotvos Lorand Geofisikai Intezet Geofizikai Kozlemenyek*, **2** 1–9 (in Hungarian).

FAJKLEWICZ, Z. J. (1976). Gravity vertical gradient measurements for the detection of small geologic and anthropogenic forms, *Geophysics*, **41**, 1016–30.

FORWARD, R. A. (1965) Rotating gravitational and inertial sensors, *AIAA Unmanned Spacecraft Meeting Los Angeles, CA*, 1–4 March, 6 p.

GILBERT, R. L. G. (1952). Gravity observations in a borehole, *Nature*, **170**, 424–5.

GOODELL, R. R. and FAY, C. H. (1964). Borehole gravity meter and its application, *Geophysics*, **29**, 774–82.

GRANT, F. S. and WEST, G. F. (1965). *Interpretation Theory in Applied Geophysics*. McGraw-Hill, New York.

HAMMER, S. (1939). Terrain corrections for gravimeter stations, *Geophysics*, **4**, 184–93.

HAMMER, S. (1950). Density determinations by underground gravity measurements, *Geophysics*, **15**, 637–52.

HAMMER, S. (1974). Approximation in gravity interpretation calculations, *Geophysics*, **39**, 205–22.

HEAD, W. J. and KOSOSKI, B. A. (1979). Borehole Gravity: a new tool for the ground-water hydrologist (abstract only), *Trans. Am. Geophys. Union*, **60**, 248.

HEALEY, D. L. (1970). Calculated *in situ* bulk densities from subsurface gravity observations and density logs, Nevada Test Site and Hot Creek Valley, Nye County, Nevada, US Geological Survey Professional Paper 700-B, p. B52–B62.

HEALEY, D. L. (1977). Personal Communication.

HEARST, J. R. (1968). Terrain corrections for borehole gravimetry, *Geophysics*, **33**, 361–2.

HEARST, J. R. (1976). Effects of mudcake and sonde angle on a simple two-detector density log, *Log Analyst*, **17**(3), 11–14.

HEARST, J. R. (1977*a*). On the range of investigation of a borehole gravimeter, *Soc. Prof. Well Log Analysts 18th Annual Logging Symp. Trans.*, 5–8 June, p. E1–E12.

HEARST, J. R. (1977*b*). Estimation of dip and lateral extent of beds with borehole gravimetry, *Geophysics*, **42**, 990–4.

HEARST, J. R. (1978*a*). BIFUR II, a program for calculating borehole gravity caused by two-dimensional structure, Lawrence Livermore Laboratory Report UCID-17 852, 22 p.

HEARST, J. R. (1978*b*). Personal communication.

HEARST, J. R. and CARLSON, R. C. (1977). The gravimetric density formula for a spherical shell, *Geophysics*, **42**, 1469.

HEARST, J. R. and MCKAGUE, H. L. (1976). Structure elucidation with borehole gravimetry, *Geophysics*, **41**, 491–505.

HEARST, J. R., KASAMEYER, P. W. and OWEN, L. B. (1978). Potential uses for a high-temperature borehole gravimeter, Lawrence Livermore Laboratory Report UCRL-52 421, 8 p.

HEARST, J. R., SCHMOKER, J. W. and CARLSON, R. C. (1980). Effects of terrain on borehole gravity data, *Geophysics*, **45**, 234–43.

HEINTZ, K. O. and ALEXANDER, M. (1979). Sulfur exploration with core hole and surface gravity (abstract only), *Geophysics*, **44**, 370.

HEISKANEN, W. A. and VENING MEINESZ, F. A. (1958). *The Earth and Its Gravity Field*. McGraw-Hill, New York.

HENDERSON, G. C. and IVERSON, R. M. (1968). Testing gravimeters for lunar surface measurements, *IEEE Trans. on Geoscience Elec.*, **GE-6**, 132–8.

HINZE, W. J., BRADLEY, J. W. and BROWN, A. R. (1978). Gravimeter survey in the Michigan deep borehole, *J. Geophys. Res.*, **83**, 5864–7.

HOWELL, L. G., HEINTZ, K. O. and BARRY, A. (1966). The development and use of a high-precision downhole gravity meter, *Geophysics*, **31**, 764–72.

JAGELER, A. H. (1976). Improved hydrocarbon reservoir evaluation through use of borehole-gravimeter data, *J. Petroleum Technol.*, **28**, 709–18.

JAGELER, A. H. (1981). Personal communication.

JONES, B. R. (1972). The use of downhole gravity data in formation evaluation, *Soc. Prof. Well Log Analysts 13th Annual Logging Symp. Trans.*, 7–10 May, p. M1-M12.

JUNG, H. (1939). Density determinations in solid rock through measurement of acceleration due to gravity at different subsurface depths (in German), *Zeit. Geoph.*, **15**, 56–65.

KUMEGAI, N., ABE, E. and YOSHIMURA, Y. (1960). Measurement of vertical gradient of gravity and its significance, *Boll. Geof.*, **2**, 607–30.

KUO, J. T., OTTAVANI, M. and SINGH. S. K. (1969). Variations of vertical gravity gradient in New York City and Alpine, New Jersey, *Geophysics*, **34**, 235–48.

LACOSTE, L. J. B. (1934). A new type long-period vertical seismograph, *Physics*, **5**, 178–82.

LACOSTE, L. J. B. (1935). A simplification in the conditions for the zero-length-spring seismograph, *Bull Seis. Soc. Am.*, **25**, 176–9.

LACOSTE, L. J. B. (1977). Method and apparatus for leveling an instrument in a well bore, US Patent No. 4,040,189.

LACOSTE, L. J. B. and ROMBERG, A. (1942). Force measuring device, US Patent No. 2 293 437.

LACOSTE, L. J. B. and ROMBERG, A. (1945). Force measuring instrument, US Patent No. 2 377 889.

LAFEHR, T. R., MERKEL, R. H. and HERRING, A. T. (1979). Evaluation and applications of new LaCoste and Romberg borehole gravity meter (abstract only), *Geophysics*, **44**, 369–70.

LONGMAN, I. M. (1959). Formulas for computing tidal acceleration, *J. Geophys. Res.*, **64**, 2351.

McCULLOH, T. H. (1965). A confirmation by gravity measurements of an underground density profile based on core densities, *Geophysics*, **30**, 1108–32.

McCULLOH, T. H. (1966a). Gravimetric effects of petroleum accumulations—a preliminary summary, US Geological Survey Circular 530, 4 p.

McCULLOH, T. H. (1966b). The promise of precise borehole gravimetry in petroleum exploration and exploitation, US Geological Survey Circular 531.

McCULLOH, T. H. (1967a). Borehole gravimetry—new developments and applications, in *New Geophysical Developments and Methods: Proc 7th World Petroleum Congress, Mexico City*, 2–8 April. Elsevier, London, pp. 85–99.

McCULLOH, T. H. (1967b). Mass properties of sedimentary rocks and gravimetric effects of petroleum and natural-gas reservoirs, US Geological Survey Professional Paper 528A, 50 p.

McCULLOH, T. H., LACOSTE, L. J. B., SCHOELLHAMER, J. E. and PAMPAYAN.

E. H. (1967a). The US Geological Survey–LaCoste and Romberg precise borehole gravimeter system—instrumentation and support equipment, US Geological Survey Professional Paper 575-D, p. D92-D100.

McCULLOH, T. H., SCHOELLHAMER, J. E., PAMPAYAN, E. H. and PARKS, H. B. (1967b). The US Geological Survey – LaCoste and Romberg precise borehole gravimeter—test results, US Geological Survey Professional Paper 575-D, p. D101-D112.

McCOLLOH, T. H., KANDLE, J. R. and SCHOELLHAMER, J. E. (1968). Application of gravity measurements in wells to problems of reservoir evaluation, Soc. Prof. Well Log Analysts 9th Annual Logging Symp. Trans., p. O1-O29.

MILLER, A. H. and INNES, M. J. S. (1953). Application of gravimeter observations to the determination of the mean density of the earth and of rock densities in mines, Public. Dominion Observatory, Ottawa, 16(4), 3–17.

NETTLETON, L. L. (1942). Gravity and magnetic calculations, Geophysics, 7, 293–310.

NETTLETON, L. L. (1976). Gravity and Magnetics in Oil Prospecting. McGraw-Hill, New York.

PLOUFF, D. (1961). Gravity profile along Roberts Tunnel, Colorado, US Geological Survey Professional Paper 424–C, p. C263-C265.

PLOUFF, D. (1966). Digital terrain corrections based on geographic coordinates, Preprint, 36th SEG Meeting, Houston.

RASMUSSEN, N. F. (1973). Borehole gravity survey planning and operations, Soc. Prof. Well Log Analysts 14th Annual Logging Symp. Trans., p. Q1-Q28.

RASMUSSEN, N. F. (1975). Borehole gravimeter finds bypassed oil, gas, Oil and Gas J., 73(39), 100–4.

ROBBINS, S. L. (1979). Description of a special logging truck built for the US Geological Survey for borehole gravity surveys, US Geological Survey Open-file Report 79-1511, 67 p.

ROBBINS, S. L. (1980). Bibliography with abridged abstracts of subsurface gravimetry (especially borehole) and corresponding in situ rock density measurements, US Geological Survey Open-file Report 80–710, 47 p.

ROBBINS, S. L. (1981). Reexamination of the values used as constants in calculating rock density from borehole gravity data, Geophysics, 46, 208–10.

ROGERS, G. R. (1952). Subsurface gravity measurements, Geophysics, 17, 365–77.

SCHMOKER, J. W. (1977a). Density variations in a quartz diorite determined from borehole gravity measurements, San Benito County, California, Log Analyst, 18(2), 32–8.

SCHMOKER, J. W. (1977b). A borehole gravity survey to determine density variations in the Devonian shale sequence of Lincoln County, West Virginia, US Dept. of Energy, Energy Research Center MERC/CR-77-7, 15 p.

SCHMOKER, J. W. (1978a) Accuracy of borehole gravity data, Geophysics, 43, 538–42.

SCHMOKER, J. W. (1978b). Personal communication.

SCHMOKER, J. W. (1979). Interpretation of borehole gravity surveys in a native-sulfur deposit, Culberson County, Texas, Economic Geol., 74, 1462–70.

SCHMOKER, J. W. (1980a). Terrain effects of cultural features upon shallow borehole gravity data, Geophysics, 45, 1869–71.

SCHMOKER, J. W. (1980b). Effect upon borehole-gravity data of salt structures typical of the WIPP site (northern Delaware Basin) Eddy County, New Mexico, US Geological Survey Oil and Gas Investigations Chart OC-109.

SHERMAN, H. and Locke, S. (1975). Depth of investigation of neutron and density sondes for 35-percent porosity sandstone, *Soc. Prof. Well Log Analysts 16th Annual Logging Symp. Trans.*, p Q1–Q25.

SMITH, N. J. (1950). The case for gravity data from boreholes, *Geophysics*, **15**, 605–36.

SNYDER, D. D. (1976). The borehole Bouguer gravity anomaly—application to interpreting borehole gravity surveys, *Soc. Prof. Well Log Analysts 17th Annual Logging Symp. Trans.*, p. AA1–AA20.

SNYDER, D. D. and MERKEL, R. H. (1977). Generalized correction for dipping beds for the borehole gravity meter, *Log Analyst*, **18**(2), 41–3.

SPAN (1980). *A Matter of Gravity*. SpaN, Standard Oil Company of Indiana, pp. 14–17.

TALWANI, M. and EWING, M. (1960). Rapid computation of gravitational attraction of two-dimensional bodies of arbitrary shape, *Geophysics*, **25**, 203–25.

THYSSEN-BORNEMISZA, S. (1963). The vertical gravity gradient in borehole exploration, *Geophysics*, **28**, 1072–3.

THYSSEN-BORNEMISZA, S. (1964). Determination of Bouguer gravity in shallow holes, *Geophysics*, **29**, 445–6.

THYSSEN-BORNEMISZA, S. (1965*a*). Determination of the vertical density gradient in a borehole, *Geophysics*, **30**, 439–40.

THYSSEN-BORNEMISZA, S. (1965*b*). The anomalous free-air vertical gradient in borehole exploration, *Geophysics*, **30**, 441–3.

TRAGESER, M. B. (1970). A gradiometer system for gravity anomaly surveying, Charles Stark Draper Laboratory Report R-588, 44 p.

TRAGESER, M. B. (1975). Feasibility model gravity gradiometer test results, Charles Stark Draper Laboratory Report P-179, 17 p.

VASCHILOV, YU. N. (1964). Allowance for the effects of the relief of the locality in gravimetric observations in underground workings and boreholes, *Razvedochnaya i Promyslovaya Geofizika*, **51**, 71–5. Trans. by. J. E. Bradley, Addis Trans. Menlo Park CA, 1965.

VENERUSO, A. F. (1978). Personal communication.

INDEX